Argument-Driven Inquiry
in
EARTH AND SPACE SCIENCE

LAB INVESTIGATIONS
for GRADES 6–10

Argument-Driven Inquiry
in
EARTH AND SPACE SCIENCE

LAB INVESTIGATIONS
for GRADES 6–10

Victor Sampson, Ashley Murphy, Kemper Lipscomb, and Todd L. Hutner

NSTApress

National Science Teachers Association
Arlington, Virginia

National Science Teachers Association

Claire Reinburg, Director
Rachel Ledbetter, Managing Editor
Deborah Siegel, Associate Editor
Andrea Silen, Associate Editor
Donna Yudkin, Book Acquisitions Manager

ART AND DESIGN
Will Thomas Jr., Director

PRINTING AND PRODUCTION
Catherine Lorrain, Director

NATIONAL SCIENCE TEACHERS ASSOCIATION
David L. Evans, Executive Director

1840 Wilson Blvd., Arlington, VA 22201
www.nsta.org/store
For customer service inquiries, please call 800-277-5300.

NSTA is committed to publishing material that promotes the best in inquiry-based science education. However, conditions of actual use may vary, and the safety procedures and practices described in this book are intended to serve only as a guide. Additional precautionary measures may be required. NSTA and the authors do not warrant or represent that the procedures and practices in this book meet any safety code or standard of federal, state, or local regulations. NSTA and the authors disclaim any liability for personal injury or damage to property arising out of or relating to the use of this book, including any of the recommendations, instructions, or materials contained therein.

PERMISSIONS
Book purchasers may photocopy, print, or e-mail up to five copies of an NSTA book chapter for personal use only; this does not include display or promotional use. Elementary, middle, and high school teachers may reproduce forms, sample documents, and single NSTA book chapters needed for classroom use only. E-book buyers may download files to multiple personal devices but are prohibited from posting the files to third-party servers or websites, or from passing files to non-buyers. For additional permission to photocopy or use material electronically from this NSTA Press book, please contact the Copyright Clearance Center (CCC) (*www.copyright.com*; 978-750-8400). Please access *www.nsta.org/permissions* for further information about NSTA's rights and permissions policies.

Library of Congress Cataloging-in-Publication Data
Names: Sampson, Victor, 1974- author. | Murphy, Ashley, 1988- author. | Lipscomb, Kemper, 1990- author. | Hutner, Todd, 1981- author.
Title: Argument-driven inquiry in earth and space science : lab investigations for grades 6-10 / by Victor Sampson, Ashley Murphy, Kemper Lipscomb, and Todd L. Hutner.
Description: Arlington, VA : National Science Teachers Association, [2018] | Includes bibliographical references and index. | Identifiers: LCCN 2017055013 (print) | LCCN 2018000680 (ebook) | ISBN 9781681403748 (e-book) | ISBN 9781681403731 (print)
Subjects: LCSH: Earth sciences--Study and teaching (Middle school)--Handbooks, manuals, etc. | Earth sciences--Study and teaching (Secondary)--Handbooks, manuals, etc. | Earth sciences--Problems, exercises, etc. | Meteorology--Study and teaching (Middle school)--Handbooks, manuals, etc. | Meteorology--Study and teaching (Secondary)--Handbooks, manuals, etc. | Outer space--Study and teaching (Middle school) | Outer space--Study and teaching (Secondary)--Handbooks, manuals, etc.
Classification: LCC QE40 (ebook) | LCC QE40 .A74 2018 (print) | DDC 520.78--dc23
LC record available at https://lccn.loc.gov/2017055013

CONTENTS

Preface xi
About the Authors xv
Introduction xvii

SECTION 1
Using Argument-Driven Inquiry

Chapter 1. Argument-Driven Inquiry 3

Chapter 2. Lab Investigations 23

SECTION 2
Space Systems

INTRODUCTION LABS

Lab 1. Moon Phases: Why Does the Appearance of the Moon Change Over Time in a Predictable Pattern?
    Teacher Notes 32
    Lab Handout 43
    Checkout Questions 49

Lab 2. Seasons: What Causes the Differences in Average Temperature and the Changes in Day Length That We Associate With the Change in Seasons on Earth?
    Teacher Notes 54
    Lab Handout 67
    Checkout Questions 74

Lab 3. Gravity and Orbits: How Does Changing the Mass and Velocity of a Satellite and the Mass of the Object That It Revolves Around Affect the Nature of the Satellite's Orbit?
    Teacher Notes 78
    Lab Handout 89
    Checkout Questions 97

APPLICATION LAB

Lab 4. Habitable Worlds: Where Should NASA Send a Probe to Look for Life?
    Teacher Notes 102
    Lab Handout 109
    Checkout Questions 114

SECTION 3
History of Earth

INTRODUCTION LABS

Lab 5. Geologic Time and the Fossil Record: Which Time Intervals in the Past 650 Million Years of Earth's History Are Associated With the Most Extinctions and Which Are Associated With the Most Diversification of Life?

Teacher Notes.. 122
Lab Handout.. 133
Checkout Questions.. 141

Lab 6. Plate Interactions: How Is the Nature of the Geologic Activity That Is Observed Near a Plate Boundary Related to the Type of Plate Interaction That Occurs at That Boundary?

Teacher Notes.. 146
Lab Handout.. 157
Checkout Questions.. 162

APPLICATION LAB

Lab 7. Formation of Geologic Features: How Can We Explain the Growth of the Hawaiian Archipelago Over the Past 100 Million Years?

Teacher Notes.. 168
Lab Handout.. 180
Checkout Questions.. 186

SECTION 4
Earth's Systems

INTRODUCTION LABS

Lab 8. Surface Erosion by Wind: Why Do Changes in Wind Speed, Wind Duration, and Soil Moisture Affect the Amount of Soil That Will Be Lost Due to Wind Erosion?

Teacher Notes.. 194
Lab Handout.. 204
Checkout Questions.. 211

Lab 9. Sediment Transport by Water: How Do Changes in Stream Flow Affect the Size and Shape of a River Delta?

Teacher Notes.. 216

Lab Handout 229

Checkout Questions 236

Lab 10. Deposition of Sediments: How Can We Explain the Deposition of Sediments in Water?

Teacher Notes 240

Lab Handout 252

Checkout Questions 259

Lab 11. Soil Texture and Soil Water Permeability: How Does Soil Texture Affect Soil Water Permeability?

Teacher Notes 262

Lab Handout 274

Checkout Questions 280

Lab 12. Cycling of Water on Earth: Why Do the Temperature and the Surface Area to Volume Ratio of a Sample of Water Affect Its Rate of Evaporation?

Teacher Notes 284

Lab Handout 296

Checkout Questions 304

APPLICATION LABS

Lab 13. Characteristics of Minerals: What Are the Identities of the Unknown Minerals?

Teacher Notes 310

Lab Handout 321

Checkout Questions 328

Lab 14. Distribution of Natural Resources: Which Proposal for a New Copper Mine Maximizes the Potential Benefits While Minimizing the Potential Costs?

Teacher Notes 332

Lab Handout 347

Checkout Questions 358

SECTION 5
Weather and Climate

INTRODUCTION LABS

Lab 15. Air Masses and Weather Conditions: How Do the Motions and Interactions of Air Masses Result in Changes in Weather Conditions?
Teacher Notes.. 364
Lab Handout.. 376
Checkout Questions... 383

Lab 16. Surface Materials and Temperature Change: How Does the Nature of the Surface Material Covering a Specific Location Affect Heating and Cooling Rates at That Location?
Teacher Notes.. 386
Lab Handout.. 398
Checkout Questions... 405

Lab 17. Factors That Affect Global Temperature: How Do Cloud Cover and Greenhouse Gas Concentration in the Atmosphere Affect the Surface Temperature of Earth?
Teacher Notes.. 408
Lab Handout.. 419
Checkout Questions... 426

Lab 18. Carbon Dioxide Levels in the Atmosphere: How Has the Concentration of Atmospheric Carbon Dioxide Changed Over Time?
Teacher Notes.. 428
Lab Handout.. 438
Checkout Questions... 444

APPLICATION LAB

Lab 19. Differences in Regional Climate: Why Do Two Cities Located at the Same Latitude and Near a Body of Water Have Such Different Climates?
Teacher Notes.. 448
Lab Handout.. 458
Checkout Questions... 465

SECTION 6
Human Impact

INTRODUCTION LABS

Lab 20. Predicting Hurricane Strength: How Can Someone Predict Changes in Hurricane Wind Speed Over Time?
Teacher Notes...472
Lab Handout...481
Checkout Questions...487

Lab 21. Forecasting Extreme Weather: When and Under What Atmospheric Conditions Are Tornadoes Likely to Develop in the Oklahoma City Area?
Teacher Notes...490
Lab Handout...501
Checkout Questions...507

APPLICATION LABS

Lab 22. Minimizing Carbon Emissions: What Type of Greenhouse Gas Emission Reduction Policy Will Different Regions of the World Need to Adopt to Prevent the Average Global Surface Temperature on Earth From Increasing by 2°C Between Now and the Year 2100?
Teacher Notes...512
Lab Handout...522
Checkout Questions...530

Lab 23. Human Use of Natural Resources: Which Combination of Water Use Policies Will Ensure That the Phoenix Metropolitan Area Water Supply Is Sustainable?
Teacher Notes...534
Lab Handout...543
Checkout Questions...552

SECTION 7
Appendixes

Appendix 1. Standards Alignment Matrixes 557

***Appendix 2. Overview of Crosscutting Concepts and Nature of Scientific
Knowledge and Scientific Inquiry Concepts*** 569

Appendix 3. Timeline Options for Implementing ADI Lab Investigations 573

Appendix 4. Investigation Proposal Options 577

***Appendix 5. Peer-Review Guides and Teacher Scoring Rubrics for
Investigation Reports*** 587

Image Credits 593
Index 599

PREFACE

A Framework for K–12 Science Education (NRC 2012; henceforth referred to as the *Framework*) and the *Next Generation Science Standards* (NGSS Lead States 2013; henceforth referred to as the *NGSS*) call for a different way of thinking about why we teach science and what we expect students to know by the time they graduate high school. As to why we teach science, these documents emphasize that schools need to

> ensure by the end of 12th grade, *all* students have some appreciation of the beauty and wonder of science; possess sufficient knowledge of science and engineering to engage in public discussions on related issues; are careful consumers of scientific and technological information related to their everyday lives; are able to continue to learn about science outside school; and have the skills to enter careers of their choice, including (but not limited to) careers in science, engineering, and technology. (NRC 2012, p. 1)

The *Framework* and the *NGSS* are based on the idea that students need to learn science because it helps them understand how the natural world works, because citizens are required to use scientific ideas to inform both individual choices and collective choices as members of a modern democratic society, and because economic opportunity is increasingly tied to the ability to use scientific ideas, processes, and habits of mind. From this perspective, it is important to learn science because it enables people to figure things out or to solve problems.

These two documents also call for a reappraisal of what students need to know and be able to do by time they graduate from high school. Instead of teaching with the goal of helping students remember facts, concepts, and terms, science teachers are now charged with the goal of helping their students become *proficient* in science. To be considered proficient in science, the *Framework* suggests that students need to understand 12 disciplinary core ideas (DCIs) in the Earth and space sciences, be able to use seven crosscutting concepts (CCs) that span the various disciplines of science, and learn how to participate in eight fundamental scientific and engineering practices (SEPs; called science and engineering practices in the *NGSS*).

The DCIs are key organizing principles that have broad explanatory power within a discipline. Scientists use these ideas to explain the natural world. The CCs are ideas that are used across disciplines. These concepts provide a framework or a lens that people can use to explore natural phenomena; thus, these concepts often influence what people focus on or pay attention to when they attempt to understand how something works or why something happens. The SEPs are the different activities that scientists engage in as they attempt to generate new concepts, models, theories, or laws that are both valid and reliable. All three of these dimensions of science are important. Students need to not only know about the DCIs, CCs, and SEPs but also must be able to use all three dimensions at the same time to figure things out or to solve problems. These important DCIs, CCs, and SEPs are summarized in Figure 1.

FIGURE 1 _____

The three dimensions of science in *A Framework for K–12 Science Education* and the *Next Generation Science Standards*

Science and engineering practices	Crosscutting concepts
1. Asking Questions and Defining Problems	1. Patterns
2. Developing and Using Models	2. Cause and Effect: Mechanism and Explanation
3. Planning and Carrying Out Investigations	3. Scale, Proportion, and Quantity
4. Analyzing and Interpreting Data	4. Systems and System Models
5. Using Mathematics and Computational Thinking	5. Energy and Matter: Flows, Cycles, and Conservation
6. Constructing Explanations and Designing Solutions	6. Structure and Function
7. Engaging in Argument From Evidence	7. Stability and Change
8. Obtaining, Evaluating, and Communicating Information	

Disciplinary core ideas in the Earth and space sciences

- ESS1.A: The Universe and Its Stars
- ESS1.B: Earth and the Solar System
- ESS1.C: The History of Planet Earth
- ESS2.A: Earth Materials and Systems
- ESS2.B: Plate Tectonics and Large-Scale System Interactions
- ESS2.C: The Roles of Water in Earth's Surface Processes
- ESS2.D: Weather and Climate
- ESS2.E: Biogeology
- ESS3.A: Natural Resources
- ESS3.B: Natural Hazards
- ESS3.C: Human Impacts on Earth Systems
- ESS3.D: Global Climate Change

Source: Adapted from NRC 2012 and NGSS Lead States 2013

To help students become proficient in science in ways described by the National Research Council in the *Framework,* teachers will need to use new instructional approaches that give students an opportunity to use the three dimensions of science to explain natural phenomena or develop novel solutions to problems. This is important because traditional instructional approaches, which were designed to help students "learn about" the concepts, theories, and laws of science rather than learn how to "figure out" how or why things work, were not created to foster the development of science proficiency inside the classroom. To help teachers make this instructional shift, this book provides 23 laboratory investigations designed using an innovative approach to lab instruction called argument-driven inquiry (ADI). This approach promotes and supports three-dimensional instruction inside classrooms because it gives students an opportunity to use DCIs, CCs, and SEPs to construct and critique claims about how things work or why things happen. The lab activities described in this book will also enable students to develop the disciplinary-based literacy skills outlined in the *Common Core State Standards* for English language arts (NGAC and CCSSO 2010) because ADI gives students an opportunity to give presentations to their peers, respond to audience questions and critiques, and then write, evaluate, and revise reports as part of each lab. Use of these labs, as a result, can help teachers align their teaching with current recommendations for improving classroom instruction in science and for making earth and space science more meaningful for students.

References

National Governors Association Center for Best Practices and Council of Chief State School Officers (NGAC and CCSSO). 2010. *Common core state standards.* Washington, DC: NGAC and CCSSO.

National Research Council (NRC). 2012. *A framework for K–12 science education: Practices, crosscutting concepts, and core ideas.* Washington, DC: National Academies Press.

NGSS Lead States. 2013. *Next Generation Science Standards: For states, by states.* Washington, DC: National Academies Press. *www.nextgenscience.org/next-generation-science-standards.*

ABOUT THE AUTHORS

Victor Sampson is an associate professor of STEM (science, technology, engineering, and mathematics) education and the director of the Center for STEM Education at The University of Texas at Austin (UT-Austin). He received a BA in zoology from the University of Washington, an MIT from Seattle University, and a PhD in curriculum and instruction with a specialization in science education from Arizona State University. Victor also taught high school biology and chemistry for nine years. He specializes in argumentation in science education, teacher learning, and assessment. To learn more about his work in science education, go to *www.vicsampson.com.*

Ashley Murphy attended Florida State University (FSU) and earned a BS with dual majors in biology and secondary science education. Ashley spent some time as a middle school biology and science teacher before entering graduate school at UT-Austin, where she is currently working toward a PhD in STEM education. Her research interests include argumentation in middle and elementary classrooms. As an educator, she frequently employed argumentation as a means to enhance student understanding of concepts and science literacy.

Kemper Lipscomb is a doctoral student studying STEM education at UT-Austin. She received a BS in biology and secondary education from FSU and taught high school biology, anatomy, and physiology for three years. She specializes in argumentation in science education and computational thinking.

Todd L. Hutner is the assistant director for teacher education and center development for the Center of STEM Education at UT-Austin. He received a BS and an MS in science education from FSU and a PhD in curriculum and instruction from UT-Austin. Todd's classroom teaching experience includes teaching chemistry, physics, and Advanced Placement (AP) physics in Texas and earth science and astronomy in Florida. His current research focuses on the impact of both teacher education and education policy on the teaching practice of secondary science teachers.

INTRODUCTION

The Importance of Helping Students Become Proficient in Science

The current aim of science education in the United States is for *all* students to become proficient in science by the time they finish high school. *Science proficiency*, as defined by Duschl, Schweingruber, and Shouse (2007), consists of four interrelated aspects. First, it requires an individual to know important scientific explanations about the natural world, to be able to use these explanations to solve problems, and to be able to understand new explanations when they are introduced to the individual. Second, it requires an individual to be able to generate and evaluate scientific explanations and scientific arguments. Third, it requires an individual to understand the nature of scientific knowledge and how scientific knowledge develops over time. Finally, and perhaps most important, an individual who is proficient in science should be able to participate in scientific practices (such as planning and carrying out investigations, analyzing and interpreting data, and arguing from evidence) and communicate in a manner that is consistent with the norms of the scientific community. These four aspects of science proficiency include the knowledge and skills that all people need to have to be able to pursue a degree in science, prepare for a science-related career, and participate in a democracy as an informed citizen.

This view of science proficiency serves as the foundation for the *Framework* (NRC 2012) and the *NGSS* (NGSS Lead States 2013). Unfortunately, our educational system was not designed to help students become proficient in science. As noted in the *Framework*,

> K–12 science education in the United States fails to [promote the development of science proficiency], in part because it is not organized systematically across multiple years of school, emphasizes discrete facts with a focus on breadth over depth, and does not provide students with engaging opportunities to experience how science is actually done. (p. 1)

Our current science education system, in other words, was not designed to give students an opportunity to learn how to use scientific explanations to solve problems, generate or evaluate scientific explanations and arguments, or participate in the practices of science. Our current system was designed to help students learn facts, vocabulary, and basic process skills because many people think that students need a strong foundation in the basics to be successful later in school or in a future career. This vision of science education defines rigor as covering more topics and learning as the simple acquisition of new ideas or skills.

Our views about what counts as rigor, therefore, must change to promote and support the development of science proficiency. Instead of using the number of different topics covered in a course as a way to measure rigor in our schools, we must start to measure rigor in terms of the number of opportunities students have to use the ideas of science as a way to make sense of the world around them. Students, in other words, should be expected to learn how to use the core ideas of science as conceptual tools to plan and carry out investigations, develop and evaluate explanations, and question how we know what we know. A rigorous course, as result, would be one where students are expected to do science, not just learn about science.

Our views about what learning is and how it happens must also change to promote and support the development of science proficiency. Rather then viewing learning as a simple process where people accumulate more information over time, learning needs to viewed as a personal and social process that involves "people entering into a different way of thinking about and explaining the natural world; becoming socialized to a greater or lesser extent into the practices of the scientific community with its particular purposes, ways of seeing, and ways of supporting its knowledge claims" (Driver et al. 1994, p. 8). Learning, from this perspective, requires a person to be exposed to the language, the concepts, and the practices of science that makes science different from other ways of knowing. This process requires input and guidance about "what counts" from people who are familiar with the goals of science, the norms of science, and the ways things are done in science. Thus, learning is dependent on supportive and informative interactions with others.

Over time, people will begin to appropriate and use the language, the concepts, and the practices of science as their own when they see how valuable they are as a way to accomplish their own goals. Learning therefore involves seeing new ideas and ways of doing things, trying out these new ideas and practices, and then adopting them when they are useful. This entire process, however, can only happen if teachers provide students with multiple opportunities to use scientific ideas to solve problems, to generate or evaluate scientific explanations and arguments, and to participate in the practices of science inside the classroom. This is important because students must have a supportive and educative environment to try out new ideas and practices, make mistakes, and refine what they know and what they do before they are able to adopt the language, the concepts, and the practices of science as their own.

A New Approach to Teaching Science

We need to use different instructional approaches to create a supportive and educative environment that will enable students to learn the knowledge and skills they need to become proficient in science. These new instructional approaches will need to give students an opportunity to learn how to "figure out" how things work or why things happen. Rather than simply encouraging students to learn about the facts, concepts, theories, and laws of science, we need to give them more opportunities to develop explanations for natural phenomena and design solutions to problems. This emphasis on "figuring things out" instead of "learning about things" represents a big change in the way we will need to teach science at all grade levels. To figure out how things work or why things happen in a way that is consistent with how science is actually done, students must do more than hands-on activities. Students must learn how to use disciplinary core ideas (DCIs), crosscutting concepts (CCs), and science and engineering practices (SEPs) to develop explanations and solve problems (NGSS Lead States 2013; NRC 2012).

A DCI is a scientific idea that is central to understanding a variety of natural phenomena. An example of a DCI in Earth and Space Sciences is that the solar system consists of the Sun and a collection of objects that are held in orbit around the Sun by its gravitational pull on them. This DCI not only explains the motion of planets around the Sun but can also explain tides, eclipses of the Sun and the Moon, and the motion of the planets in the sky relative to the stars.

CCs are those concepts that are important across the disciplines of science; there are similarities and differences in the treatment of the CCs in each discipline. The CCs can be used as a lens to help people think about what to focus on or pay attention to during an investigation. For example, one of the CCs from the *Framework* is Energy and Matter: Flows, Cycles, and Conservation. This CC is important in many different fields of study, including astronomy, geology, and meteorology. This CC is equally important in physics and biology. Physicists use this CC to study mechanics, thermodynamics, electricity, and magnetism. Biologists use this CC to study cells, growth and development, and ecosystems. It is important to highlight the centrality of this idea, and other CCs, for students as we teach the subject-specific DCIs.

SEPs describe what scientists do to investigate the natural world. The practices outlined in the *Framework* and the *NGSS* explain and extend what is meant by *inquiry* in science and the wide range of activities that scientists engage in as they attempt to generate and validate new ideas. Students engage in practices to build, deepen, and apply their knowledge of DCIs and CCs. The SEPs include familiar aspects of inquiry, such as Asking Questions and Defining Problems, Planning and Carrying Out Investigations, and Analyzing and Interpreting Data. More important, however, the SEPs include other activities that are at the core of doing science: Developing and Using Models, Constructing Explanations and Designing Solutions, Engaging in Argument From Evidence, and Obtaining, Evaluating, and Communicating Information. All of these SEPs are important to learn, because there is no single scientific method that all scientists must follow; scientists engage in different practices, at different times, and in different orders depending on what they are studying and what they are trying to accomplish at that point in time.

This focus on students using DCIs, CCs, and SEPs during a lesson is called *three-dimensional instruction* because students have an opportunity to use all three dimensions of science to understand how something works, to explain why something happens, or to develop a novel solution to a problem. When teachers use three-dimensional instruction inside their classrooms, they encourage students to develop or use conceptual models, design investigations, develop explanations, share and critique ideas, and argue from evidence, all of which allow students to develop the knowledge and skills they need to be proficient in science (NRC 2012). Current research suggests that all students benefit from three-dimensional instruction because it gives all students more voice and choice during a lesson and it makes the learning process inside the classroom more active and inclusive (NRC 2012).

We think the school science laboratory is the perfect place to integrate three-dimensional instruction into the science curriculum. Well-designed lab activities can provide opportunities for students to participate in an extended investigation where they can not only use one or more DCIs to understand how something works, to explain why something happens, or to develop a novel solution to a problem but also use several different CCs and SEPs during the same lesson. A teacher, for example, can give his or her students an opportunity to develop and use a model of the Earth-Sun-Moon system to describe the cyclic patterns of lunar phases, eclipses of the Sun and Moon, and the seasons. The teacher can then encourage them to use what they know about Earth and the Solar System (a DCI) and their understanding of Patterns and of Scale, Proportion, and Quantity (two different CCs) to plan and carry out an investigation to figure out how Earth, the Sun, and the Moon move relative to each other. During this investigation they must ask questions, analyze and interpret data, use mathematics, develop a model, argue from evidence, and obtain, evaluate, and communicate information (six different SEPs). Using multiple DCIs, CCs, and SEPs at the same time is important because it creates a classroom experience that parallels how science is done. This, in turn, gives all students who participate in a school science lab activity an opportunity to deepen their understanding of what it means to do science and to develop science-related identities. In the following section, we will describe how to promote and support the development of science proficiency during school science labs through three-dimensional instruction.

How School Science Labs Can Help Foster the Development of Science Proficiency Through Three-Dimensional Instruction

As defined by the NRC (2005, p. 3), "[l]aboratory experiences provide opportunities for students to interact directly with the material world … using the tools, data collection techniques, models, and theories of science." School science laboratory experiences tend to follow a similar format in most U.S. science classrooms (Hofstein and Lunetta 2004; NRC 2005). This format begins with the teacher introducing students to an important concept or principle through direct instruction, usually by giving a lecture about it or by assigning a chapter from a textbook to read. This portion of instruction often takes several class periods. Next, the students will complete a hands-on lab activity. The purpose of the hands-on activity is help students understand a concept or principle that was introduced to the students earlier. To ensure that students "get the right result" during the lab and that the lab actually illustrates, confirms, or verifies the target concept or principle, the teacher usually provides students with a step-by-step procedure to follow and a data table to fill out. Students are then asked to answer a set of analysis questions to ensure that everyone "reaches the right conclusion" based on the data they collected during the lab. The lab experience ends with the teacher going over what the students should have done during the lab, what they should have observed, and what answers they should have given in response to the analysis questions; this review step is done to ensure that the students "learned what they were supposed to have learned" from the hands-on activity and is usually done, once again, through whole-class direct instruction.

Classroom-based research, however, suggests that this type of approach to lab instruction does little to help students learn key concepts. The National Research Council (2005, p. 5), for example, conducted a synthesis of several different studies that examined what students learn from lab instruction and found that "research focused on the goal of student mastery of subject matter indicates that typical laboratory experiences are no more or less effective than other forms of science instruction (such as reading, lectures, or discussion)." This finding is troubling because, as noted earlier, the main goal of this type of lab experience is to help students understand an important concept or principle by giving them a hands-on and concrete experience with it. In addition, this type of lab experience does little to help students learn how to plan and carry out investigations or analyze and interpret data because students have no voice or choice during the activity. Students are expected to simply follow a set of directions rather than having to think about what data they will collect, how they will collect it, and what they will need to do to analyze it once they have it. These types of activities also can lead to misunderstanding about the nature of scientific knowledge and how this knowledge is developed over time due to the emphasis on following procedure and getting the right results. These "cookbook" labs, as a result, do not reflect how science is done at all.

Over the past decade, many teachers have changed their labs to be inquiry-based in order to address the many shortcomings of more traditional cookbook lab activities. Inquiry-based lab experiences that are consistent with the definition of *inquiry* found in the *National Science Education Standards* (NRC 1996) and *Inquiry and the National Science Education Standards* (NRC 2000) share five key features:

1. Students need to answer a scientifically oriented question.

2. Students must collect data or use data collected by someone else.

3. Students formulate an answer to the question based on their analysis of the data.

4. Students connect their answer to some theory, model, or law.

5. Students communicate their answer to the question to someone else.

Teachers tend to use inquiry-based labs as a way to introduce students to new concepts and to give them an opportunity to learn how to collect and analyze data in science (NRC 2012).

Although inquiry-based approaches give students much more voice and choice during a lab, especially when compared with more traditional cookbook approaches, they do not do as much as they could do to promote the development of science proficiency. Teachers tend to use inquiry-based labs as a way to help students learn about a new idea rather than as a way to help students learn how to figure out how things work or why they happen. Students, as a result, rarely have an opportunity to learn how to use DCIs, CCs, and SEPs to develop explanations or solve problems. In addition, inquiry-based approaches rarely give students an opportunity to participate in the full range of scientific practices. Inquiry-based labs tend to be designed so students have many opportunities to learn how to ask questions, plan and carry out investigations, and analyze and interpret data but few opportunities to learn how to participate in the practices that focus on how new ideas are developed, shared, refined, and eventually validated within the scientific community. These important practices include developing and using models, constructing explanations, arguing from evidence, and obtaining, evaluating, and communicating information (Duschl, Schweingruber, and Shouse 2007; NRC 2005). Inquiry-based labs also do not give students an opportunity to improve their science-specific literacy skills. Students, as a result, are rarely expected to read, write, and speak in scientific manner because the focus of these labs is learning about content and how to collect and analyze data in science, not how to propose, critique, and revise ideas.

Changing the focus and nature of inquiry-based labs so they are more consistent with three-dimensional instruction can help address these issues. To implement such a change, teachers will not only have to focus on using DCIs, CCs, and SEPs during a lab but will also need to emphasize "how we know" in the different Earth and space science disciplines (i.e., how new knowledge is generated and validated) equally with "what we know" about plate tectonics, climate, stars, and energy (i.e., the theories, laws, and unifying concepts). We have found that this shift in focus is best accomplished by making the practice of arguing from evidence or scientific argumentation the central feature of all lab activities. We define *scientific argumentation* as the process of proposing, supporting, evaluating, and refining claims based on evidence (Sampson, Grooms, and Walker 2011). The *Framework* (NRC 2012) provides a good description of the role argumentation plays in science:

> Scientists and engineers use evidence-based argumentation to make the case for their ideas, whether involving new theories or designs, novel ways of collecting data, or interpretations of evidence. They and their peers then attempt to identify weaknesses and limitations in the argument, with the ultimate goal of refining and improving the explanation or design. (p. 46)

When teachers make the practice of arguing from evidence the central focus of lab activities students have more opportunities to learn how to construct and support scientific knowledge claims through argument (NRC 2012). Students also have more opportunities to learn how to evaluate the claims and arguments made by others. Students, as a result, learn how to read, write, and speak in a scientific manner because they need to be able to propose and support their claims when they share them and evaluate, challenge, and refine the claims made by others.

We developed the argument-driven inquiry (ADI) instructional model (Sampson and Gleim 2009; Sampson, Grooms, and Walker 2009, 2011) as a way to change the focus and nature of labs so they are consistent with three-dimensional instruction. ADI gives students an opportunity to learn how to use DCIs, CCs, and SEPs to figure out how things work or why things happen. This instructional approach also places scientific argumentation as the central feature of all lab activities. ADI lab investigations, as a result, make lab activities more authentic and educative for students and thus help teachers promote and support the development of science proficiency. This instructional model reflects current theories about how people learn science (NRC 1999, 2005, 2008, 2012) and is also based on what is known about how to engage students in argumentation and other important scientific practices (Erduran and Jimenez-Aleixandre 2008; McNeill and Krajcik 2008; Osborne, Erduran, and Simon 2004; Sampson and Clark 2008; Sampson, Enderle, and Grooms, 2013). We will explain the stages of ADI and how each stage works in Chapter 1.

Organization of This Book

This book is divided into seven sections. Section 1 includes two chapters: the first chapter describes the ADI instructional model, and the second chapter describes the development of the ADI lab investigations and provides an overview of what is included with each investigation. Sections 2–6 contain the 23 lab investigations. Each investigation includes three components:

- Teacher Notes, which provides information about the purpose of the lab and what teachers need to do to guide students through it.
- Lab Handout, which can be photocopied and given to students at the beginning of the lab. It provides the students with a phenomenon to investigate, a guiding question to answer, and an overview of the DCIs and CCs that students can use during the investigation.
- Checkout Questions, which can be photocopied and given to students at the conclusion of the lab activity. The Checkout Questions consist of items that target students' understanding of the DCIs and CCs and the concepts of the nature of scientific knowledge (NOSK) and the nature of scientific inquiry (NOSI) addressed during the lab.

Section 7 consists of five appendixes:

- Appendix 1 contains several standards alignment matrixes that can be used to assist with curriculum or lesson planning.
- Appendix 2 provides an overview of the CCs and the NOSK and NOSI concepts that are a focus of the lab investigations. This information about the CCs and the NOSK and NOSI are included as a reference for teachers.
- Appendix 3 provides several options (in tabular format) for implementing an ADI investigation over multiple 50-minute class periods.
- Appendix 4 provides options for investigation proposals, which students can use as graphic organizers to plan an investigation. The proposals can be photocopied and given to students during the lab.
- Appendix 5 provides two versions of a peer-review guide and teacher scoring rubric (one for middle school and one for high school), which can also be photocopied and given to students.

Safety Practices in the Science Laboratory

It is important for all of us to do what we can to make school science laboratory experiences safer for everyone in the classroom. We recommend four important guidelines to follow. First, we need to have proper safety equipment such as, but not limited to, fume hoods, fire extinguishers, eye wash, and showers in the classroom or laboratory. Second, we need to ensure that students use appropriate personal protective equipment (PPE; e.g., sanitized indirectly vented chemical-splash goggles, chemical-resistant aprons, and nonlatex gloves) during all components of lab activities (i.e., setup, hands-on investigation, and takedown). At a minimum, the PPE we provide for students to use must meet the ANSI/ISEA Z87.1D3 standard. Third, we must review and comply with all safety policies and procedures, including but not limited to appropriate chemical management, that have been established by our place of employment. Finally, and perhaps most important, we all need to adopt safety standards and better professional safety practices and enforce them inside the classroom or laboratory.

We provide safety precautions for each investigation and recommend that all teachers follow these safety precautions to provide a safer learning experience inside the classroom. The safety precautions associated with each lab investigation are based, in part, on the use of the recommended materials and instructions, legal safety standards, and better professional safety practices. Selection of alternative materials or procedures for these activities may jeopardize the level of safety and therefore is at the user's own risk.

We also recommend that you encourage students to read the National Science Teacher Association's document *Safety in the Science Classroom, Laboratory, or Field Sites* before allowing them to work in the laboratory for the first time. This document is available online at *www.nsta.org/docs/SafetyInTheScienceClassroomLabAndField.pdf*. Your students and their parent(s) or guardian(s) should then sign the document to acknowledge that they understand the safety procedures that must be followed during a school science laboratory experience.

Remember that a lab includes three parts: (1) setup, which includes setting up the lab and preparing the materials; (2) conducting the actual investigation; and (3) the cleanup, also called the takedown. The safety procedures and PPE we recommend for each investigation apply to all three parts.

References

Duschl, R. A., H. A. Schweingruber, and A. W. Shouse, eds. 2007. *Taking science to school: Learning and teaching science in grades K–8*. Washington, DC: National Academies Press.

Driver, R., Asoko, H., Leach, J., Mortimer, E., and Scott, P. 1994. Constructing scientific knowledge in the classroom. *Educational Researcher* 23 (7): 5–12.

Erduran, S., and M. Jimenez-Aleixandre, eds. 2008. *Argumentation in science education: Perspectives from classroom-based research*. Dordrecht, The Netherlands: Springer.

Hofstein, A., and V. Lunetta. 2004. The laboratory in science education: Foundations for the twenty-first century. *Science Education* 88 (1): 28–54.

McNeill, K., and J. Krajcik. 2008. Assessing middle school students' content knowledge and reasoning through written scientific explanations. In *Assessing science learning:*

Perspectives from research and practice, eds. J. Coffey, R. Douglas, and C. Stearns, 101–116. Arlington, VA: NSTA Press.

National Research Council (NRC). 1999. *How people learn: Brain, mind, experience, and school.* Washington, DC: National Academies Press.

National Research Council (NRC). 2000. *Inquiry and the National Science Education Standards.* Washington, DC: National Academies Press.

National Research Council (NRC). 2005. *America's lab report: Investigations in high school science.* Washington, DC: National Academies Press.

National Research Council (NRC). 2008. *Ready, set, science: Putting research to work in K–8 science classrooms.* Washington, DC: National Academies Press.

National Research Council (NRC). 2012. *A framework for K–12 science education: Practices, crosscutting concepts, and core ideas.* Washington, DC: National Academies Press.

NGSS Lead States. 2013. *Next Generation Science Standards: For states, by states.* Washington, DC: National Academies Press. *www.nextgenscience.org/next-generation-science-standards.*

Osborne, J., S. Erduran, and S. Simon. 2004. Enhancing the quality of argumentation in science classrooms. *Journal of Research in Science Teaching* 41 (10): 994–1020.

Sampson, V., and D. Clark. 2008. Assessment of the ways students generate arguments in science education: Current perspectives and recommendations for future directions. *Science Education* 92 (3): 447–472.

Sampson, V., and L. Gleim. 2009. Argument-driven inquiry to promote the understanding of important concepts and practices in biology. *American Biology Teacher* 71 (8): 471–477.

Sampson, V., Enderle, P., and Grooms, J. 2013. Argumentation in science and science education. *The Science Teacher* 80 (5): 30–33.

Sampson, V., J. Grooms, and J. Walker. 2009. Argument-driven inquiry: A way to promote learning during laboratory activities. *The Science Teacher* 76 (7): 42–47.

Sampson, V., J. Grooms, and J. Walker. 2011. Argument-driven inquiry as a way to help students learn how to participate in scientific argumentation and craft written arguments: An exploratory study. *Science Education* 95 (2): 217–257.

SECTION 1
Using Argument-Driven Inquiry

CHAPTER 1
Argument-Driven Inquiry

Stages of Argument-Driven Inquiry

The argument-driven inquiry (ADI) instructional model was designed to change the focus and nature of labs so they are consistent with three-dimensional science instruction. ADI therefore gives students an opportunity to learn how to use disciplinary core ideas (DCIs), crosscutting concepts (CCs), and science and engineering practices (SEPs) (NGSS Lead States 2013; NRC 2012) to figure out how things work or why things happen. This instructional approach also places scientific argumentation as the central feature of all lab activities. ADI lab investigations, as a result, make lab activities more authentic (students have an opportunity to use the three dimensions of science) *and* educative (students receive the feedback and explicit guidance that they need to improve on each aspect of science proficiency).

In this chapter, we will explain what happens during each of the eight stages of ADI. These eight stages are the same for every ADI lab experience. Students, as a result, quickly learn what is expected of them during each stage of an ADI lab and can focus on learning how to use DCIs, CCs, and SEPs to develop explanations or solve problems. Figure 2 summarizes the eight stages of the ADI instructional model.

FIGURE 2

Stages of the argument-driven inquiry instructional model

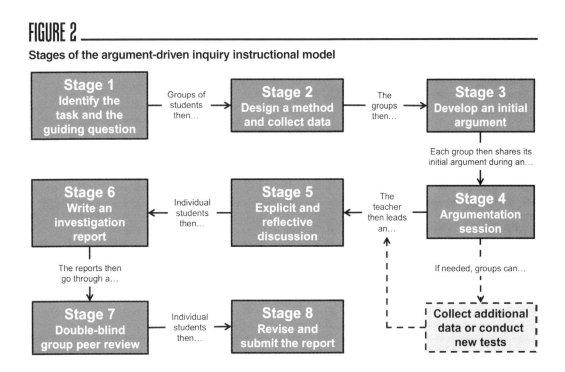

Stage 1: Identify the Task and the Guiding Question

An ADI lab activity begins with the teacher identifying a phenomenon to investigate and offering a guiding question for the students to answer. The goal of the teacher at this stage of the model is to capture the students' interest and provide them with a reason to complete the investigation. To aid in this, teachers should provide each student with a copy of the Lab Handout. This handout includes a brief introduction that provides a description of a puzzling phenomenon or a problem to solve, the DCIs and CCs that students can use during the investigation, a reason to investigate, and the task the students will need to complete. This handout also includes information about the nature of the argument they will need to produce, some helpful tips on how to get started, and criteria that will be used to judge argument quality (e.g., the sufficiency of the claim and the quality of the evidence).

Teachers often begin an ADI investigation by selecting a different student to read each section of the Lab Handout out loud while the other students follow along. As the students read, they can annotate the text to identify important or useful ideas and information or terms that may be unfamiliar or confusing. After each section is read, the teacher can pause to clarify expectations, answer questions, and provide additional information as needed. Teachers can also spark student interest by giving a demonstration or showing a video of the phenomenon.

It is also important for the teacher to hold a "tool talk" during this stage, taking a few minutes to explain how to use the available lab equipment, how to use a computer simulation, or even how to use software to analyze data. Teachers need to hold a tool talk because students are often unfamiliar with specialized lab equipment, simulations, or software. Even if the students are familiar with the available tools, they will often use them incorrectly or in an unsafe manner unless they are reminded about how the tools work and the proper way to use them. The teacher should therefore review specific safety protocols and precautions as part of the tool talk.

Including a tool talk during this stage is useful because students often find it difficult to design a method to collect the data needed to answer the guiding question (the task of stage 2) when they do not understand how to use the available materials. We also recommend that teachers give students a few minutes to tinker with the equipment, simulation, or software they will be using to collect data as part of the tool talk. We have found that students can quickly figure out how the equipment, simulation, or software works and what they can and cannot do with it simply by tinkering with the available materials for 5-10 minutes. When students are given this opportunity to tinker with the equipment, simulation, or software as part of the tool talk, they end up designing much better investigations (the task of stage 2) because they understand what they can and cannot do with the tools they will use to collect data.

Once all the students understand the goal of the activity and how to use the available materials, the teacher should divide the students into small groups (we recommend three or four students per group) and move on to the second stage of the instructional model.

Stage 2: Design a Method and Collect Data

In stage 2, small groups of students develop a method to gather the data they need to answer the guiding question and carry out that method. How students complete this stage depends on the nature of the investigation. Some investigations call for groups to answer the guiding question by designing a controlled experiment, whereas others require students to analyze an existing data set (e.g., a database or information sheets).

If students need assistance in designing their method, teachers can have students complete an investigation proposal. These proposals guide students through the process of developing a method by encouraging them to think about what type of data they will need to collect, how to collect it, and how to analyze it. We have included six different investigation proposals in Appendix 4 (p. 577) of this book that students can use to design their investigations. Investigation Proposal A (long or short version) can be used when students need to collect systematic observations for a descriptive investigation. Investigation Proposal B (long or short version) or Investigation Proposal C (long or short version) can be used when students need to design a comparative or experimental study to test potential explanations or relationships as part of their investigation. Investigation Proposal B requires students to design a test of two alternative hypotheses, and Investigation Proposal C requires students to design a test of three alternative hypotheses.

The overall intent of this stage is to provide students with an opportunity to interact directly with the natural world (or in some cases with data drawn from the natural world) using appropriate tools and data collection techniques and to learn how to deal with the uncertainties of empirical work. This stage of the model also gives students a chance to learn why some approaches to data collection or analysis work better than others and how the method used during a scientific investigation is based on the nature of the question and the phenomenon under investigation. At the end of this stage, students should have collected all the data they need to answer the guiding question.

Stage 3: Develop an Initial Argument

The next stage of the instructional model calls for students to develop an initial argument in response to the guiding question. To do this, each group needs to be encouraged to first analyze the measurements (e.g., temperature and mass) and/or observations (e.g., appearance and location) collected during stage 2 of the model. Once the groups have analyzed and interpreted the results of their analysis, they can create an initial argument. The argument consists of a claim, the evidence they are using to support their claim, and a justification of their evidence. The *claim* is their answer to the guiding question. The *evidence* consists of an analysis of the data they collected and an interpretation of the analysis. The *justification of the evidence* is a statement that defends their choice of evidence by explaining why it is important and relevant, making the concepts or assumptions underlying the analysis and interpretation explicit. The components of a scientific argument are illustrated in Figure 3.

To illustrate each of the three components of a scientific argument, consider the following example. This argument was made in response to the guiding question, "How does soil texture affect soil water permeability?"

> *Claim:* Soil that is composed of mostly small particles is less permeable than soil that is composed of mostly large particles.
>
> *Evidence:* Sand has a soil water permeability rate of 32.3 ml/minute. Water moves through sandy loam at a rate of 3.7 ml/minute. The soil water permeability rate of sandy clay loam is 1.1 ml/minute. These results suggest that water moves faster through soils that are made up of sand and slower through soils that made up of mostly clay.
>
> *Justification of the evidence:* Our evidence is based on four important assumptions. First, soil water permeability is determined by how much water can flow through a soil in a given amount of time. Second, there are three classes of particles that make up soil. These particles are classified by size. Sand is the largest, silt is smaller than sand, and clay is smaller than silt. Third, soil classification is based on the amount of each type of particle in it. Fourth, it is important to examine soil particle size because it affects how much empty space is between the particles.

The claim in this argument provides an answer to the guiding question. The author then uses genuine evidence to support the claim by providing an analysis of the data collected (measure of soil water permeability for three different types of soil) and an interpretation of the analysis (water moves faster through soils that are made up of sand and slower through soils that are made up of mostly clay). Finally, the author provides a justification of the evidence in the argument by making explicit the underlying concepts (the definition of soil water permeability, the types of particles found in soil, and how soil is classified) and assumptions (soil particle size affects how much empty space there is between the particles in different types of soil) that guided the analysis of the data and the interpretation of the analysis.

It is important for students to understand that, in science, some arguments are better than others. An important aspect of science and scientific argumentation involves the evaluation of the various components of the arguments put forward by others. Therefore, the framework provided in Figure 3 also highlights two types of criteria that students can and should be encouraged to use to evaluate an argument in science: empirical criteria and theoretical criteria. *Empirical criteria* include

- how well the claim fits with all available evidence,
- the sufficiency of the evidence,
- the relevance of the evidence,
- the appropriateness and rigor of the method used to collect the data, and
- the appropriateness and soundness of the method used to analyze the data.

FIGURE 3

The components of a scientific argument and some criteria for evaluating the merits of the argument

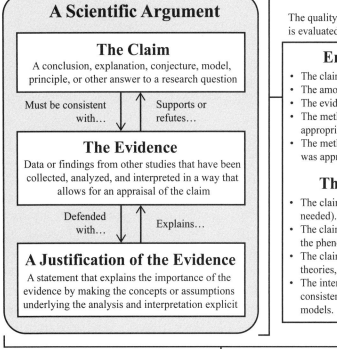

Theoretical criteria refer to standards that are important in science but are not empirical in nature; examples of these criteria are

- the sufficiency of the claim (i.e., Does it include everything needed?);

- the usefulness of the claim (i.e., Does it help us understand the phenomenon we are studying?);

- how consistent the claim is with accepted theories, laws, or models (i.e., Does it fit with our current understanding of the behavior of different Earth systems or what we know about Earth's place in the universe); and

- how consistent the interpretation of the results of the analysis is with accepted theories, laws, or models (i.e., Is the interpretation based on what is known about materials found on Earth, the behavior of the different Earth systems, or how Earth moves through space relative to other objects?).

What counts as quality within these different components, however, varies from discipline to discipline (e.g., geology, meteorology, astronomy, physics, chemistry, biology) and within the specific fields of each discipline (e.g., paleontology, volcanology, semiology, hydrogeology, structural geology). This variation is due to differences in the types of phenomena investigated, what counts as an accepted mode of inquiry (e.g., descriptive studies, experimentation, computer modeling), and the theory-laden nature of scientific inquiry. It is important to keep in mind that "what counts" as a quality argument in science is discipline and field dependent.

To allow for the critique and refinement of the initial arguments during the next stage of ADI, each group of students should create their initial argument in a medium that can easily be viewed by the other groups. We recommend using a 2' × 3' whiteboard. Students should include the guiding question of the lab and the three main components of the argument on the board. Figure 4 shows the general layout for a presentation of an argument, and Figure 5 provides an example of an argument crafted by students. Students can also create their initial arguments using presentation software such as Microsoft's PowerPoint or Apple's Keynote and devote one slide to each component of an argument. The choice of medium is not important as long as students are able to easily modify the content of their argument as they work and it enables others to easily view their argument.

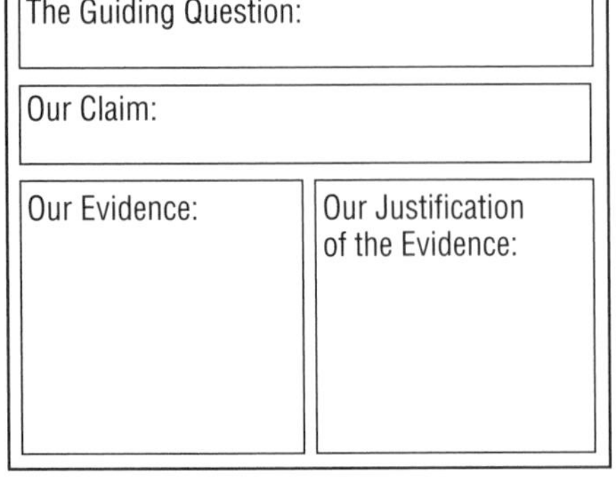

The intention of this stage of the model is to provide the student groups with an opportunity to make sense of what they are seeing or doing during the investigation. As students work together to create an initial argument, they must talk with each other and determine if their analysis is useful or not and how to best interpret the trends, differences, or relationships that they identify as a result of the analysis. They must also decide if the evidence (data that have been analyzed and interpreted) that they chose to include in their argument is relevant, sufficient, and an acceptable way to support their claim. This process, in turn, enables the groups of students to evaluate competing ideas and

FIGURE 4 _____

The components of an argument that should be included on a whiteboard (outline)

The Guiding Question:	
Our Claim:	
Our Evidence:	Our Justification of the Evidence:

FIGURE 5

An example of a student-generated argument on a whiteboard

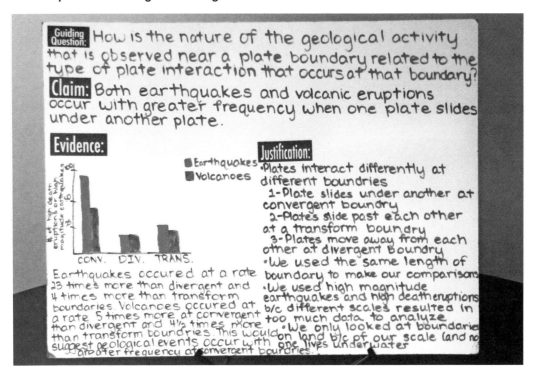

weed out any claim that is inaccurate, does not fit with all the available data, or contains contradictions.

This stage of the model is challenging for students because they are rarely asked to make sense of a phenomenon based on raw data, so it is important for teachers to actively work to support their "sense-making." In this stage, teachers should circulate from group to group to act as a resource person for the students, asking questions urging them to think about what they are doing and why. To help students remember the goal of the activity, you can ask questions such as "What are you trying to figure out?" You can ask them questions such as "Why is that information important?" or "Why is that analysis useful?" to encourage them to think about whether or not the data they are analyzing are relevant or the analysis is informative. To help them remember to use rigorous criteria to evaluate the merits of a tentative claim, you can ask, "Does that fit with all the data?" or "Is that consistent with what we know about plate tectonics?"

It is important to remember that at the beginning of the school year, students will struggle to develop arguments and will often rely on inappropriate criteria such as plausibility (e.g., "That sounds good to me") or fit with personal experience (e.g., "But that is what I saw on TV once") as they attempt to make sense of their data. However, as students learn why it is useful to use evidence in an argument, what makes evidence valid or acceptable in science, and why it is important to justify why they used a particular type of evidence through practice, *students will improve their ability to argue from evidence* (Grooms, Enderle, and Sampson 2015). This is an important principle underlying the ADI instructional model.

Stage 4: Argumentation Session

The fourth stage of ADI is the argumentation session. In this stage, each group is given an opportunity to share, evaluate, and revise their initial arguments by interacting with members from the other groups (see Figure 6). This stage is included in the model for three reasons:

1. Scientific argumentation (i.e., arguing from evidence) is an important practice in science because critique and revision lead to better outcomes.

2. Research indicates that students learn more about the content and develop better critical-thinking skills when they are exposed to alternative ideas, respond to the questions and challenges of other students, and evaluate the merits of competing ideas (Duschl, Schweingruber, and Shouse 2007; NRC 2012).

3. During the argumentation sessions, students learn how to distinguish between ideas using rigorous scientific criteria and are able to develop scientific habits of mind such as treating ideas with initial skepticism, insisting that the reasoning and assumptions be made explicit, and insisting that claims be supported by valid evidence.

This stage, as a result, provides the students with an opportunity to learn from and about scientific argumentation.

It is important to note, however, that supporting and promoting productive interactions between students inside the classroom can be difficult because the practice of arguing from evidence is foreign to most students when they first begin participating in ADI. To aid these interactions, students are required to generate their arguments in a medium that can be seen by others. By looking at whiteboards, paper, or slides, students tend to focus their attention on evaluating evidence and the DCIs or CCs that were used to justify the evidence rather than attacking the source of the ideas. This strategy often makes the discussion more productive and makes it easier for students to identify and weed out faulty ideas. It is also important for the students to view the argumentation session as an opportunity to learn. The teacher, therefore, should describe the argumentation session as an opportunity for students to collaborate with their peers and as a chance to give each other feedback so the quality of all the arguments can be improved, rather than as an opportunity determine who is right or wrong.

To ensure that all students remain engaged during the argumentation session, we recommend that teachers use a modified "gallery walk" format rather than a whole-class presentation format. In the modified gallery walk format, one or two members of each group stay at their workstation to share their group's ideas while the other group members go to different groups one at a time to listen to and critique the arguments developed by their classmates (see Figure 7). This type of format ensures that all ideas are heard and that more students are actively involved in the process. We recommend that the students who are responsible for critiquing

FIGURE 6

A student presents her group's argument to students from other groups during the argumentation session

arguments visit at least three different groups during the argumentation session. We also recommend that the presenters keep a record of the critiques made by their classmates and any suggestions for improvement. The students who are responsible for critiquing the

FIGURE 7

A modified gallery walk format is used during the argumentation session to allow multiple groups to share their arguments at the same time

arguments should also be encouraged to keep a record of good ideas or potential ways to improve their own arguments as they travel from group to group.

Just as is the case in earlier stages of ADI, it is important for the classroom teacher to be involved in (without leading) the discussions during the argumentation session. Once again, the teacher should move from group to group to keep students on task and model good scientific argumentation. The teacher can ask the presenter(s) questions such as "Why did you decide to analyze the available data like that?" or "Were there any data that did not fit with your claim?" to encourage students to use empirical criteria to evaluate the quality of the arguments. The teacher can also ask the presenter(s) to explain how the claim they are presenting fits with the theories, laws, or models of science or to explain why the evidence they used is important. In addition, the teacher can also ask the students who are listening to the presentation questions such as "Do you think their analysis is accurate?" or "Do you think their interpretation is sound?" or even "Do you think their claim fits with what we know about forces and motion?" These questions can serve to remind students to use empirical and theoretical criteria to evaluate an argument during the discussions. Overall, it is the goal of the teacher at this stage of the lesson to encourage students to think about how they know what they know and why some claims are more valid or acceptable in science. This stage of the model, however, is not the time to tell the students that they are right or wrong.

At the end of the argumentation session, it is important to give the students time to meet with their original group so they can discuss what they learned by interacting with individuals from the other groups and revise their initial arguments. This process can begin with the presenters sharing the critiques and the suggestions for improvement that they heard during the modified gallery walk. The students who visited the other groups during the argumentation can then share their ideas for making the arguments better based on what they observed and discussed at other stations. Students often realize that the way they collected or analyzed data was flawed in some way at this point in the process. The teacher should therefore encourage students to collect new data or reanalyze the data they collected as needed. Teachers can also give students time to conduct additional tests of ideas or claims. At the end of this stage, each group should have a final argument that is much better than their initial one.

Stage 5: Explicit and Reflective Discussion

The teacher should lead a whole-class explicit and reflective discussion during stage 5 of ADI. The intent of this discussion is to give students an opportunity to think about and share what they know and how they know it. This stage enables the classroom teacher to ensure that all students understand the DCIs and CCs they used during the investigation. It also encourages students to think about ways to improve their participation in scientific practices such as planning and carrying out investigations, analyzing and interpreting data, and arguing from evidence. At this point in the instructional sequence, the teacher should also encourage students to think about one or two nature of scientific knowledge (NOSK) or nature of scientific inquiry (NOSI) concepts.

It is important to emphasize that an explicit and reflective discussion is not a lecture; it is an opportunity for students to think about important ideas and practices and to share what they know or do not understand. The more students talk during this stage, the more meaningful the experience will be for them and the more a teacher can learn about student thinking.

Teachers should begin the discussion by asking students to share what they know about the DCIs and the CCs they used to figure things out during the lab (the DCIs and CCs can be found in the "Your Task" section of the Lab Handout). The teacher can give several images as prompts and then ask students questions to encourage students to think about how these ideas or concepts helped them explain the phenomenon under investigation and how they used these ideas or concepts to provide a justification of the evidence in their arguments. The teacher should not tell the students what results they should have obtained or what information should be included in each argument. Instead, the teacher should focus on the students' thoughts about the DCIs and CCs by providing a context for students to share their views and explain their thinking. Remember, this stage of ADI is a *discussion,* not a presentation about what the students "should have seen" or "should have learned." We provide recommendations about what teachers can do and the types of questions that teachers can ask to facilitate a productive discussion about the DCIs and CCs during this stage as part of the Teacher Notes for each lab investigation.

Next, the teacher should encourage the students to think about what they learned about the practices of science and how to design better investigations in the future. This is important because students are expected to design their own investigations, decide how to analyze and interpret data, and support their claims with evidence in every ADI lab investigation. These practices are complex, and students cannot be expected to master them without being given opportunities to try, fail, and then learn from their mistakes. To encourage students to learn from their mistakes during a lab, students must have an opportunity to reflect on what went well and what went wrong during their investigation. The teacher should therefore encourage the students to think about what they did during their investigation, how they chose to analyze and interpret data, and how they decided to argue from evidence and what they could have done better. The teacher can then use the students' ideas to highlight what does and does not count as quality or rigor in science and to offer advice about ways to improve in the future. Over time, students will gradually improve their abilities to participate in the practices of science as they learn what works and what does not. To help facilitate this process, in the Teacher Notes for each ADI lab investigation we provide questions that teachers can ask students to help elicit their ideas about the practices of science and set goals for future investigations.

The teacher should end this stage with an explicit discussion of one or two aspects of NOSK or NOSI, using what the students did during the investigation to help illustrate these important concepts (NGSS Lead States 2013). This stage provides a golden opportunity for explicit instruction about NOSK and how this knowledge develops over time in a context that is meaningful to the students. For example, teachers can use the lab as a way to illustrate the differences between

- observations and inferences,
- data and evidence, or
- theories and laws.

Teachers can also use the lab investigation as a way to illustrate NOSI. For example, teachers might discuss

- how the culture of science, societal needs, and current events influence the work of scientists;
- the wide range of methods that scientists can use to collect data;
- what does and does not count as an experiment in science; or
- the role that creativity and imagination play during an investigation.

Research in science education suggests that students only develop an appropriate understanding of the NOSK and NOSI when teachers *explicitly* discuss these specific concepts as part of a lesson (Abd-El-Khalick and Lederman 2000; Lederman and Lederman 2004; Schwartz, Lederman, and Crawford 2004). In addition, by embedding a discussion of NOSK and NOSI into each lab investigation, teachers can highlight these important concepts over and over again throughout the school year rather than just focusing on them during a single unit. This type of approach makes it easier for students to learn these abstract and sometimes counterintuitive concepts. As part of the Teacher Notes for each lab investigation, we provide recommendations about which concepts to focus on and examples of questions that teachers can ask to facilitate a productive discussion about these concepts during this stage of the instructional sequence.

Stage 6: Write an Investigation Report

Stage 6 is included in the ADI model because writing is an important part of doing science. Scientists must be able to read and understand the writing of others as well as evaluate its worth. They also must be able to share the results of their own research through writing. In addition, writing helps students learn how to articulate their thinking in a clear and concise manner, encourages metacognition, and improves student understanding of the content (Wallace, Hand, and Prain 2004). Finally, and perhaps most important, writing makes each student's thinking visible to the teacher (which facilitates assessment) and enables the teacher to provide students with the educative feedback they need to improve.

In stage 6, each student is required to write an individual investigation report using his or her group's argument as a starting point. The report should be centered on three fundamental questions:

1. What question were you trying to answer and why?

2. What did you do to answer your question and why?

3. What is your argument?

Teachers should encourage students to use tables or graphs to help organize their evidence and require them to reference this information in the body of the report.

Stage 6 of ADI is important because it allows students to learn how to construct an explanation, argue from evidence, and communicate information. It also enables students to master the disciplinary-based writing skills outlined in the *Common Core State Standards* in English Language Arts (*CCSS ELA*; NGAC and CCSSO 2010). The report can be written during class or can be assigned as homework.

The format of the report is designed to emphasize the persuasive nature of science writing and to help students learn how to communicate in multiple modes (words, figures, tables, and equations). The three-question format is well aligned with the components of a traditional laboratory report (i.e., introduction, procedure, results and discussion) but allows students to see the important role argument plays in science. We strongly recommend that teachers *limit the length of the investigation report* to two double-spaced pages or one single-spaced page. This limitation encourages students to write in a clear and concise manner, because there is little room for extraneous information. This limitation also makes the assignment less intimidating than a lengthier report requirement, and it lessens the work required in the subsequent stages.

Stage 7: Double-Blind Group Peer Review

During stage 7, each student is required to submit to the teacher one or more copies of his or her investigation report. We recommend that students bring in multiple copies of their report to make it easier for a group of students to review it at the same time; however, this is not a requirement if students are unable to bring in multiple copies of their reports. Instead of reading multiple copies of the same report as they review it, the group of reviewers can simply share a single copy of a report. Students should not place their names on the report before they turn it in to the teacher at the beginning of this stage; instead they should use an identification number to maintain anonymity—to ensure that reviews are based on the ideas presented and not the person presenting the ideas.

We recommend that teachers place students into groups of three to review the reports (these groups can be different from the groups that students worked in during stages 1–4). The teacher should give each group a report written by a single student (or the multiple copies of the report submitted by a single student) and a peer-review guide and teacher scoring rubric (PRG/TSR). Two versions of the PRG/TSR are included in Appendix 5 (p. 587): one version is designed for middle school students, and the other version is designed for high school students.

The students in each group are asked to review the report (or copies of the report) as a team using the PRG/TSR (see Figure 8). The PRG/TSR contains specific criteria that are to be used by the group as they evaluate the quality of each section of the investigation report as well as quality of the writing. There is also space for the reviewers to provide the author with feedback about how to improve the report. Once a group finishes reviewing a report as a team, they are given another report to review. When students are grouped together in threes, they only need to review three different reports. Be sure to give students only 15 minutes to review each set of reports (we recommend setting a timer to help manage time). When students are grouped into three and given 15 minutes to complete each review, the entire peer-review process can be completed in one 50-minute class period (3 different reports × 15 minutes = 45 minutes).

Reviewing each report as a group using the PRG/TSR is an important component of the peer-review process because it provides students with a forum to discuss "what counts" as high quality or acceptable and, in so doing, forces them to reach a consensus during the process. This method also helps prevent students from checking off "yes" for each criterion on the PRG/TSR without thorough consideration of the merits of the paper. It is also important for students to provide constructive and specific feedback to the author when areas of the paper are found to not meet the standards established by the PRG/TSR. The peer-review process provides students with an opportunity to read good and bad examples of the reports. This helps the students learn new ways to organize and present information, which in turn will help them write better on subsequent reports.

This stage of the model also gives students more opportunities to develop reading skills that are needed to be successful in science. Students must be able to determine the central

FIGURE 8

A group of students review a report written by a classmate using the peer-review guide and teacher scoring rubric

ideas or conclusions of a text and determine the meaning of symbols, key terms, and other domain-specific words. In addition, students must be able to assess the reasoning and evidence that an author includes in a text to support his or her claim and compare or contrast findings presented in a text with those from other sources when they read a scientific text. Students can develop all these skills, as well as the other discipline-based reading standards found in the *CCSS ELA*, when they are required to read and critically review reports written by their classmates.

Stage 8: Revise and Submit the Report

The final stage in the ADI instructional model is to revise the report based on the suggestions given during the peer review. If the report met all the criteria, the student may simply submit the paper to the teacher with the original peer-reviewed "rough draft" and PRG/TSR attached, ensuring that his or her name replaces the identification number. Students whose reports are found by the peer-review group to be acceptable can maintain the option to revise it if they so desire after reviewing the work of other students. If a report was found to be unacceptable by the reviewers during the peer-review stage, the author is required to rewrite his or her report using the reviewers' comments and suggestions as a guideline. The author is also required to explain what he or she did to improve each section of the report in response to the reviewers' suggestions (or explain why he or she decided to ignore the reviewers' suggestions) in the author response section of the PRG/TSR.

Once the report is revised, it is turned in to the teacher for evaluation with the original rough draft and the PRG/TSR attached. The teacher can then provide a score on the PRG/TSR in the column labeled "Teacher Score" and use these ratings to assign an overall grade for the report. This approach provides students with a chance to improve their writing mechanics and develop their reasoning and understanding of the content. This process also offers students the added benefit of reducing academic pressure by providing support in obtaining the highest possible grade for their final product.

The PRG/TSR is designed to be used with any ADI lab investigation, thus allowing teachers to use the same scoring rubric throughout the entire year. This is beneficial for several reasons. First, the criteria for what counts as a high-quality report do not change from lab to lab. Students therefore quickly learn what is expected from them when they write a report, and teachers do not have to spend valuable class time explaining the various components of the PRG/TSR each time they assign a report. Second, the PRG/TSR makes it clear which components of a report need to be improved next time, because the grade is not based on a holistic evaluation of the report. Students, as a result, can see which aspects of their writing are strong and which aspects need improvement. Finally, and perhaps most important, the PRG/TSR provides teachers with a standardized measure of student performance that can be compared over multiple reports across semesters, thus allowing teachers to track improvement over time.

The Role of the Teacher During Argument-Driven Inquiry

If the ADI instructional model is to be successful and student learning is to be optimized, the role of the teacher during a lab activity designed using this model must be different from the teacher's role during a more traditional lab. The teacher *must* act as a resource for the students, rather than as a director, as students work through each stage of the activity; the teacher must encourage students to think about *what they are doing* and *why they made that decision* throughout the process. This encouragement should take the form of probing questions that teachers ask as they walk around the classroom, such as "Why do you want to set up your equipment that way?" or "What type of data will you need to collect to be able to answer that question?"

Teachers must restrain themselves from telling or showing students how to "properly" conduct the investigation. However, teachers must emphasize the need to maintain high standards for a scientific investigation by requiring students to use rigorous standards for "what counts" as a good method or a strong argument in the context of science.

Finally, and perhaps most important for the success of an ADI activity, teachers must be willing to let students try and fail, and then help them learn from their mistakes. Teachers should not try to make the lab investigations included in this book "student-proof" by providing additional directions to ensure that students do everything right the first time. We have found that students often learn more from an ADI lab activity when they design a flawed method to collect data during stage 2 or analyze their results in an inappropriate manner during stage 3, because their classmates quickly point out these mistakes during the argumentation session (stage 4) and it leads to more teachable moments.

Because the teacher's role in an ADI lab is different from what typically happens in lab, we've provided a chart describing teacher behaviors that are consistent and inconsistent with each stage of the instructional model (see Table 1). This table is organized by stage because what the students and the teacher need to accomplish during each stage is different. It might be helpful to keep this table handy as a guide when you are first attempting to implement the lab activities found in the book.

TABLE 1

Teacher behaviors during the stages of the ADI instructional model

Stage	What the teacher does that is ...	
	Consistent with ADI model	**Inconsistent with ADI model**
1: Identify the task and the guiding question	• Sparks students' curiosity • "Creates a need" for students to design and carry out an investigation • Organizes students into collaborative groups of three or four • Supplies students with the materials they will need • Holds a "tool talk" to show students how to use equipment or to illustrate proper technique • Reviews relevant safety precautions and protocols • Provides students with hints • Allows students to tinker with the equipment they will be using later	• Provides students with possible answers to the research question • Tells students that there is one correct answer • Provides a list of vocabulary terms or explicitly describes the content addressed in the lab
2: Design a method and collect data	• Encourages students to ask questions as they design their investigations • Asks groups questions about their method (e.g., "Why did you do it this way?") and the type of data they expect from that design • Reminds students of the importance of specificity when completing their investigation proposals	• Gives students a procedure to follow • Does not question students about the method they design or the type of data they expect to collect • Approves vague or incomplete investigation proposals
3: Develop an initial argument	• Reminds students of the research question and what counts as appropriate evidence in science • Requires students to generate an argument that provides and supports a claim with genuine evidence • Asks students what opposing ideas or rebuttals they might anticipate • Provides related theories and reference materials as tools	• Requires only one student to be prepared to discuss the argument • Moves to groups to check on progress without asking students questions about why they are doing what they are doing • Does not interact with students (uses the time to catch up on other responsibilities) • Tells students the right answer
4: Argumentation session	• Reminds students of appropriate behaviors in the learning community • Encourages students to ask questions of peers • Keeps the discussion focused on the elements of the argument • Encourages students to use appropriate criteria for determining what does and does not count	• Allows students to negatively respond to others • Asks questions about students' claims before other students can ask • Allows students to discuss ideas that are not supported by evidence • Allows students to use inappropriate criteria for determining what does and does not count

Continued

TABLE 1 (*continued*)

Stage	What the teacher does that is ...	
	Consistent with ADI model	**Inconsistent with ADI model**
5: Explicit and reflective discussion	• Encourages students to discuss what they learned about the content and how they know what they know • Encourages students to discuss what they learned about the nature of scientific knowledge and the nature of scientific inquiry • Encourages students to think of ways to be more productive next time	• Provides a lecture on the content • Skips over the discussion about the nature of scientific knowledge and the nature of scientific inquiry to save time • Tells students "what they should have learned" or "this is what you all should have figured out"
6: Write an investigation report	• Reminds students about the audience, topic, and purpose of the report • Provides the peer-review guide in advance • Provides an example of a good report and an example of a bad report	• Has students write only a portion of the report • Moves on to the next activity/topic without providing feedback
7: Double-blind group peer review	• Reminds students of appropriate behaviors for the review process • Ensures that all groups are giving a quality and fair peer review to the best of their ability • Encourages students to remember that while grammar and punctuation are important, the main goal is an acceptable scientific claim with supporting evidence and justification • Holds the reviewers accountable	• Allows students to make critical comments about the author (e.g., "This person is stupid") rather than their work (e.g., "This claim needs to be supported by evidence") • Allows students to just check off "Yes" on each item without providing a critical evaluation of the report
8: Revise and submit the report	• Requires students to edit their reports based on the reviewers' comments • Requires students to respond to the reviewers' ratings and comments • Has students complete the Checkout Questions after they have turned in their report	• Allows students to turn in a report without a completed peer-review guide • Allows students to turn in a report without revising it first

References

Abd-El-Khalick, F., and N. G. Lederman. 2000. Improving science teachers' conceptions of nature of science: A critical review of the literature. *International Journal of Science Education* 22 (7): 665–701.

Duschl, R. A., H. A. Schweingruber, and A. W. Shouse, eds. 2007. *Taking science to school: Learning and teaching science in grades K–8*. Washington, DC: National Academies Press.

Lederman, N. G., and J. S. Lederman. 2004. Revising instruction to teach the nature of science. *The Science Teacher* 71 (9): 36–39.

National Governors Association Center for Best Practices and Council of Chief State School Officers (NGAC and CCSSO). 2010. *Common core state standards.* Washington, DC: NGAC and CCSSO.

National Research Council (NRC). 2012. *A framework for K–12 science education: Practices, crosscutting concepts, and core ideas*. Washington, DC: National Academies Press.

NGSS Lead States. 2013. *Next Generation Science Standards: For states, by states.* Washington, DC: National Academies Press. *www.nextgenscience.org/next-generation-science-standards*.

Schwartz, R. S., N. Lederman, and B. Crawford. 2004. Developing views of nature of science in an authentic context: An explicit approach to bridging the gap between nature of science and scientific inquiry. *Science Education* 88 (4): 610–645.

Wallace, C., B. Hand, and V. Prain, eds. 2004. *Writing and learning in the science classroom*. Boston: Kluwer Academic Publishers.

CHAPTER 2
Lab Investigations

This book includes 23 Earth and space science lab investigations designed around the argument-driven inquiry (ADI) instructional model. Please note that these investigations are not designed to replace an existing curriculum, but as a way to change the nature of the labs that are included in the curriculum. These investigations are designed to function as stand-alone lessons, which gives teachers the flexibility they need to decide which ones to use and when to use them during the academic year. We do not expect teachers to use every lab included in this book. We do, however, recommend that teachers attempt to incorporate between 8 and 12 of these labs into their science curriculum to give students an opportunity to learn how to use disciplinary core ideas (DCIs), crosscutting concepts (CCs), and science and engineering practices (SEPs) to figure things out over time.

A teacher can use these investigations as a way to introduce students to a new concept related to a DCI at the beginning of a unit (introduction labs) or as a way to give students an opportunity to apply a specific concept related to a DCI that they learned about earlier in class in a novel situation (application labs). All of the labs, however, were designed to give students an opportunity to learn how to use at least one DCI and multiple CCs and SEPs to figure out how or why things happen during each investigation. Each lab is labeled as being either an introduction lab or an application lab, and labs are grouped together into sections by the middle school and high school DCIs found in the *NGSS*. The different sections included in the book (such as Section 2: Space Systems or Section 4: Earth's Systems) can be integrated into a curriculum in any order, but teachers should not assign an application lab before an introduction lab from the same section if the two labs focus on the same DCI (such Labs 6 and 7 in Section 3).

The 23 lab investigations have been aligned with the following sources to facilitate curriculum and lesson planning:

- *A Framework for K–12 Science Education* (see Standards Matrix A in Appendix 1);
- Aspects of the nature of scientific knowledge (NOSK) and the nature of scientific inquiry (NOSI) (see Standards Matrix B in Appendix 1);
- The *Next Generation Science Standards* (see Standards Matrix C in Appendix 1);
- The *Common Core State Standards* in English language arts (*CCSS ELA*; see Standards Matrix D in Appendix 1)

We wrote all the investigations included in this book to align with a specific performance expectation found in the *NGSS*. Many of the ideas for the investigations in this book came from existing resources; however, we modified these existing activities to target the DCI and CC found within a specific performance expectation and to fit with the ADI instructional model. Once we finished writing the labs, we had them reviewed for content accuracy. Several different middle school science teachers then piloted the labs in their courses (including general and honors sections). We then revised each investigation based on their feedback. The revised version of each lab is included in this book.

Research that has been conducted on ADI in classrooms indicates that students have much better inquiry and writing skills after participating in at least eight ADI investigations over the course of an academic year and that they make substantial gains in their understanding of DCIs, CCs, SEPs, the NOSK, and the NOSI (Grooms, Enderle, and Sampson 2015; Sampson et al. 2013; Strimaitis et al. 2017). To learn more about the research on the ADI instructional model, visit *www.argumentdriveninquiry.com*.

Teacher Notes

Each teacher must decide when and how to use a laboratory experience to best support student learning. To help with this decision making, we have included Teacher Notes for each investigation. These notes include information about the purpose of the lab, the time needed to implement each stage of the model for that lab, the materials needed, and hints for implementation. We have also included a "Connections to Standards" section showing how each ADI lab activity is aligned with the *NGSS* performance expectations and the *CCSS ELA*. In the sections that follow, we will describe the information provided in each section of the Teacher Notes.

Purpose

This section describes the content of the lab and indicates whether the activity is designed to help students think about a new idea or think with a new idea. Labs that are designed to help students *think about* a new idea are called introduction labs. Introduction labs require students to explore potential cause-and-effect relationships or how things change over time. These labs are best used at the beginning of a unit of study. Labs that are designed to help students learn to *think with* a new idea are called application labs. Application labs require students to use an idea they are already familiar with to develop an explanation or to solve a problem. These labs are best used at the end of the unit of study.

Please note that because of the nature of the ADI approach, in both introduction labs and application labs very *little* emphasis needs to be placed on making sure the students "learn the vocabulary first" or "know the content" before the lab investigation begins. Instead, with the combination of the information provided in the Lab Handout and your students' evolving understanding of the DCIs, CCs, and SEPs, they will develop a better understanding of the content *as they work through the eight stages of ADI*. The "Purpose" section also highlights the NOSK or NOSI concepts that should be emphasized during the explicit and reflective discussion stage of the activity.

Important Earth and Space Science Content

This section of the Teacher Notes provides a basic overview of the major concepts that the students will explore and or use during the investigation.

Timeline

Unlike most traditional labs, ADI labs typically take four or five days to complete. The amount of time it will take to complete each lab will vary depending on how long it takes to collect data and whether or not the students write in class or at home. The time associated with each ADI lab investigation may be longer in the first few labs your students conduct, but the time will be reduced as your students become familiar with the practices used in the model (argumentation, designing investigations, writing reports). We therefore provide suggestions about which stages of ADI you should be able to complete in a 50-minute class period (see Appendix 3 [p. 573]).

Materials and Preparation

This section describes the lab supplies (i.e., consumables and equipment) and instructional materials (e.g., Lab Handout, Investigation Proposal, and peer-review guide and teacher scoring rubric [PRG/TSR]) needed to implement the lab activity. The lab supplies listed are designed for one group; however, multiple groups can share if resources are scarce. We have also included specific suggestions for some lab supplies, based on our finding that these supplies worked best during the field tests. However, if needed, substitutions can be made. Always be sure to test all lab supplies before conducting the lab with the students, because using new materials often has unexpected consequences.

This section also describes the setup that needs to be done *before* students can do the investigation. Please note that some of the labs may require some preparation up to 24 hours in advance. Make sure to read this section at least two days before you plan to have the students in your class start the investigation.

Safety Precautions and Laboratory Waste Disposal

This section provides an overview of potential safety hazards as well as safety protocols that should be followed to make the laboratory safer for students. These are based on legal safety standards and current better professional safety practices. Teachers should also review and follow all local policies and protocols used within their school district and/or school (e.g., the district chemical hygiene plan, Board of Education safety policies).

Topics for the Explicit and Reflective Discussion

This section begins with an overview of some the DCIs and CCs that students use to figure things out during the lab. We provide advice about ways to encourage students to think about how these ideas or concepts helped them explain the phenomenon under investigation and how they used these ideas or concepts to provide a justification of the evidence in their arguments. The section also provides some advice for teachers about how to encourage students to reflect on ways to improve the design of their investigation in the future. This section concludes with an overview of the relevant NOSK and NOSI concepts to discuss during the explicit and reflective discussion and some sample questions that teachers can pose to help students be reflective about what they know about these concepts.

Hints for Implementing the Lab

Many teachers have tested these labs many different times. As a result, we have collected hints from the teachers for each stage of the ADI process. These hints are designed to help you avoid some of the pitfalls earlier teachers have experienced and make the investigation run smoothly. The section also includes tips for making the investigation safer.

Connections to Standards

This section is designed to inform curriculum and lesson planning by highlighting how the investigation can be used to address specific performance expectations from the *NGSS* and *CCSS ELA*.

Instructional Materials

The instructional materials included in this book are reproducible copy masters that are designed to support students as they participate in an ADI lab activity. The materials needed for each lab include a Lab Handout, the PRG/TSR, and a set of Checkout Questions. Some labs also require an investigation proposal and supplementary materials.

Lab Handout

At the beginning of each lab activity, each student should be given a copy of the Lab Handout. This handout provides information about the phenomenon that they will investigate and a guiding question for the students to answer. The handout also provides hints for students to help them design their investigation in the "Getting Started" section, information about what to include in their initial argument, and the requirements for the investigation report. The last part of the Lab Handout provides space for students to keep track of critiques, suggestions for improvement, and good ideas that arise during the argumentation session.

Peer-Review Guide and Teacher Scoring Rubric

The PRG/TSR is designed to make the criteria that are used to judge the quality of an investigation report explicit. Appendix 5 (p. 587) includes two versions of the PRG/TSR: a middle school version and a high school version. We recommend that teachers make one copy of the appropriate version for each student and provide it to the students before they begin writing their investigation report. This will ensure that students understand how they will be evaluated. Then during the double-blind group peer-review stage of the model, each group should fill out the peer-review guide as they review the reports of their classmates. (Each group will need to review at least three reports.) The reviewers should rate the report on each criterion and then provide advice to the author about ways to improve. Once the review is complete, the author needs to revise his or her report and respond to the reviewers' rating and comments in the appropriate sections in the PRG/TSR.

The PRG/TSR should be submitted to the instructor along with the first draft and the final report for a final evaluation. To score the report, the teacher can simply fill out the "Teacher Score" column of the PRG/TSR and then total the scores.

Checkout Questions

To facilitate formative assessment inside the classroom, we have included a set of Checkout Questions for each lab investigation. The questions target the key ideas and the NOSK and NOSI concepts that are addressed in the lab. Students should complete the Checkout Questions on the same day they turn in their final report. One handout is needed for each student. The students should complete these questions on their own. The teacher can use the students' responses, along with the report, to determine if the students learned what they needed to during the lab, and then reteach as needed.

Investigation Proposal

To help students design better investigations, we have developed and included three different types of investigation proposals in this book, with a short version and a long version of each type (see Appendix 4 [p. 577]). These investigation proposals are optional, but we have found that students design and carry out much better investigations when they are required to fill out a proposal and then get teacher feedback about their method before they begin. We provide recommendations about which investigation proposal (A, B, or C) to use for a particular lab as part of the Teacher Notes. If a teacher decides to use an investigation proposal as part of a lab activity, we recommend providing one copy for each group. The Lab Handout for students also has a heading labeled "Investigation Proposal Required?" that is followed by "Yes" and "No" check boxes. The teacher should be sure to have students check the appropriate box on the Lab Handout when introducing the lab activity.

Supplementary Materials

Some lab activities include supplementary materials such as data files or images. Students will need to be able to use these materials during their investigation. These materials can be downloaded from the book's Extras page at *www.nsta.org/adi-ess*.

References

Grooms, J., P. Enderle, and V. Sampson. 2015. Coordinating scientific argumentation and the *Next Generation Science Standards* through argument driven inquiry. *Science Educator* 24 (1): 45–50.

National Governors Association Center for Best Practices and Council of Chief State School Officers (NGAC and CCSSO). 2010. *Common core state standards*. Washington, DC: NGAC and CCSSO.

National Research Council (NRC). 2012. *A framework for K–12 science education: Practices, crosscutting concepts, and core ideas*. Washington, DC: National Academies Press.

NGSS Lead States. 2013. *Next Generation Science Standards: For states, by states*. Washington, DC: National Academies Press. *www.nextgenscience.org/next-generation-science-standards*.

Sampson, V., P. Enderle, J. Grooms, and S. Witte. 2013. Writing to learn and learning to write during the school science laboratory: Helping middle and high school students develop argumentative writing skills as they learn core ideas. *Science Education* 97 (5): 643–670.

Strimaitis, A., S. Southerland, V. Sampson, P. Enderle, and J. Grooms. 2017. Promoting equitable biology lab instruction by engaging all students in a broad range of science practices: An exploratory study. *School Science and Mathematics* 117 (3–4): 92–103.

SECTION 2
Space Systems

Introduction Labs

Teacher Notes

Lab 1. Moon Phases: Why Does the Appearance of the Moon Change Over Time in a Predictable Pattern?

Purpose

The purpose of this lab is to *introduce* students to the disciplinary core ideas (DCIs) of (a) The Universe and Its Stars and (b) Earth and the Solar System by having them develop a model that explains the underlying cause of the lunar cycle. In addition, students have an opportunity to learn about the crosscutting concepts (CCs) of (a) Patterns and (b) Systems and System Models. During the explicit and reflective discussion, students will also learn about (a) the use of models as tools for reasoning about natural phenomena and (b) how scientists use different methods to answer different types of questions.

Important Earth and Space Science Content

The appearance of the Moon phases is due to how the Sun illuminates the Moon and how the positions of the Earth, Moon, and Sun relative to each other change over time. The Moon is *always* half illuminated by sunlight, and the side that is illuminated is *always* facing the Sun. The amount of the illuminated side of the Moon *that is visible from Earth,* however,

changes over time as the Moon orbits Earth. The Moon takes 27.3 days to orbit Earth. A full lunar cycle (from new Moon to new Moon), however, takes 29.5 days to complete. The lunar cycle takes more time to complete because Earth travels about 45 million miles in its orbit around the Sun during the time it takes the Moon to completes one orbit around Earth. The Moon, as a result, takes an additional 2.2 days to reach the same relative position.

Figure 1.1, which is not drawn to scale, shows the relative position of the Earth, Moon, and Sun as viewed from high above Earth. The inner circle around the diagram of Earth shows the half-illuminated moon *as viewed from high above the Earth.* The outer ring of Moon pictures shows what the Moon

FIGURE 1.1

The location of the Moon in relation to Earth and the Sun during each major Moon phase

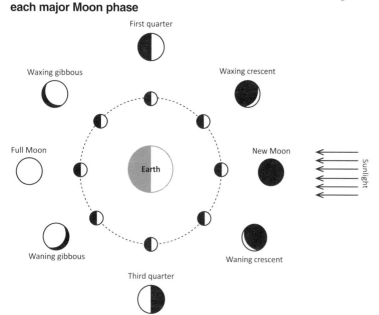

looks like *as viewed from the surface of Earth* at that point in time. At a full Moon, the Earth, Moon, and Sun are in approximate alignment. The Earth, Moon, and Sun are also in approximate alignment during the new Moon. During a new Moon, however, the Moon is between the Sun and Earth, so the entire sunlit part of the Moon is facing away from us. The illuminated portion is entirely hidden from view. The first quarter Moon and third quarter Moon happen when the Moon is at a 90-degree angle with respect to Earth and the Sun.

The Moon appears to rise and set in the sky because Earth rotates on its axis in a counterclockwise direction. The Moon rises in the east and sets in the west because Earth spins toward the east. As Earth rotates, it carries a person standing on the surface of Earth eastward as it turns, so whatever lies beyond that eastern horizon eventually comes up over the horizon and the observer will see it. The point at which the Moon becomes visible to an observer on Earth is dependent on the Moon's position in its own orbit. We can assume that when an observer on Earth is directly in line with the Sun, the Sun appears to be at its highest in the sky, and the time is solar noon. Using this point in time as a reference, we can estimate the times that the Moon would appear to rise and set. A new Moon will rise with the Sun at dawn and set with the Sun at sunset. A full Moon, in contrast, would appear to rise at sunset and set at dawn. These exact rise and set times change based on an observer's location on Earth and the current season because Earth's tilted axis causes an unequal distribution of light across the globe.

Solar eclipses occur when the Moon moves directly in front of the Sun. Lunar eclipses occur when Earth is between the Sun and the Moon. Solar and lunar eclipses are rare because the Moon orbits Earth at an angle that is tilted to the ecliptic by about 5 degrees. The *ecliptic* is the path the Sun takes as it moves through the sky as seen from Earth. This means that on each orbit around Earth, the Moon only crosses the ecliptic twice and these two instances are the only two opportunities for an either a solar or a lunar eclipse to happen. The Moon, as a result, usually crosses the ecliptic when it is not in alignment with the Sun and Earth.

Timeline

The instructional time needed to complete this lab investigation is 270–330 minutes. Appendix 3 (p. 573) provides options for implementing this lab investigation over several class periods. Option G (330 minutes) should be used if students are unfamiliar with scientific writing, because this option provides extra instructional time for scaffolding the writing process. You can scaffold the writing process by modeling, providing examples, and providing hints as students write each section of the report. Option H (270 minutes) should be used if students are familiar with scientific writing and have developed the skills needed to write an investigation report on their own. In option H, students complete stage 6 (writing the investigation report) and stage 8 (revising the investigation report) as homework.

Materials and Preparation

The materials needed to implement this investigation are listed in Table 1.1. Most of the equipment can be purchased from a big-box retail store such as Wal-Mart or Target or through an online retailer such as Amazon. The wood blocks, dowels and foam balls can be purchased from a craft store such as Michaels or Hobby Lobby. Moon phase calendars for June 2016 (A) and July 2016 (B) can be downloaded from the book's Extras page at *www.nsta.org/adi-ess*. However, you can use different Moon phase calendars if you prefer.

TABLE 1.1

Materials list for Lab 1

Item	Quantity
Indirectly vented chemical-splash goggles	1 per student
Physical model of Earth (large foam ball on a stand)	1 per group
Physical model of the Moon A (small foam ball on a stick)	1 per group
Physical model of the Moon B (small foam ball on stand)	2 per group
Light source (lamp with lightbulb)	1 per group
Moon phase calendar A	1 per group
Moon phase calendar B	1 per group
Investigation Proposal A (optional)	1 per group
Whiteboard, 2' × 3'*	1 per group
Lab Handout	1 per student
Peer-review guide and teacher scoring rubric	1 per student
Checkout Questions	1 per student

*As an alternative, students can use computer and presentation software such as Microsoft PowerPoint or Apple Keynote to create their arguments.

Before the lab begins, you need to make the physical models of Earth and the Moon for students to use; directions for making the physical models are as follows:

Physical model of Earth (large foam ball on a stand—see Figure 1.2; one per group)

- Drill a hole in a wood block. The hole should be the same diameter as the dowel. Do not drill all the way through the block.
- Add some glue to the hole.
- Place a 4-inch piece of dowel in the hole. Set aside and let the glue dry.
- Drill a hole all the way through the center of a large foam ball. The diameter of the hole should be slightly larger than the dowel.

- Insert a thumbtack into one side of the ball. The thumbtack will represent an observer standing on Earth.
- Use a marker to label the directions around the thumbtack. This will enable students to see the direction that the "observer" is looking when they use the model.
- Place the foam ball onto the dowel. The foam ball should be able to spin.

FIGURE 1.2 _____

How to assemble the physical models of Earth (large foam ball on a stand)

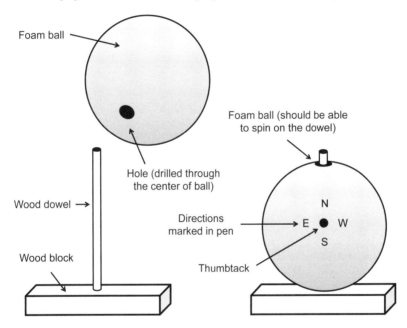

Physical model of the Moon A (small foam ball on a stick—see Figure 1.3 (p. 36); two per group)

- Drill a hole into a small foam ball but not all the way through the ball. The hole should be the same diameter as the wood dowel.
- Insert a 12-inch piece of dowel into the hole.
- Add glue around the edge of the hole. Set aside to dry.
- Use a marker to label "X" on one side of the small foam ball.

Physical model of the Moon B (small foam ball on a stand—see Figure 1.3; one per group)

- Glue a small foam ball to a wood block.
- Use a marker to label "X" on one side of the small foam ball.

LAB 1

FIGURE 1.3

How to assemble the physical models of moon A (small foam ball on a stick) and moon B (small foam ball on a stand)

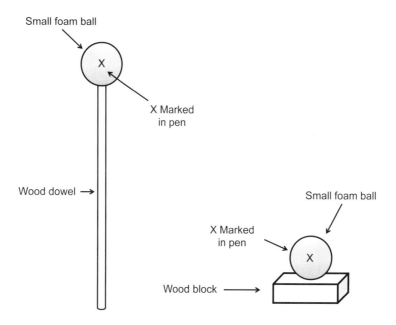

   Be sure to use a set routine for distributing and collecting the materials during the lab investigation. One option is to set up the materials for each group at each group's lab station before class begins. This option works well when there is a dedicated section of the classroom for lab work and the materials are large and difficult to move (such as a stream table). A second option is to have all the materials on a table or cart at a central location. You can then assign a member of each group to be the "materials manager." This individual is responsible for collecting all the materials his or her group needs from the table or a cart during class and for returning all the materials at the end of the class. This option works well when the materials are small and easy to move (such as stopwatches, meter sticks, or thermometers). It also makes it easy to inventory the materials at the end of the class before students leave for the day.

Safety Precautions

Remind students to follow all normal lab safety rules. In addition, tell them to take the following safety precautions:

- Wear sanitized indirectly vented chemical-splash goggles throughout the entire investigation (which includes setup and cleanup).

- Use only GFCI-protected electrical receptacles for lamps to prevent or reduce risk of shock.

- Handle the lamps with care; they can get hot when left on for long periods of time.

- Wash hands with soap and water when done collecting the data and after completing the lab.

Topics for the Explicit and Reflective Discussion

Reflecting on the Use of Core Ideas and Crosscutting Concepts During the Investigation

Teachers should begin the explicit and reflective discussion by asking students to discuss what they know about the DCIs they used during the investigation. The following are some important concepts related to the DCIs of (a) The Universe and Its Stars and (b) Earth and the Solar System that students need in order to develop a conceptual model that can be used to explain the phases of the Moon:

- Planets and dwarf planets orbit the Sun in a counterclockwise direction.

- Moons orbit planets.

- Earth spins in a counterclockwise direction.

- The Moon's orbital plane is not the same as Earth's.

- The Moon does not emit its own light; it only reflects light from the Sun.

To help students reflect on what they know about these concepts, we recommend showing them two or three images using presentation software that help illustrate these important ideas. You can then ask the students the following questions to encourage students to share how they are thinking about these important concepts:

1. What do we see going on in this image?

2. Does anyone have anything else to add?

3. What might be going on that we can't see?

4. What are some things that we are not sure about here?

You can then encourage students to think about how CCs played a role in their investigation. There are at least two CCs that students need to use to develop a conceptual model that can be used to explain the phases of the Moon: (1) Patterns and (2) Systems and System Models (see Appendix 2 [p. 569] for a brief description of these CCs). To help students reflect on what they know about these CCs, we recommend asking the following questions:

1. Why is it important to look for, and attempt to explain, patterns in nature during an investigation?

2. What patterns did you identify during your investigation? What did the identification of these patterns allow you to do?

3. Why do scientists often define a system and then develop a model of it as part of an investigation?

4. How did you use a model to understand why we see the phases of the Moon in a predictable pattern? Why was that useful?

You can then encourage the students to think about how they used all these different concepts to help answer the guiding question and why it is important to use these ideas to help justify their evidence for their final arguments. Be sure to remind your students to explain why they included the evidence in their arguments and make the assumptions underlying their analysis and interpretation of the data explicit in order to provide an adequate justification of their evidence.

Reflecting on Ways to Design Better Investigations

It is important for students to reflect on the strengths and weaknesses of the investigation they designed during the explicit and reflective discussion. Students should therefore be encouraged to discuss ways to eliminate potential flaws, measurement errors, or sources of uncertainty in their investigations. To help students be more reflective about the design of their investigation and what they can do to make their investigations more rigorous in the future, you can ask the following questions:

1. What were some of the strengths of the way you planned and carried out your investigation? In other words, what made it scientific?

2. What were some of the weaknesses of the way you planned and carried out your investigation? In other words, what made it less scientific?

3. What rules can we make, as a class, to ensure that our next investigation is more scientific?

Reflecting on the Nature of Scientific Knowledge and Scientific Inquiry

This investigation can be used to illustrate two important concepts related to the nature of scientific knowledge and the nature of scientific inquiry: (a) the use of models as tools for reasoning about natural phenomena and (b) how scientists use different methods to answer different types of questions (see Appendix 2 [p. 569] for a brief description of these two concepts). Be sure to review these concepts during and at the end of the explicit and reflective discussion. To help students think about these concepts in relation to what they did during the lab, you can ask the following questions:

1. I asked you to develop a model of the Earth-Sun-Moon system as part of your investigation. Why is it useful to develop models in science?

2. Can you work with your group to come up with a rule that you can use to decide what a model is and what a model is not in science? Be ready to share in a few minutes.

3. There is no universal step-by step scientific method that all scientists follow. Why do you think there is no universal scientific method?

4. Think about what you did during this investigation. How would you describe the method you used to develop your model? Why would you call it that?

You can also use presentation software or other techniques to encourage students to think about these concepts. You can show examples and non-examples of scientific models and ask students to classify each one and explain their thinking. You can also show one or more images of a "universal scientific method" that misrepresent the nature of scientific inquiry (see, e.g., *https://commons.wikimedia.org/wiki/File:The_Scientific_Method_as_an_Ongoing_Process.svg*) and ask students why each image is *not* a good representation of what scientists do to develop scientific knowledge. You can also ask students to suggest revisions to the image that would make it more consistent with the way scientists develop scientific knowledge.

Remind your students that, to be proficient in science, it is important that they understand what counts as scientific knowledge and how that knowledge develops over time.

Hints for Implementing the Lab

- Be sure to turn off the overhead lights in the classroom and have students set up the lamps so they are facing away from the center of the room. This will reduce light pollution and make it easier for students to see the different phases of the Moon as they work with their physical models.

- Students should only be given Moon phase calendar A when they are developing their conceptual model. Do not give the students Moon phase calendar B until they are ready to test their conceptual model. If you give students both calendars at the same time, the students will use both calendars to develop their models and then will have no way to test their model using new data.

- Many students have alternative conceptions about Moon phases and the Moon's orbit that will affect their thinking during the lab or how they go about attempting to answer the guiding question. For example, many students believe that Moon phases are the result of Earth's casting a shadow on the Moon. Many of the students in your class, as a result, will think Earth must be at an angle that will produce a shadow on the Moon.

- The claim in the students' argument should be a conceptual model. This model should show how the components of the Earth-Sun- Moon system move over time in relationship to each other. The evidence should include information about how well the model is able to predict the Moon phases, moonrise times, and moonset times included in Moon phase calendar B. The justification of the evidence should include their underlying assumptions about the components of the Earth-Sun-Moon system (e.g., moons orbit planets, Earth spins in a counterclockwise direction, the Moon does not emit its own light)

- Students often make mistakes when developing their conceptual models and/or initial arguments, but they should quickly realize these mistakes during the argumentation session. Be sure to allow students to revise their models and arguments at the end of the argumentation session. The explicit and reflective discussion will also give students an opportunity to reflect on and identify ways to improve how they develop and test models. This also offers an opportunity to discuss what scientists do when they realize a mistake is made.

Connections to Standards

Table 1.2 highlights how the investigation can be used to address specific (a) performance expectations from the *NGSS* and (b) *Common Core State Standards* in English language arts (*CCSS ELA*).

TABLE 1.2 _____

Lab 1 alignment with standards

***NGSS* performance expectation**	Space systems • MS-ESS1-1: Develop and use a model of the Earth-Sun-Moon system to describe the cyclic patterns of lunar phases, eclipses of the Sun and Moon, and seasons.
***CCSS ELA*—Reading in Science and Technical Subjects**	Key ideas and details • CCSS.ELA-LITERACY.RST.6-8.1: Cite specific textual evidence to support analysis of science and technical texts. • CCSS.ELA-LITERACY.RST.6-8.2: Determine the central ideas or conclusions of a text; provide an accurate summary of the text distinct from prior knowledge or opinions. Craft and structure • CCSS.ELA-LITERACY.RST.6-8.4: Determine the meaning of symbols, key terms, and other domain-specific words and phrases as they are used in a specific scientific or technical context relevant to *grade 6–8 texts and topics*. • CCSS.ELA-LITERACY.RST.6-8.5: Analyze the structure an author uses to organize a text, including how the major sections contribute to the whole and to an understanding of the topic. • CCSS.ELA-LITERACY.RST.6-8.6: Analyze the author's purpose in providing an explanation, describing a procedure, or discussing an experiment in a text. • Integration of knowledge and ideas • CCSS.ELA-LITERACY.RST.6-8.7: Integrate quantitative or technical information expressed in words in a text with a version of that information expressed visually (e.g., in a flowchart, diagram, model, graph, or table). • CCSS.ELA-LITERACY.RST.6-8.8: Distinguish among facts, reasoned judgment based on research findings, and speculation in a text. • CCSS.ELA-LITERACY.RST.6-8.9: Compare and contrast the information gained from experiments, simulations, video, or multimedia sources with that gained from reading a text on the same topic.

Continued

TABLE 1.2 (*continued*)

***CCSS ELA*—Writing in Science and Technical Subjects**	Text types and purposes • CCSS.ELA-LITERACY.WHST.6-8.1: Write arguments focused on *discipline-specific content.* • CCSS.ELA-LITERACY.WHST.6-8.2: Write informative or explanatory texts, including the narration of historical events, scientific procedures/experiments, or technical processes. Production and distribution of writing • CCSS.ELA-LITERACY.WHST.6-8.4: Produce clear and coherent writing in which the development, organization, and style are appropriate to task, purpose, and audience. • CCSS.ELA-LITERACY.WHST.6-8.5: With some guidance and support from peers and adults, develop and strengthen writing as needed by planning, revising, editing, rewriting, or trying a new approach, focusing on how well purpose and audience have been addressed. • CCSS.ELA-LITERACY.WHST.6-8.6: Use technology, including the internet, to produce and publish writing and present the relationships between information and ideas clearly and efficiently. Range of writing • CCSS.ELA-LITERACY.WHST.6-8.10: Write routinely over extended time frames (time for reflection and revision) and shorter time frames (a single sitting or a day or two) for a range of discipline-specific tasks, purposes, and audiences.
***CCSS ELA*—Speaking and Listening**	Comprehension and collaboration • CCSS.ELA-LITERACY.SL.6-8.1: Engage effectively in a range of collaborative discussions (one-on-one, in groups, and teacher-led) with diverse partners on grade 6–8 topics, texts, and issues, building on others' ideas and expressing their own clearly. • CCSS.ELA-LITERACY.SL.6-8.2:* Interpret information presented in diverse media and formats (e.g., visually, quantitatively, orally) and explain how it contributes to a topic, text, or issue under study. • CCSS.ELA-LITERACY.SL.6-8.3:* Delineate a speaker's argument and specific claims, distinguishing claims that are supported by reasons and evidence from claims that are not. Presentation of knowledge and ideas • CCSS.ELA-LITERACY.SL.6-8.4:* Present claims and findings, sequencing ideas logically and using pertinent descriptions, facts, and details to accentuate main ideas or themes; use appropriate eye contact, adequate volume, and clear pronunciation. • CCSS.ELA-LITERACY.SL.6-8.5:* Include multimedia components (e.g., graphics, images, music, sound) and visual displays in presentations to clarify information. • CCSS.ELA-LITERACY.SL.6-8.6: Adapt speech to a variety of contexts and tasks, demonstrating command of formal English when indicated or appropriate.

* Only the standard for grade 6 is provided because the standards for grades 7 and 8 are similar. Please see *www.corestandards.org/ELA-Literacy/SL* for the exact wording of the standards for grades 7 and 8.

Lab Handout

Lab 1. Moon Phases: Why Does the Appearance of the Moon Change Over Time in a Predictable Pattern?

Introduction

We have all seen the Moon in the sky and how it looks different at various times of the month. In fact, differences in the appearance of the Moon over time were the basis for the Chinese, Islamic, Hindu, and Judaic calendars, as well as most of the other calendar systems that were used in ancient times. People can use the appearance of the Moon to mark the passage of time because the Moon's appearance changes in a predictable pattern over a period of 29.5 days. Figure L1.1 shows the pattern that the appearance of the Moon follows. As can be seen in this figure, the portion of the Moon that is illuminated gradually increases until the Moon is full, and then the portion of the Moon that is illuminated gradually decreases until it is completely dark. People often describe this pattern as a lunar cycle. Each phase, or how the Moon looks at a given point in the lunar cycle, has a specific name (see Figure L1.1).

FIGURE L1.1

The phases of the Moon follow a predictable pattern over a period of 29.5 days

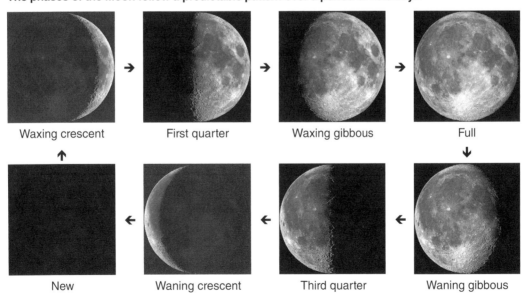

Waxing crescent	First quarter	Waxing gibbous	Full
New	Waning crescent	Third quarter	Waning gibbous

There are some other important facts that we know about Moon in addition to the fact that it goes through a series of phases over the course of a lunar cycle. First, the Moon rises in the east and sets in the west once every 24 hours. The Moon, therefore, travels from east to west across the sky just like the Sun. Second, the time that the Moon rises and sets in the sky changes each day. Sometimes the Moon will rise at dusk and set at dawn, and other times it will rise late at night and set in the morning. The Moon can even rise at dawn and set at dusk just like the Sun. The times that we can see the Moon in the sky therefore change over the course of a lunar cycle. Third, we always see the same light and dark regions on the surface of Moon regardless of its current phase (see Figure L1.1). We always see the same surface features when we look at the Moon because the same side of the Moon is always facing Earth. Finally, we see solar and lunar eclipses from time to time. A solar eclipse occurs during the day. A solar eclipse results in the light from the Sun being blocked for about 5 to 10 minutes (see Figure L1.2). A lunar eclipse, in contrast, occurs at night. A lunar eclipse causes the full Moon to get darker and turn red for a few minutes (see Figure L1.3). All of these different facts about the Moon can be explained if you understand what causes the lunar cycle.

FIGURE L1.2 _____

A solar eclipse

FIGURE L1.3 _____

A lunar eclipse

To explain the lunar cycle and all these different facts about the Moon, it is important to know a little about the types of objects that are found in our solar system and how all these objects move over time in relation to each other. The solar system consists of the Sun, the eight official planets, at least five dwarf planets, more than 130 moons, and numerous small bodies (including comets and asteroids). At the center of the solar system is the Sun. The inner solar system includes the planets Mercury, Venus, Earth, and Mars; the dwarf planet Ceres; and three moons. The outer solar system includes the planets Jupiter, Saturn, Uranus, and Neptune; the four other dwarf planets; and the remaining moons. In our solar system, all the planets and dwarf planets orbit (revolve around) the Sun, and all the moons orbit planets or dwarf planets. All the planets in our solar system travel around the Sun in a counterclockwise direction (when looking down from above the Sun's north pole). All of the planets and dwarf plants, with the exception of Venus, Uranus and Pluto, also spin (or rotate) in a counterclockwise direction.

You can use this information about our solar system to develop a physical model of the Earth-Sun-Moon system. You can then use your physical model to explore how Earth, the Sun, and the Moon move in relation to each other and how the light from the Sun illuminates the Moon as it orbits Earth. You can also use your physical model to determine how different positions of Earth, the Sun, and the Moon in relation to each other affect the appearance of the Moon over time (as seen from Earth). You will then be able to use what you learned about how the Moon and Earth move in relation to each other over time by working with a physical model to create a conceptual model that you can use to explain the lunar cycle.

Your Task

Develop a conceptual model that you can use to explain the phases of the Moon. Your conceptual model must be based on what we know about system and system models, patterns, the objects that are found in our solar system, and how these objects move in relationship to each other. You should be able to use your conceptual model to predict when and where you will be able to see the Moon in the sky during a lunar cycle.

The guiding question of this investigation is, *Why does the appearance of the Moon change over time in a predictable pattern?*

Materials

You may use any of the following materials during your investigation:

Equipment
- Safety glasses or goggles (required)
- Physical model of Earth (large ball on a stand)
- Physical model of the Moon A (small ball on a stand)
- Physical model of the Moon B (small ball on a stick)
- Lamp and lightbulb

Other Resources
- Moon phase calendar A (use to develop your conceptual model)
- Moon phase calendar B (use to test your conceptual model)

Safety Precautions

Follow all normal lab safety rules. In addition, take the following safety precautions:

- Wear sanitized indirectly vented chemical-splash goggles throughout the entire investigation (which includes setup and cleanup).
- Use only GFCI-protected electrical receptacles for lamps to prevent or reduce risk of shock.
- Handle the lamps with care; they can get hot when left on for long periods of time.
- Wash hands with soap and water when done collecting the data and after completing the lab.

LAB 1

Investigation Proposal Required? ☐ Yes ☐ No

Getting Started

The first step in developing a conceptual model is to design and carry out an investigation to determine how movement of Earth, the Sun, and the Moon over time results in the Moon looking different from our perspective on Earth. To accomplish this task, you will need to create a physical model of the Earth-Sun-Moon system using the available materials. You can then use this physical model to see how light shines on Earth and the Moon when they are in different positions relative to each other. You can also use this model to test your different ideas about the underlying cause of the Moon phases. As you develop your physical model, be sure to consider the following questions:

- What are the boundaries of the system you are studying?
- What are the components of this system?
- How can you quantitatively describe changes within the system over time?
- What could be causing the pattern that we observe?

Once you have used your physical model to test your ideas about the underlying cause of the Moon phases, your group can use what you learned to develop your conceptual model. A conceptual model is an idea or set of ideas that explains what causes a particular phenomenon in nature. People often use words, images, and arrows to describe a conceptual model. Your conceptual model needs to be able to explain why we see the phases of the Moon in the same pattern. It also needs to be able to explain

- why we see the Moon rise in the east and set in the west,
- why the Moon rises and sets at different times of the day,
- why we see the same side of the Moon regardless of its current phase, and
- why there are occasional solar and lunar eclipses.

The last step in your investigation will be to generate the evidence that you need to convince others that your conceptual model is valid or acceptable. To accomplish this goal, you can use your model to predict when and where the Moon will be in the night sky over the next month. You can also attempt to show how using a different version of your model or making a specific change to a portion of your model would make your model inconsistent with data you have or the facts we know about the Moon. Scientists often make comparisons between different versions of a model in this manner to show that a model is valid or acceptable. If you are able to use your conceptual model to make accurate predictions about the behavior of the Moon over time or you are able show how your conceptual model explains the behavior of the Moon better than other models, then you should be able to convince others that it is valid or acceptable.

Connections to the Nature of Scientific Knowledge and Scientific Inquiry

As you work through your investigation, be sure to think about

- the use of models as tools for reasoning about natural phenomena, and
- how scientists use different methods to answer different types of questions.

Initial Argument

Once your group has finished collecting and analyzing your data, your group will need to develop an initial argument. Your initial argument needs to include a claim, evidence to support your claim, and a justification of the evidence. The claim is your group's answer to the guiding question. The evidence is an analysis and interpretation of your data. Finally, the justification of the evidence is why your group thinks the evidence matters. The justification of the evidence is important because scientists can use different kinds of evidence to support their claims. Your group will create your initial argument on a whiteboard. Your whiteboard should include all the information shown in Figure L1.4.

FIGURE L1.4

Argument presentation on a whiteboard

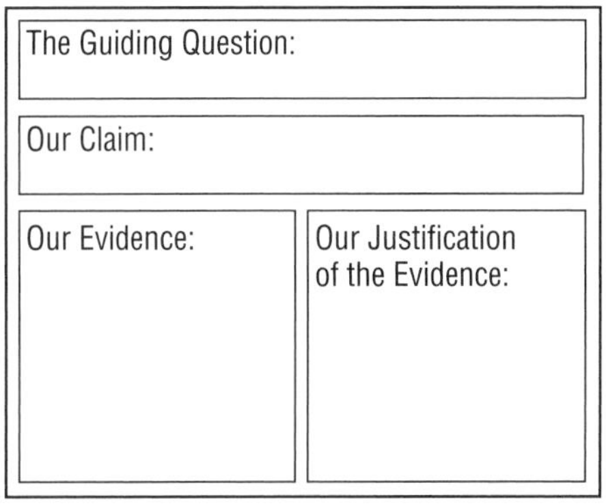

The Guiding Question:	
Our Claim:	
Our Evidence:	Our Justification of the Evidence:

Argumentation Session

The argumentation session allows all of the groups to share their arguments. One or two members of each group will stay at the lab station to share that group's argument, while the other members of the group go to the other lab stations to listen to and critique the other arguments. This is similar to what scientists do when they propose, support, evaluate, and refine new ideas during a poster session at a conference. If you are presenting your group's argument, your goal is to share your ideas and answer questions. You should also keep a record of the critiques and suggestions made by your classmates so you can use this feedback to make your initial argument stronger. You can keep track of specific critiques and suggestions for improvement that your classmates mention in the space below.

Critiques of our initial argument and suggestions for improvement:

If you are critiquing your classmates' arguments, your goal is to look for mistakes in their arguments and offer suggestions for improvement so these mistakes can be fixed. You should look for ways to make your initial argument stronger by looking for things that the other groups did well. You can keep track of interesting ideas that you see and hear during the argumentation session in the space below. You can also use this space to keep track of any questions that you will need to discuss with your team.

Interesting ideas from other groups or questions to take back to my group:

Once the argumentation session is complete, you will have a chance to meet with your group and revise your initial argument. Your group might need to gather more data or design a way to test one or more alternative claims as part of this process. Remember, your goal at this stage of the investigation is to develop the best argument possible.

Report

Once you have completed your research, you will need to prepare an investigation report that consists of three sections. Each section should provide an answer for the following questions:

1. What question were you trying to answer and why?

2. What did you do to answer your question and why?

3. What is your argument?

Your report should answer these questions in two pages or less. You should write your report using a word processing application (such as Word, Pages, or Google Docs), if possible, to make it easier for you to edit and revise it later. You should embed any diagrams, figures, or tables into the document. Be sure to write in a persuasive style; you are trying to convince others that your claim is acceptable or valid.

Checkout Questions

Lab 1. Moon Phases: Why Does the Appearance of the Moon Change Over Time in a Predictable Pattern?

1. The diagram below shows Earth and the Sun, as well as four different possible positions for the Moon (A, B, C, and D).

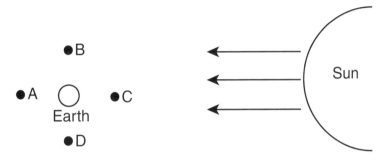

   a. Which position of the Moon (A, B, C, or D) would cause it to appear like the picture below when viewed from Earth? Circle the letter below.

   A   B   C   D

   b. How do you know?

2. You observe a crescent Moon rising in the east.

    a. Which of the following pictures illustrates how it will appear in six hours?

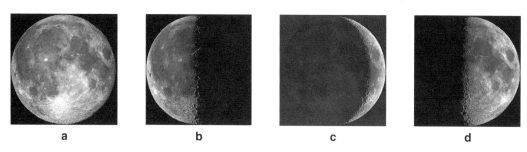

        a   b   c   d

    b. How do you know?

3. One night you look at the Moon and see this:

    a. Which of the following pictures illustrates how it will appear one week later?

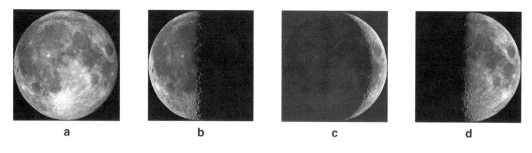

        a   b   c   d

    b. How do you know?

4. The image below is a model of the Earth-Sun-Moon system.

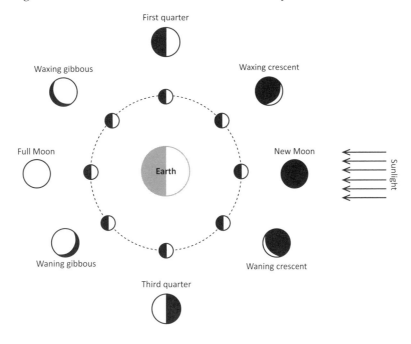

a. What are some strengths of this model?

b. Based on what you know about the Earth-Sun-Moon system from your investigation, what are some limitations or incorrect portions of this model?

5. Scientists often look at patterns to identify relationships in nature. Give an example of a pattern from your investigation on Moon phases.

6. Models are pictures of things that we cannot see.

    a. I agree with this statement.

    b. I disagree with this statement.

    Explain your answer, using an example from your investigation about the phases of the Moon.

7. There is a single scientific method that all scientists follow when conducting an investigation.

   a. I agree with this statement.

   b. I disagree with this statement.

   Explain your answer, using an example from your investigation about phases of the Moon.

LAB 2

Teacher Notes

Lab 2. Seasons: What Causes the Differences in Average Temperature and the Changes in Day Length That We Associate With the Change in Seasons on Earth?

Purpose

The purpose of this lab is to *introduce* students to the disciplinary core ideas (DCIs) of (a) The Universe and Its Stars and (b) Earth and the Solar System by having them develop a model that explains seasonal average temperature change. In addition, students have an opportunity to learn about the crosscutting concepts (CCs) of (a) Patterns and (b) Systems and System Models. During the explicit and reflective discussion, students will also learn about (a) the use of models as tools for reasoning about natural phenomena and (b) how scientists use different methods to answer different types of questions.

Important Earth and Space Science Content

A *season* is a period of the year that is marked by different climate conditions. Temperate and subpolar regions have four calendar-based seasons: winter, spring, summer, and fall. Tropical regions near the equator, in contrast, experience only two seasons: a rainy (monsoon) season and a dry season.

The difference in average temperature that is associated with a change in the seasons, especially in the temperate and subpolar regions on Earth, is due to changes in the length of day and the altitude of the Sun in the sky over the course of a year. These two factors affect the amount of energy that different regions of Earth's surface receive from the Sun. Longer days during a season expose a region to more solar radiation, and shorter days in a different season expose a region to less. The change in the altitude of the Sun in the sky influences the amount of solar radiation received by Earth's surface in two ways:

1. When the Sun is high in the sky, sunlight is most concentrated (see Figure 2.1a). The lower the angle, the more spread out and less intense the solar radiation reaching the surface (see Figures 2.1b and 2.1c).

2. Sunlight has to travel through more or less atmosphere depending on the angle of the Sun. When the Sun is directly overhead, sunlight only needs to pass through a thickness of 1 atmosphere (atm) to reach the surface of the Earth. In contrast, when the Sun is at a 30° angle, sunlight must pass through a thickness of 2 atm before it reaches the surface of Earth. The more atmosphere sunlight must travel through, the greater the chance for absorption, reflection, and scattering, all of which reduce

Seasons

What Causes the Differences in Average Temperature and the Changes in Day Length That We Associate With the Change in Seasons on Earth?

the intensity of solar radiation at the surface. The same effect accounts for the fact that it is difficult to look directly at the midday Sun, but it is easier to watch the Sun rise or set.

FIGURE 2.1

Changes in the angle of incoming sunlight affect the concentration and intensity of solar radiation. When light strikes the surface at a 90° angle (a), it is the most concentrated. When sunlight strikes the surface at a 45° angle (b), it is spread out over a larger surface area and is less intense. Sunlight is spread out over an even larger surface area when it strikes the surface at a 30° angle (c).

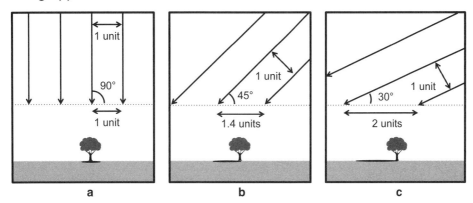

The seasonal change in the altitude of the Sun in the sky that takes place in the Northern and Southern Hemispheres can be can quite noticeable. Figure 2.2, for example, shows the daily paths of the Sun for a spot located at 40° N latitude for (a) the June solstice, (b) the March or September equinox, and (c) the December solstice. In this location, the angle of the Sun will decrease from 73.5° to 26.5°, which is a difference of 47°. In addition, the location of sunrise (east) and sunset (west) will change over the course of a year.

FIGURE 2.2

Daily paths of the Sun for a spot located at 40° N latitude during (a) the June solstice, (b) the March and September equinox, and (c) the December solstice

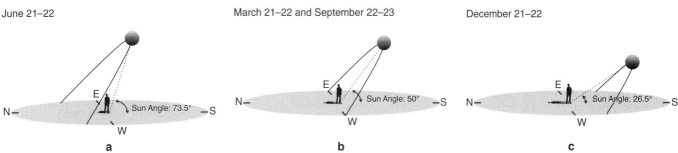

The length of day and altitude of the Sun in the sky changes over the course of a year because Earth's axis is not perpendicular to the plane of its orbit around the Sun. Instead, it is tilted at an angle of 23.4° from the perpendicular to orbit (see Figure 2.3). Earth's axis is an invisible line that runs from pole to pole through the center of the planet. Earth rotates around its axis. Earth's axis also remains pointed in the same direction (toward the North Star) as Earth moves around the Sun. Earth's orientation in relation to the Sun, as a result, changes as it travels along its orbit.

FIGURE 2.3

The axial tilt is the angle between a planet's rotational axis at its north pole and a line perpendicular to the orbital plane of the planet. Earth's axial tilt is currently 23.4°.

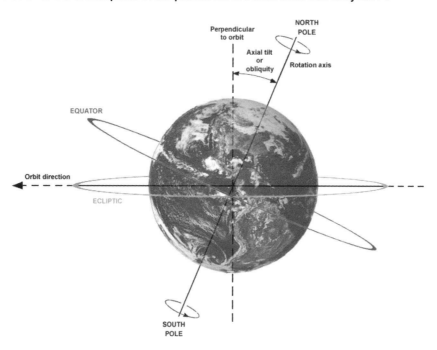

The Northern Hemisphere is tilted 23.4° toward the Sun for one day in June each year (see Figure 2.4). This day is called the June solstice. Six months later, when Earth has moved to the opposite side of its orbit, the Northern Hemisphere is tilted 23.4° away from the Sun. This day is called the December solstice. On days between these two extremes, the Northern Hemisphere is tilted at an amount less than 23.4° toward or away from the Sun. This change in orientation causes the spot on Earth where sunlight strikes Earth at a 90° angle to make an annual migration from 23.4° north of the equator to 23.4° south of the equator. Sunlight strikes Earth's surface at an ever-decreasing angle as one moves more north or more south from this location. Thus, the farther away a location is from the latitude where sunlight strikes Earth's surface at 90°, the lower the noonday Sun will be in the sky.

Seasons

*What Causes the Differences in Average Temperature and the Changes in
Day Length That We Associate With the Change in Seasons on Earth?*

FIGURE 2.4

The relative positions of the solstices and the equinox in relation to Earth's orbit around the Sun (illustration is not to scale)

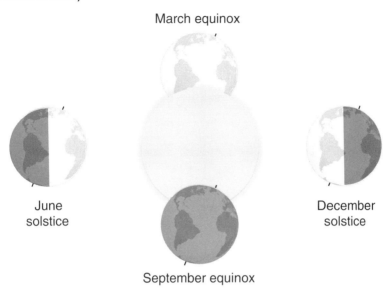

The annual migration of the location where sunlight strikes Earth at a 90° angle causes the maximum altitude of the Sun to change by up to 46.8° (23.4° + 23.4°) for many locations over the course of the year. For example, in a mid-latitude city such as New York, which is located at about 40° N latitude, the maximum altitude of the Sun is 73.5° in June (when the Sun reaches its farthest northward location in the sky) and the minimum altitude of the Sun is 26.5° in December (see Figure 2.2).

The length of daylight at a location is also determined by Earth's position in orbit. The length of daylight on the June solstice is greater than the length of night in the Northern Hemisphere and shorter than the length of night in the Southern Hemisphere. The opposite is true for the December solstice, when the nights are longer than days in the Northern Hemisphere and nights are shorter than days in the Southern Hemisphere. Table 2.1 (p. 58) provides the length of daylight observed at different latitudes in the Northern Hemisphere during the solstices and equinoxes. As can be seen in this table, the farther north a location is from the equator, the more daylight that it will receive during the June solstice. At locations above 66.5° N latitude, which marks the edge of the Arctic Circle, the length of daylight is 24 hours and the Sun does not set. Locations that are farther north from the equator also have less daylight during the December solstice. During the March and September equinoxes, however, the length of daylight is 12 hours at every location on Earth. The length of daylight is 12 hours everywhere on Earth on these two dates because the circle of illumination passes directly though the North and South Poles.

TABLE 2.1

Length of daylight at different latitudes in the Northern Hemisphere during solstices and equinoxes

Latitude	June solstice	December solstice	Equinoxes
0°	12 hr	12 hr	12 hr
10° N	12 hr 35 min	11 hr 25 min	12 hr
20° N	13 hr 12 min	10 hr 48 min	12 hr
30° N	13 hr 56 min	10 hr 04 min	12 hr
40° N	14 hr 52 min	9 hr 08 min	12 hr
50° N	16 hr 18 min	7 hr 42 min	12 hr
60° N	18 hr 27 min	5 hr 33 min	12 hr
70° N	24 hr	0 hr 00 min	12 hr
80° N	24 hr	0 hr 00 min	12 hr
90° N	24 hr	0 hr 00 min	12 hr

In summary, seasonal differences in the amount of solar energy reaching different places on Earth's surface is caused by the annual migration of the location where sunlight strikes Earth at a 90° angle and the corresponding changes in the Sun's angle and the length of daylight. The changes occur because Earth's orientation in relation to the Sun continually changes as it travels along its orbit. These changes over the year cause the month-to-month changes in the temperature that is observed in temperate and subpolar regions on Earth. Figure 2.5 shows the mean monthly temperatures for selected cities located at different latitudes. Notice that the cities located farther from the equator, such as Point Barrow and Winnipeg, experience greater temperature differences from June to December than cities that are located closer to the equator.

It is important to note that all places at the same latitude will have identical Sun angles and lengths of daylight during specific times of the year. If the changing orientation of Earth in relation to the Sun over the course of the year was the only factor that affected the temperature at specific locations, we would expect all places on Earth that are located at the same latitude to be the same temperature at the same time of year. In fact, places that are located at the same latitude have different average temperatures during the same season. As a result, the altitude of the Sun in the sky and length of daylight explains only why there are seasons in a given region; it does not explain why different regions have different climates.

Seasons

What Causes the Differences in Average Temperature and the Changes in Day Length That We Associate With the Change in Seasons on Earth?

FIGURE 2.5

Mean monthly temperatures for four cities located at different latitudes

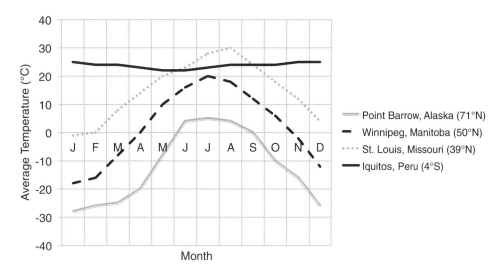

Point Barrow, Alaska (71°N)
Winnipeg, Manitoba (50°N)
St. Louis, Missouri (39°N)
Iquitos, Peru (4°S)

Timeline

The instructional time needed to complete this lab investigation is 170–230 minutes. Appendix 3 (p. 573) provides options for implementing this lab investigation over several class periods. Option C (230 minutes) should be used if students are unfamiliar with scientific writing, because this option provides extra instructional time for scaffolding the writing process. You can scaffold the writing process by modeling, providing examples, and providing hints as students write each section of the report. Option D (170 minutes) should be used if students are familiar with scientific writing and have developed the skills needed to write an investigation report on their own. In option D, students complete stage 6 (writing the investigation report) and stage 8 (revising the investigation report) as homework.

Materials and Preparation

The materials needed to implement this investigation are listed in Table 2.2 (p. 60). The *Seasons and Ecliptic Simulator* simulation, which was developed by the Astronomy Education Department at the University of Nebraska–Lincoln, is available at *http://astro. unl.edu/naap/motion1/animations/seasons_ecliptic.html*. Information about the location (latitude and longitude), weather, and hours of daylight for most major cities around the world can be found at *www.climate-charts.com/world-index.html*. Both of these resources are free to use and can be run online using an internet browser. You should access the *Seasons and Ecliptic Simulator* simulation and learn how the simulation works before beginning the lab investigation. In addition, it is important to check if students can access and use the online resources from a school computer or tablet, because some schools have set up firewalls and other restrictions on web browsing.

LAB 2

TABLE 2.2

Materials list for Lab 2

Item	Quantity
Computer or tablet with internet access	1 per group
Investigation Proposal A (optional)	1 per group
Whiteboard, 2' × 3'*	1 per group
Lab Handout	1 per student
Peer-review guide and teacher scoring rubric	1 per student
Checkout Questions	1 per students

* As an alternative, students can use computer and presentation software such as Microsoft PowerPoint or Apple Keynote to create their arguments.

Safety Precautions

Remind students to follow all normal lab safety rules.

Topics for the Explicit and Reflective Discussion

Reflecting on the Use of Core Ideas and Crosscutting Concepts During the Investigation

Teachers should begin the explicit and reflective discussion by asking students to discuss what they know about the DCIs they used during the investigation. The following are some important concepts related to the DCIs of (a) The Universe and Its Stars and (b) Earth and the Solar System that students need to develop a conceptual model that can be used to explain the seasons:

- The Sun is at the center of the solar system, and all other objects in this system orbit the Sun.
- Earth is tilted on its axis 23.4° from perpendicular to its orbit.
- The angle that sunlight strikes Earth's surface is different at different latitudes.
- The lower the angle that sunlight strikes Earth's surface, the more spread out and less intense the solar radiation reaching the surface.
- Energy cannot be created or destroyed.
- Some of the energy that reaches Earth from the Sun is transferred into heat.

To help students reflect on what they know about these concepts, we recommend showing them two or three images using presentation software that help illustrate these important ideas. You can then ask the students the following questions to encourage students to share how they are thinking about these important concepts:

1. What do we see going on in this image?

Seasons

What Causes the Differences in Average Temperature and the Changes in Day Length That We Associate With the Change in Seasons on Earth?

2. Does anyone have anything else to add?

3. What might be going on that we can't see?

4. What are some things that we are not sure about here?

You can then encourage students to think about how CCs played a role in their investigation. There are at least two CCs that students need to use to develop a conceptual model that can be used to explain the seasons: (a) Patterns and (b) Systems and System Models (see Appendix 2 [p. 569] for a brief description of these CCs). To help students reflect on what they know about these CCs, we recommend asking them the following questions:

1. Why is it important to look for, and attempt to explain, patterns in nature during an investigation?

2. What patterns did you identify during your investigation? What did the identification of these patterns allow you to do?

3. Why do scientists often define a system and then develop a model of it as part of an investigation?

4. How did you use a model to understand what causes the difference in average temperature we associate with the change in seasons on Earth? Why was that useful?

You can then encourage students to think about how they used all these different concepts to help answer the guiding question and why it is important to use these ideas to help justify their evidence for their final arguments. Be sure to remind your students to explain why they included the evidence in their arguments and make the assumptions underlying their analysis and interpretation of the data explicit in order to provide an adequate justification of their evidence.

Reflecting on Ways to Design Better Investigations

It is important for students to reflect on the strengths and weaknesses of the investigation they designed during the explicit and reflective discussion. Students should therefore be encouraged to discuss ways to eliminate potential flaws, measurement errors, or sources of uncertainty in their investigations. To help students be more reflective about the design of their investigation and what they can do to make their investigations more rigorous in the future, you can ask the following questions:

1. What were some of the strengths of the way you planned and carried out your investigation? In other words, what made it scientific?

2. What were some of the weaknesses of the way you planned and carried out your investigation? In other words, what made it less scientific?

3. What rules can we make, as a class, to ensure that our next investigation is more scientific?

Reflecting on the Nature of Scientific Knowledge and Scientific Inquiry

This investigation can be used to illustrate two important concepts related to the nature of scientific knowledge and the nature of scientific inquiry: (a) the use of models as tools for reasoning about natural phenomena and (b) how scientists use different methods to answer different types of questions (see Appendix 2 [p. 569] for a brief description of these two concepts). Be sure to review these concepts during and at the end of the explicit and reflective discussion. To encourage students, help students think about these concepts in relation to what they did during the lab, you can ask the following questions:

1. I asked you to develop a conceptual model to explain the seasons as part of your investigation. Why is it useful to develop models in science?

2. Can you work with your group to come up with a rule that you can use to decide what a model is and what a model is not in science? Be ready to share in a few minutes.

3. There is no universal step-by step scientific method that all scientists follow. Why do you think there is no universal scientific method?

4. Think about what you did during this investigation. How would you describe the method you used to develop you model? Why would you call it that?

You can also use presentation software or other techniques to encourage your students to think about these concepts. You can show examples and non-examples of scientific models and ask students to classify each one and explain their thinking. You can also show one or more images of a "universal scientific method" that misrepresent the nature of scientific inquiry (see, e.g., *https://commons.wikimedia.org/wiki/File:The_Scientific_Method_as_an_Ongoing_Process.svg*) and ask students why each image is *not* a good representation of what scientists do to develop scientific knowledge. You can also ask students to suggest revisions to the image that would make it more consistent with the way scientists develop scientific knowledge

Remind your students that, to be proficient in science, it is important that they understand what counts as scientific knowledge and how that knowledge develops over time.

Hints for Implementing the Lab

- Learn how to use the simulation before the lab begins. It is important for you to know how to use the simulation so you can help students when they get stuck or confused.

Seasons

What Causes the Differences in Average Temperature and the Changes in Day Length That We Associate With the Change in Seasons on Earth?

- A group of three students per computer or tablet tends to work well.

- Allow the students to play with the simulation as part of the tool talk before they begin to design their investigation. This gives students a chance to see what they can and cannot do with the simulation.

- Allowing students to design their own procedures for collecting data gives students an opportunity to try, to fail, and to learn from their mistakes. However, you can scaffold students as they develop their procedure by having them fill out an investigation proposal. These proposals provide a way for you to offer students hints and suggestions without telling them how to do it. You can also check the proposals quickly during a class period. We recommend using Investigation Proposal A for this lab.

- Be sure that students record actual values (e.g., latitude, Sun's altitude) and are not just attempting to hand draw what they see on the computer screen.

- The claim in the students' argument should be a conceptual model. This model should show how the components of the Earth-Sun system move over time in relationship to each other. The evidence should include information about how well it predicts the length of day and average temperature at different times of the year in several additional cities or how making a specific change to a portion of your model will make the model inconsistent with the available data or the facts we know about seasons. The justification of the evidence should include their underlying assumptions about the components of the Earth-Sun system (e.g., the Sun is at the center of the solar system, Earth is tilted on its axis 23.4° from perpendicular to its orbit, sunlight that strikes Earth's surface at a 90° angle is more concentrated and intense).

- Students often make mistakes when developing their conceptual models and/ or initial arguments, but they should quickly realize these mistakes during the argumentation session. Be sure to allow students to revise their models and arguments at the end of the argumentation session. The explicit and reflective discussion will also give students an opportunity to reflect on and identify ways to improve how they develop and test models. This also offers an opportunity to discuss what scientists do when they realize a mistake is made.

- Students often make mistakes during the data collection stage, but they should quickly realize these mistakes during the argumentation session. It will only take them a short period of time to re-collect data, and they should be allowed to do so. During the explicit and reflective discussion, students will also have the opportunity to reflect on and identify ways to improve the way they design investigations (especially how they attempt to control variables as part of an experiment). This offers another opportunity to discuss what scientists do when they realize that a mistake is made during a study.

Connections to Standards

Table 2.3 highlights how the investigation can be used to address specific (a) performance expectations from the *NGSS* and (b) *Common Core State Standards* in English language arts (*CCSS ELA*).

TABLE 2.3

Lab 2 alignment with standards

***NGSS* performance expectation**	Space systems • MS-ESS1-1: Develop and use a model of the Earth-Sun-Moon system to describe the cyclic patterns of lunar phases, eclipses of the Sun and Moon, and seasons.
***CCSS ELA*—Reading in Science and Technical Subjects**	Key ideas and details • CCSS.ELA-LITERACY.RST.6-8.1: Cite specific textual evidence to support analysis of science and technical texts. • CCSS.ELA-LITERACY.RST.6-8.2: Determine the central ideas or conclusions of a text; provide an accurate summary of the text distinct from prior knowledge or opinions. Craft and structure • CCSS.ELA-LITERACY.RST.6-8.4: Determine the meaning of symbols, key terms, and other domain-specific words and phrases as they are used in a specific scientific or technical context relevant to *grade 6–8 texts and topics*. • CCSS.ELA-LITERACY.RST.6-8.5: Analyze the structure an author uses to organize a text, including how the major sections contribute to the whole and to an understanding of the topic. • CCSS.ELA-LITERACY.RST.6-8.6: Analyze the author's purpose in providing an explanation, describing a procedure, or discussing an experiment in a text. Integration of knowledge and ideas • CCSS.ELA-LITERACY.RST.6-8.7: Integrate quantitative or technical information expressed in words in a text with a version of that information expressed visually (e.g., in a flowchart, diagram, model, graph, or table). • CCSS.ELA-LITERACY.RST.6-8.8: Distinguish among facts, reasoned judgment based on research findings, and speculation in a text. • CCSS.ELA-LITERACY.RST.6-8.9: Compare and contrast the information gained from experiments, simulations, video, or multimedia sources with that gained from reading a text on the same topic.

Continued

Seasons

*What Causes the Differences in Average Temperature and the Changes in
Day Length That We Associate With the Change in Seasons on Earth?*

TABLE 2.3 (*continued*)

***CCSS ELA*—Writing in Science and Technical Subjects**	Text types and purposes • CCSS.ELA-LITERACY.WHST.6-8.1: Write arguments focused on *discipline-specific content.* • CCSS.ELA-LITERACY.WHST.6-8.2: Write informative or explanatory texts, including the narration of historical events, scientific procedures/experiments, or technical processes. Production and distribution of writing • CCSS.ELA-LITERACY.WHST.6-8.4: Produce clear and coherent writing in which the development, organization, and style are appropriate to task, purpose, and audience. • CCSS.ELA-LITERACY.WHST.6-8.5: With some guidance and support from peers and adults, develop and strengthen writing as needed by planning, revising, editing, rewriting, or trying a new approach, focusing on how well purpose and audience have been addressed. • CCSS.ELA-LITERACY.WHST.6-8.6: Use technology, including the internet, to produce and publish writing and present the relationships between information and ideas clearly and efficiently. Range of writing • CCSS.ELA-LITERACY.WHST.6-8.10: Write routinely over extended time frames (time for reflection and revision) and shorter time frames (a single sitting or a day or two) for a range of discipline-specific tasks, purposes, and audiences.
***CCSS ELA*—Speaking and Listening**	Comprehension and collaboration • CCSS.ELA-LITERACY.SL.6-8.1: Engage effectively in a range of collaborative discussions (one-on-one, in groups, and teacher-led) with diverse partners on *grade 6–8 topics,* texts, and issues, building on others' ideas and expressing their own clearly. • CCSS.ELA-LITERACY.SL.6-8.2:* Interpret information presented in diverse media and formats (e.g., visually, quantitatively, orally) and explain how it contributes to a topic, text, or issue under study. • CCSS.ELA-LITERACY.SL.6-8.3:* Delineate a speaker's argument and specific claims, distinguishing claims that are supported by reasons and evidence from claims that are not.

Continued

TABLE 2.3 (*continued*)

CCSS ELA—**Speaking and Listening** (*continued*)	Presentation of knowledge and ideas • CCSS.ELA-LITERACY.SL.6-8.4:* Present claims and findings, sequencing ideas logically and using pertinent descriptions, facts, and details to accentuate main ideas or themes; use appropriate eye contact, adequate volume, and clear pronunciation. • CCSS.ELA-LITERACY.SL.6-8.5:* Include multimedia components (e.g., graphics, images, music, sound) and visual displays in presentations to clarify information. • CCSS.ELA-LITERACY.SL.6-8.6: Adapt speech to a variety of contexts and tasks, demonstrating command of formal English when indicated or appropriate.

* Only the standard for grade 6 is provided because the standards for grades 7 and 8 are similar. Please see *www.corestandards.org/ELA-Literacy/SL* for the exact wording of the standards for grades 7 and 8.

Seasons
What Causes the Differences in Average Temperature and the Changes in
Day Length That We Associate With the Change in Seasons on Earth?

Lab Handout

Lab 2. Seasons: What Causes the Differences in Average Temperature and the Changes in Day Length That We Associate With the Change in Seasons on Earth?

Introduction

A season is a subdivision of a year, which is often marked by changes in average daily temperature, amount of precipitation, and hours of daylight. People who live in temperate and subpolar regions around the globe experience four calendar-based seasons: spring, summer, fall, and winter. People who live in regions near the equator, in contrast, only experience two seasons: a rainy (or monsoon) season and a dry season.

Figure L2.1 shows four satellite images of Lake George in New York in February, April, July, and October. These images illustrate how the surface of the Earth looks different during different seasons.

FIGURE L2.1 ─────────────────────────

The change of seasons as seen in four satellite images of Lake George, New York, from the Advanced Spaceborne Thermal Emission and Reflection Radiometer instrument on NASA's Terra spacecraft

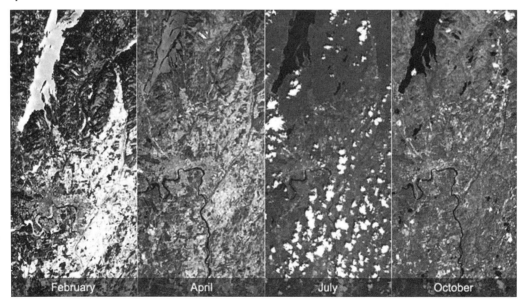

Note: A full-color version of this figure is available on the book's Extras page at *www.nsta.org/adi-ess*.

LAB 2

To understand why we experience different seasons in different locations on Earth, we must first think about the objects that are found in our solar system and how all these objects move over time in relation to each other. The Sun is at the center of our solar system. All the other objects in the solar system, which include planets, dwarf plants, asteroids, and comets, revolve (orbit) around it. All the planets in our solar system travel around the Sun in a counterclockwise direction (when looking down from above the Sun's north pole).

Earth takes 365.25 days to orbit the Sun. The distance that Earth must travel to complete one full revolution around the Sun is 940 million km. Earth, as a result, travels around the Sun at a speed of about 30 km/s. Earth is closest to the Sun in early January due to its slightly elliptical orbit. At this time, the Earth is about 146 million km away from the Sun. Earth is farthest from the Sun in early July, when the distance between Earth and the Sun is about 152 million km.

Earth also spins (or rotates) on its axis as it travels around the Sun. Earth spins on its axis in a counterclockwise direction (when looking down from above Earth's North Pole; see Figure L2.2). It takes 23 hours and 56 minutes for Earth to complete one full rotation. The rotation of Earth on its axis is what gives us day and night. During the day, we are facing the Sun and during the night we are facing away from it. Earth's axis, however, is not perpendicular to its orbit (or straight up if we were able to look down at it from above the solar system). Earth currently has an axial tilt of 23.4° (see Figure L2.2). Earth remains tilted in the same direction regardless of where it is in its orbit. This means that Earth's North Pole is directed toward the Sun in June but directed away from the Sun in December. In contrast, Earth's South Pole is directed toward the Sun in December and is directed away from the Sun in June.

These facts are useful and can help us understand the change in seasons. Yet, these facts do not provide us with all the information that we need to develop a complete conceptual model that explains the cause of the seasons. You will therefore need to learn more about how the average temperature and the hours of daylight change over the course of year at different locations on Earth. Next, you will need to use an online simulation called the Seasons and Ecliptic Simulator to explore how the tilt of Earth affects the amount of sunlight and the angle that sunlight strikes Earth at various locations over time. Finally, you will have an opportunity to put all these pieces of information together to develop a conceptual model that explains the cause of the seasons.

Seasons

What Causes the Differences in Average Temperature and the Changes in Day Length That We Associate With the Change in Seasons on Earth?

FIGURE L2.2

The axial tilt is the angle between a planet's rotational axis at its north pole and a line perpendicular to the orbital plane of the planet. Earth's axial tilt is currently 23.4°

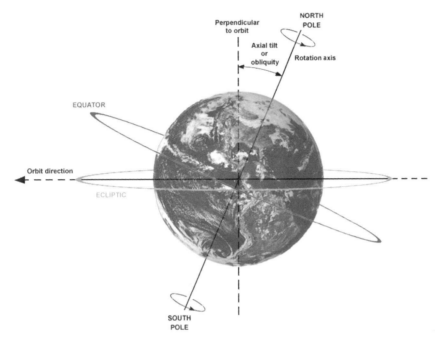

Your Task

Develop a conceptual model that you can use to explain the cause of the seasons. You must base your conceptual model on what we know about how Earth revolves around the Sun and spins on its axis. You will also need to use what you know about systems and system models and the importance of looking for patterns in nature to develop your conceptual model. To be considered valid or acceptable, your conceptual model should not only explain the underlying cause of the seasons but also predict the changes in average daily temperature and hours of daylight at several different locations on Earth.

The guiding question of this investigation is, *What causes the differences in average temperature and the changes in day length that we associate with the change in seasons on Earth?*

Materials

You will use an online simulation called Seasons and Ecliptic Simulator to conduct your investigation; the simulation is available at *http://astro.unl.edu/naap/motion1/animations/seasons_ecliptic.html.*

Information about the location (latitude and longitude), weather, and hours of daylight for most major cities around the world can be found at *www.climate-charts.com/world-index.html.*

LAB 2

Safety Precautions

Follow all normal lab safety rules.

Investigation Proposal Required?   ☐ Yes   ☐ No

Getting Started

The first step in developing a conceptual model that explains the cause of the seasons is to collect information about the changes in average daily temperature and hours of daylight over a year at several different locations on Earth. This information can be found for cities in 149 countries at the World Climate website, which contains the largest set of accessible climate data on the web. Be sure to collect information from cities at a wide range of latitudes and longitudes. Once you collect this information, look for any patterns that you can use to help develop your conceptual model.

FIGURE L2.3 _____

A screenshot from the Seasons and Ecliptic Simulator simulation

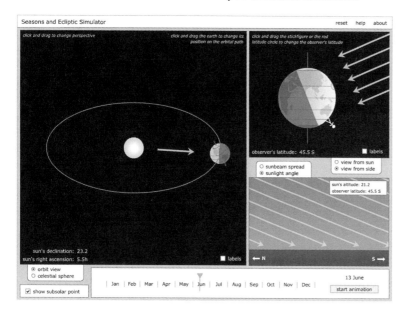

Next, you can use the *Seasons and Ecliptic Simulato*r to learn more about how light from the Sun strikes Earth over the course of the year (see Figure L2.3). This simulation allows you to move an observer to different latitudes on Earth and to track the Sun's altitude in the sky and the sunlight angle over the course of the year for that observer. Given the nature of the simulation, you must determine what type of data you need to collect, how you will collect it, and how you will analyze it to learn more about how Earth's tilt affects the amount of sunlight and the angle at which sunlight strikes Earth over time.

To determine *what type of data you need to collect,* think about the following questions:

- What are the boundaries and components of the system you are studying?
- How do the components of the system interact with each other?
- What type of measurements or observations will you need to record to determine how Earth's tilt affects the amount of sunlight that strikes Earth over time?
- What type of measurements or observations will you need to record to determine how Earth's tilt affects the angle that sunlight strikes Earth over time?

To determine *how you will collect the data,* think about the following questions:

- How often will you need to make the measurements or observations?
- What scale or scales should you use when you take your measurements?
- What types of comparisons will you need to make?

To determine *how you will analyze the data,* think about the following questions:

- What types of calculations will you need to make?
- What types of patterns could you look for as you analyze your data?
- How could you use mathematics to describe a change over time?
- How could you use mathematics to describe a relationship between variables?

Once you have finished using the *Seasons and Ecliptic Simulator,* your group can develop a conceptual model that can be used to explain the cause of the seasons. Be sure to incorporate the information you collected from the World Climate website. To be valid or acceptable, your conceptual model must be able to explain (a) why the length of day changes by different amounts in different locations and (b) why the average temperature for each month changes by different amounts in different locations.

The last step in your investigation will be to generate the evidence that you need to convince others that your conceptual model is valid or acceptable. To accomplish this goal, you can use your model to predict the length of day and average temperature at different times of the year in several additional cities. These cities should be ones that you have not looked up before. You can also attempt to show how using a different version of your model or making a specific change to a portion of your model will make your model inconsistent with data you have or the facts we know about seasons. Scientists often make comparisons between different versions of a model in this manner to show that a model is valid or acceptable. If you are able to use your conceptual model to make accurate predictions about the changes in average daily temperature and hours of daylight at several locations on Earth or you are able show how your conceptual model explains the cause of the seasons better than other models, then you should be able to convince others that it is valid or acceptable.

Connections to the Nature of Scientific Knowledge and Scientific Inquiry

As you work through your investigation, be sure to think about

- the use of models as tools for reasoning about natural phenomena, and
- how scientists use different methods to answer different types of questions.

LAB 2

Initial Argument

Once your group has finished collecting and analyzing your data, your group will need to develop an initial argument. Your initial argument needs to include a claim, evidence to support your claim, and a justification of the evidence. The *claim* is your group's answer to the guiding question. The *evidence* is an analysis and interpretation of your data. Finally, the *justification* of the evidence is why your group thinks the evidence matters. The justification of the evidence is important because scientists can use different kinds of evidence to support their claims. Your group will create your initial argument on a whiteboard. Your whiteboard should include all the information shown in Figure L2.4.

FIGURE L2.4

Argument presentation on a whiteboard

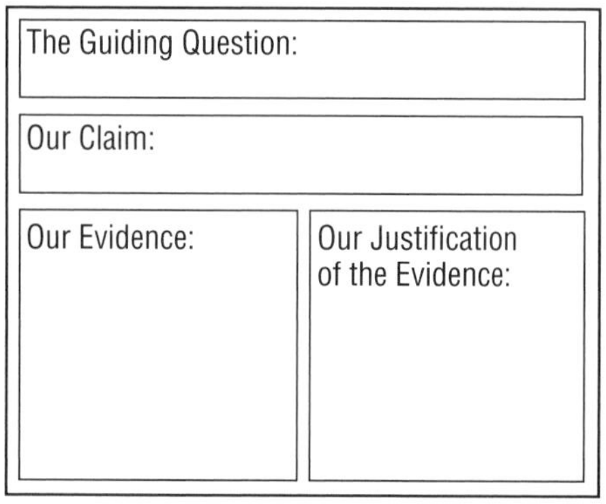

The Guiding Question:	
Our Claim:	
Our Evidence:	Our Justification of the Evidence:

Argumentation Session

The argumentation session allows all of the groups to share their arguments. One or two members of each group will stay at the lab station to share that group's argument, while the other members of the group go to the other lab stations to listen to and critique the other arguments. This is similar to what scientists do when they propose, support, evaluate, and refine new ideas during a poster session at a conference. If you are presenting your group's argument, your goal is to share your ideas and answer questions. You should also keep a record of the critiques and suggestions made by your classmates so you can use this feedback to make your initial argument stronger. You can keep track of specific critiques and suggestions for improvement that your classmates mention in the space below.

Critiques of our initial argument and suggestions for improvement:

National Science Teachers Association

Seasons

What Causes the Differences in Average Temperature and the Changes in Day Length That We Associate With the Change in Seasons on Earth?

If you are critiquing your classmates' arguments, your goal is to look for mistakes in their arguments and offer suggestions for improvement so these mistakes can be fixed. You should look for ways to make your initial argument stronger by looking for things that the other groups did well. You can keep track of interesting ideas that you see and hear during the argumentation in the space below. You can also use this space to keep track of any questions that you will need to discuss with your team.

Interesting ideas from other groups or questions to take back to my group:

Once the argumentation session is complete, you will have a chance to meet with your group and revise your initial argument. Your group might need to gather more data or design a way to test one or more alternative claims as part of this process. Remember, your goal at this stage of the investigation is to develop the best argument possible.

Report

Once you have completed your research, you will need to prepare an *investigation report* that consists of three sections. Each section should provide an answer for the following questions:

1. What question were you trying to answer and why?

2. What did you do to answer your question and why?

3. What is your argument?

Your report should answer these questions in two pages or less. You should write your report using a word processing application (such as Word, Pages, or Google Docs), if possible, to make it easier for you to edit and revise it later. You should embed any diagrams, figures, or tables into the document. Be sure to write in a persuasive style; you are trying to convince others that your claim is acceptable or valid.

Checkout Questions

Lab 2. Seasons: What Causes the Differences in Average Temperature and the Changes in Day Length That We Associate With the Change in Seasons on Earth?

1. Pictured below are five different pictures (A–E) of Earth as it orbits the Sun. Use numbers to rank the pictures from highest to lowest temperature at the locations indicated by the black circle. If you think two or more pictures show locations with equal temperatures, give them the same number.

Picture	Relative position of Earth as it orbits the Sun*	Rank
A		_____
B		_____
C		_____
D		_____
E		_____

* Not to scale

Seasons

What Causes the Differences in Average Temperature and the Changes in Day Length That We Associate With the Change in Seasons on Earth?

Explain your answer. Why do you think the order that you chose is correct?

2. The December solstice occurs when Earth is positioned as it is in the picture below. Sometimes people refer to this as the winter solstice—is this an appropriate name for this solstice?

Explain your answer, using an example from your investigation about the cause of the seasons.

3. Scientists create pictures of things to teach people about them. These pictures are models.

    a. I agree with this statement.

    b. I disagree with this statement.

Explain your answer, using an example from your investigation about the cause of the seasons.

4. All scientific investigations are experiments.

    a. I agree with this statement.

    b. I disagree with this statement.

Explain your answer, using an example from your investigation about the cause of the seasons.

Seasons

What Causes the Differences in Average Temperature and the Changes in Day Length That We Associate With the Change in Seasons on Earth?

5. Scientists often work to understand the cause of patterns we observe in nature. Provide an example of a pattern and its cause from your investigation about the cause of the seasons.

6. Scientists often develop or use models. Explain why scientists might use a model to study natural phenomena, using an example from your investigation about the cause of the seasons.

LAB 3

Teacher Notes

Lab 3. Gravity and Orbits: How Does Changing the Mass and Velocity of a Satellite and the Mass of the Object That It Revolves Around Affect the Nature of the Satellite's Orbit?

Purpose

The purpose of this lab is to *introduce* students to the disciplinary core ideas (DCI) of (a) The Universe and Its Stars and (b) Earth and the Solar System by having them explore the role that gravity plays in the motion of objects within the solar system. In addition, students have an opportunity to learn about the crosscutting concepts (CCs) of (a) Scale, Proportion, and Quantity; and (b) Systems and System Models. During the explicit and reflective discussion, students will also learn about (a) the difference between laws and theories in science and (b) the assumptions made by scientists about order and consistency in nature.

Important Earth and Space Science Content

An *orbit* is a regular, repeating path that one object in space takes around another one. Any object in an orbit around a planet or a star is called a satellite. A satellite can be natural, like planets (which orbit a star), moons (which orbit planets), and comets (which orbit stars). Or a satellite can be artificial, created by engineers and scientists and sent into space, such as the International Space Station (which orbits Earth), the Hubble Space Telescope (which orbits Earth), and the Cassini space probe (which orbits Saturn).

The time it takes a satellite to make one full orbit is called its period. All orbits are elliptical. An elliptical orbit means that the satellite follows a path that is round but can range in shape from a perfect circle to a long, thin oval. Space scientists use the term *eccentricity* to describe how much an orbit deviates from a perfect circle. Eccentricities for a satellite are always between zero and one. The orbits of the planets are almost circular, but the orbits of comets are highly eccentric. Satellites that orbit Earth, including the Moon, are not always the same distance from Earth because they have elliptical orbits. The closest point a satellite comes to Earth is called its perigee, and the farthest point is called its apogee. For planets, the point in their orbit closest to the Sun is perihelion, and the point in their orbit farthest from the Sun is called aphelion. Earth reaches its aphelion during summer in the Northern Hemisphere.

When sending an artificial satellite into orbit, it is important to consider Newton's first law: an object in motion will stay in motion unless something pushes or pulls on it. Without the force of gravity acting on it, an artificial satellite therefore moves in a straight line at a constant velocity. With the force of gravity acting on it, however, the artificial satellite is pulled toward the center of the object it is orbiting (usually a planet, but occasionally a

Gravity and Orbits

How Does Changing the Mass and Velocity of a Satellite and the Mass of the Object
That It Revolves Around Affect the Nature of the Satellite's Orbit?

moon or the Sun). This force causes the satellite to change direction. A satellite's velocity and the force of gravity acting on it therefore have to be balanced for the satellite to stay in orbit around a planet. If the satellite is moving too fast, it will not enter into an orbit at all and will fly off into space; but if it is moving too slow, it will be pulled into the planet and crash. When the velocity of the satellite and the force of gravity are balanced, the satellite will enter a stable orbit at a specific distance above the planet. Scientists use the term *orbital velocity* to describe how fast a satellite has to be moving in order to enter a stable orbit at a specific altitude. On Earth, for example, the orbital velocity for a satellite located at an altitude of 150 miles above the surface is about 17,000 miles per hour. Satellites that have higher orbits have slower orbital velocities.

The shape of an orbit, the distance of an orbit, and the period of an orbit depend on three factors: (1) is the mass of the satellite, (2) the initial velocity of the satellite, and (3) the mass of the object that the satellite revolves around. In this investigation, students can use the *My Solar System* simulation to change each of these three factors, one at a time, to determine how they affect the nature of an orbit. This kind of investigation is not possible without a simulation because we cannot change the mass of planets, stars, moons, or comets in our actual solar system; we can only observe how they interact with each other.

The initial preset for the *My Solar System* simulation includes

- a sun (body 1) with a mass of 200 units,
- a planet (body 2) with a mass of 10 units,
- a *y*-velocity for body 2 (which is the initial velocity of the object in the *y* direction) of 140 units/s, and
- an *x*-velocity for body 2 (which is the initial velocity of the object in the *x* direction) of 0 units/s

Under these conditions, the planet (body 2) will enter a stable orbit with a major axis of 206 units, a minor axis of 190 units, and an orbital period of about 14 seconds. Increasing the mass of the planet or the star will decrease the size of the major axis, the size of the minor axis, and the length of the orbital period. Increasing the *y*-velocity of the planet will increase the size of the major axis, the size of the minor axis, and the length of the orbital period.

Timeline

The instructional time needed to complete this lab investigation is 170–230 minutes. Appendix 3 (p. 573) provides options for implementing this lab investigation over several class periods. Option C (230 minutes) should be used if students are unfamiliar with scientific writing, because this option provides extra instructional time for scaffolding the writing process. You can scaffold the writing process by modeling, providing examples, and providing hints as students write each section of the report. Option D (170 minutes) should be used if students are familiar with scientific writing and have developed the skills needed to write an investigation report on their own. In option D, students complete stage 6 (writing the investigation report) and stage 8 (revising the investigation report) as homework.

Materials and Preparation

The materials needed to implement this investigation are listed in Table 3.1. The *My Solar System* simulation, which was developed by PhET Interactive Simulations, University of Colorado (*http://phet.colorado.edu*), is available at *https://phet.colorado.edu/en/simulation/legacy/my-solar-system*. It is free to use and can be run online using an internet browser. You should access the website and learn how the simulation works before beginning the lab investigation. In addition, it is important to check if students can access and use the simulation from a school computer or tablet, because some schools have set up firewalls and other restrictions on web browsing.

You can also download an app-based version of PhET that includes this simulation. The app-based version currently only works on Apple products and is available at the Apple App Store.

TABLE 3.1

Materials list for Lab 3

Item	Quantity
Computer with internet access (or tablet with PhET app)	1 per group
Investigation Proposal C (optional)	3 per group
Whiteboard, 2' × 3'*	1 per group
Lab Handout	1 per student
Peer-review guide and teacher scoring rubric	1 per student
Checkout Questions	1 per student

* As an alternative, students can use computer and presentation software such as Microsoft PowerPoint or Apple Keynote to create their arguments.

Safety Precautions

Remind students to follow all normal lab safety rules.

Topics for the Explicit and Reflective Discussion

Reflecting on the Use of Core Ideas and Crosscutting Concepts During the Investigation

Teachers should begin the explicit and reflective discussion by asking students to discuss what they know about DCIs they used during the investigation. The following are some important concepts related to the DCIs of (a) The Universe and Its Stars and (b) Earth and the Solar System that students need to know to explore the role gravity plays in the motion of objects within the solar system:

Gravity and Orbits
How Does Changing the Mass and Velocity of a Satellite and the Mass of the Object
That It Revolves Around Affect the Nature of the Satellite's Orbit?

- *Gravity* is an attractive force between two objects that have mass.

- *Orbital period* is the time taken for a given object to make one complete orbit around another object.

- *Eccentricity* describes how much an orbit deviates from a perfect circle. Eccentricities for a satellite are always between zero and one.

- *Velocity* is the rate of change in position over a period of time. *Acceleration* is the rate of change in velocity over a period of time.

To help students reflect on what they know about these concepts, we recommend showing them two or three images using presentation software that help illustrate these important ideas. You can then ask the students the following questions to encourage students to share how they are thinking about these important concepts:

1. What do we see going on in this image?

2. Does anyone have anything else to add?

3. What might be going on that we can't see?

4. What are some things that we are not sure about here?

You can then encourage students to think about how CCs played a role in their investigation. There are at least two CCs that students need to know to explore the role gravity plays in the motion of objects within the solar system: (a) Scale, Proportion, and Quantity; and (b) Systems and System Models (see Appendix 2 [p. 569] for a brief description of these CCs). To help students reflect on what they know about these CCs, we recommend asking them the following questions:

1. Why is it important to keep track of changes in a system quantitatively during an investigation?

2. What did you keep track of quantitatively during your investigation? What did that allow you to do?

3. Why do scientists often define a system and then develop a model of it as part of an investigation?

4. How did you use a model to understand the role gravity plays in the motion of objects within the solar system? Why was that useful?

You can then encourage students to think about how they used all these different concepts to help answer the guiding question and why it is important to use these ideas to help justify their evidence for their final arguments. Be sure to remind your students to explain why they included the evidence in their arguments and make the assumptions underlying their analysis and interpretation of the data explicit in order to provide an adequate justification of their evidence.

Reflecting on Ways to Design Better Investigations

It is important for students to reflect on the strengths and weaknesses of the investigation they designed during the explicit and reflective discussion. Students should therefore be encouraged to discuss ways to eliminate potential flaws, measurement errors, or sources of uncertainty in their investigations. To help students be more reflective about the design of their investigation and what they can do to make their investigations more rigorous in the future, you can ask the following questions:

1. What were some of the strengths of the way you planned and carried out your investigation? In other words, what made it scientific?

2. What were some of the weaknesses of the way you planned and carried out your investigation? In other words, what made it less scientific?

3. What rules can we make, as a class, to ensure that our next investigation is more scientific?

Reflecting on the Nature of Scientific Knowledge and Scientific Inquiry

This investigation can be used to illustrate two important concepts related to the nature of scientific knowledge and the nature of scientific inquiry: (a) the difference between laws and theories in science and (b) the assumptions made by scientists about order and consistency in nature (see Appendix 2 [p. 569] for a brief description of these two concepts). Be sure to review these concepts during and at the end of the explicit and reflective discussion. To help students think about these concepts in relation to what they did during the lab, you can ask the following questions:

1. Laws and theories are different in science. Is gravity an example of a theory or a law? Why?

2. Can you work with your group to come up with a rule that you can use to decide if something is a law or a theory? Be ready to share in a few minutes.

3. Scientists assume that the universe is a vast single system in which basic laws are consistent. Why do you think this assumption is important?

4. Think about the simulation you used during this investigation. How would it need to be different if the universe was not a single system in which basic laws are consistent?

Gravity and Orbits

How Does Changing the Mass and Velocity of a Satellite and the Mass of the Object That It Revolves Around Affect the Nature of the Satellite's Orbit?

You can also use presentation software or other techniques to encourage the students to think about these concepts. You can show examples of laws (such as $\mathbf{g} = GM/\mathbf{r}^2$) or theories (such as *gravity is the curvature of four-dimensional space-time due to the presence of mass*) and ask students to indicate if they think each example is a law or a theory and explain their thinking. You can also show images of different scientific laws and ask students if these laws would be the same everywhere if the universe was not a single system. Then ask them to think about what scientists would need to do to be able to study the universe if it was made up of many different systems.

Remind your students that, to be proficient in science, it is important that they understand what counts as scientific knowledge and how that knowledge develops over time.

Hints for Implementing the Lab

- Learn how to use the simulation before the lab begins. It is important for you to know how to use the simulation so you can help students when they get stuck or confused.

- A group of three students per computer or tablet tends to work well.

- Allow the students to play with the simulation as part of the tool talk before they begin to design their investigation. This gives students a chance to see what they can and cannot do with the simulation.

- Allowing students to design their own procedures for collecting data gives students an opportunity to try, to fail, and to learn from their mistakes. However, you can scaffold students as they develop their procedure by having them fill out an investigation proposal. These proposals provide a way for you to offer students hints and suggestions without telling them how to do it. You can also check the proposals quickly during a class period. We recommend using Investigation Proposal C for this lab.

- Investigation Proposal C allows students to generate three alternative hypotheses about how the orbit will change. For example, one hypothesis might be that the size of the orbit will increase as the velocity of the satellite increases. A second hypothesis might be that the size of the orbit will decrease as the velocity of the satellite increases. A third hypothesis might be that the size of the orbit will stay the same as the velocity of the satellite increases. We recommend that students fill out an investigation proposal for each experiment they do. Each group will therefore need three copies of the investigation proposal because each group will be responsible for planning and carrying out three different experiments.

- Encourage students to sketch graphs for each prediction based on each hypothesis. You can then encourage students to think about how well their result fits with each prediction as they analyze their data.

- Encourage student to focus on changing only one factor at time (mass of star, mass of the planet, initial y-velocity of the planet) so it is easier to determine a relationship.

- Tell the students to use the initial conditions for the sun and planet preset as a baseline or control condition. The students can measure the major axis and minor axis of the planet's orbit and how long the planet takes to complete one orbit (its orbital period) for this condition and then see how the orbit of the planet changes when the mass of the sun, the mass of the planet, and the y-velocity of the planet are changed.

- Be sure that students record actual values (e.g., size of the orbit, major axis of the orbit, minor axis of the orbit, eccentricities of the orbit, y-velocity of the planet, mass of the star, mass of the planet) when they use the simulation, rather than just attempting to describe what they see on the computer screen (e.g., "the orbit was bigger," "it looked more like an oval," "the velocity was greater"). The simulation contains a number of tools that will allow the students to collect quantitative data during this investigation.

- Students often make mistakes during the data collection stage, but they should quickly realize these mistakes during the argumentation session. It will only take them a short period of time to re-collect data, and they should be allowed to do so. During the explicit and reflective discussion, students will also have the opportunity to reflect on and identify ways to improve the way they design investigations (especially how they attempt to control variables as part of an experiment). This also offers an opportunity to discuss what scientists do when they realize that a mistake is made during a study.

- This lab also provides an excellent opportunity to discuss how scientists identify a signal (a pattern or trend) from the noise (measurement error) in their data. Be sure to use this activity as a concrete example during the explicit and reflective discussion.

Connections to Standards

Table 3.2 highlights how the investigation can be used to address specific (a) performance expectations from the *NGSS* and (b) *Common Core State Standards* in English language arts (*CCSS ELA*).

Gravity and Orbits

*How Does Changing the Mass and Velocity of a Satellite and the Mass of the Object
That It Revolves Around Affect the Nature of the Satellite's Orbit?*

TABLE 3.2

Lab 3 alignment with standards

***NGSS* performance expectations**	Space systems • MS-ESS1-2: Develop and use a model to describe the role of gravity in the motions within galaxies and the solar system. • HS-ESS1-4: Use mathematical or computational representations to predict the motion of orbiting objects in the solar system.
***CCSS ELA*—Reading in Science and Technical Subjects**	Key ideas and details • CCSS.ELA-LITERACY.RST.6-8.1: Cite specific textual evidence to support analysis of science and technical texts. • CCSS.ELA-LITERACY.RST.6-8.2: Determine the central ideas or conclusions of a text; provide an accurate summary of the text distinct from prior knowledge or opinions. • CCSS.ELA-LITERACY.RST.9-10.1: Cite specific textual evidence to support analysis of science and technical texts, attending to the precise details of explanations or descriptions. • CCSS.ELA-LITERACY.RST.9-10.2: Determine the central ideas or conclusions of a text; trace the text's explanation or depiction of a complex process, phenomenon, or concept; provide an accurate summary of the text. • CCSS.ELA-LITERACY.RST.9-10.3: Follow precisely a complex multistep procedure when carrying out experiments, taking measurements, or performing technical tasks, attending to special cases or exceptions defined in the text. Craft and structure • CCSS.ELA-LITERACY.RST.6-8.4: Determine the meaning of symbols, key terms, and other domain-specific words and phrases as they are used in a specific scientific or technical context relevant to *grade 6–8 texts and topics*. • CCSS.ELA-LITERACY.RST.6-8.5: Analyze the structure an author uses to organize a text, including how the major sections contribute to the whole and to an understanding of the topic. • CCSS.ELA-LITERACY.RST.6-8.6: Analyze the author's purpose in providing an explanation, describing a procedure, or discussing an experiment in a text. • CCSS.ELA-LITERACY.RST.9-10.4: Determine the meaning of symbols, key terms, and other domain-specific words and phrases as they are used in a specific scientific or technical context relevant to *grade 9–10 texts and topics*. • CCSS.ELA-LITERACY.RST.9-10.5: Analyze the structure of the relationships among concepts in a text, including relationships among key terms (e.g., *force, friction, reaction force, energy*).

Continued

TABLE 3.2 (*continued*)

CCSS ELA—**Reading in Science and Technical Subjects** (*continued*)	Craft and structure (*continued*) • CCSS.ELA-LITERACY.RST.9-10.6: Analyze the author's purpose in providing an explanation, describing a procedure, or discussing an experiment in a text, defining the question the author seeks to address. Integration of knowledge and ideas • CCSS.ELA-LITERACY.RST.6-8.7: Integrate quantitative or technical information expressed in words in a text with a version of that information expressed visually (e.g., in a flowchart, diagram, model, graph, or table). • CCSS.ELA-LITERACY.RST.6-8.8: Distinguish among facts, reasoned judgment based on research findings, and speculation in a text. • CCSS.ELA-LITERACY.RST.6-8.9: Compare and contrast the information gained from experiments, simulations, video, or multimedia sources with that gained from reading a text on the same topic. • CCSS.ELA-LITERACY.RST.9-10.7: Translate quantitative or technical information expressed in words in a text into visual form (e.g., a table or chart) and translate information expressed visually or mathematically (e.g., in an equation) into words. • CCSS.ELA-LITERACY.RST.9-10.8: Assess the extent to which the reasoning and evidence in a text support the author's claim or a recommendation for solving a scientific or technical problem. • CCSS.ELA-LITERACY.RST.9-10.9: Compare and contrast findings presented in a text to those from other sources (including their own experiments), noting when the findings support or contradict previous explanations or accounts.
CCSS ELA—**Writing in Science and Technical Subjects**	Text types and purposes • CCSS.ELA-LITERACY.WHST.6-10.1: Write arguments focused on *discipline-specific content.* • CCSS.ELA-LITERACY.WHST.6-10.2: Write informative or explanatory texts, including the narration of historical events, scientific procedures/experiments, or technical processes. Production and distribution of writing • CCSS.ELA-LITERACY.WHST.6-8.4: Produce clear and coherent writing in which the development, organization, and style are appropriate to task, purpose, and audience. • CCSS.ELA-LITERACY.WHST.6-10.5: With some guidance and support from peers and adults, develop and strengthen writing as needed by planning, revising, editing, rewriting, or trying a new approach, focusing on how well purpose and audience have been addressed.

Continued

TABLE 3.2 (*continued*)

CCSS ELA—**Writing in Science and Technical Subjects** (*continued*)	Production and distribution of writing (*continued*) • CCSS.ELA-LITERACY.WHST.6-8.6: Use technology, including the Internet, to produce and publish writing and present the relationships between information and ideas clearly and efficiently. • CCSS.ELA-LITERACY.WHST.9-10.5: Develop and strengthen writing as needed by planning, revising, editing, rewriting, or trying a new approach, focusing on addressing what is most significant for a specific purpose and audience. • CCSS.ELA-LITERACY.WHST.9-10.6: Use technology, including the internet, to produce, publish, and update individual or shared writing products, taking advantage of technology's capacity to link to other information and to display information flexibly and dynamically. Range of writing • CCSS.ELA-LITERACY.WHST.6-10.10: Write routinely over extended time frames (time for reflection and revision) and shorter time frames (a single sitting or a day or two) for a range of discipline-specific tasks, purposes, and audiences.
CCSS ELA—**Speaking and Listening**	Comprehension and collaboration • CCSS.ELA-LITERACY.SL.6-8.1: Engage effectively in a range of collaborative discussions (one-on-one, in groups, and teacher-led) with diverse partners on grade 6–8 topics, texts, and issues, building on others' ideas and expressing their own clearly. • CCSS.ELA-LITERACY.SL.6-8.2:* Interpret information presented in diverse media and formats (e.g., visually, quantitatively, orally) and explain how it contributes to a topic, text, or issue under study. • CCSS.ELA-LITERACY.SL.6-8.3:* Delineate a speaker's argument and specific claims, distinguishing claims that are supported by reasons and evidence from claims that are not. • CCSS.ELA-LITERACY.SL.9-10.1: Initiate and participate effectively in a range of collaborative discussions (one-on-one, in groups, and teacher-led) with diverse partners on grade 9–10 topics, texts, and issues, building on others' ideas and expressing their own clearly and persuasively. • CCSS.ELA-LITERACY.SL.9-10.2: Integrate multiple sources of information presented in diverse media or formats (e.g., visually, quantitatively, orally) evaluating the credibility and accuracy of each source. • CCSS.ELA-LITERACY.SL.9-10.3: Evaluate a speaker's point of view, reasoning, and use of evidence and rhetoric, identifying any fallacious reasoning or exaggerated or distorted evidence.

Continued

TABLE 3.2 (*continued*)

*CCSS ELA—***Speaking and Listening** (*continued*)	Presentation of knowledge and ideas • CCSS.ELA-LITERACY.SL.6-8.4:* Present claims and findings, sequencing ideas logically and using pertinent descriptions, facts, and details to accentuate main ideas or themes; use appropriate eye contact, adequate volume, and clear pronunciation. • CCSS.ELA-LITERACY.SL.6-8.5:* Include multimedia components (e.g., graphics, images, music, sound) and visual displays in presentations to clarify information. • CCSS.ELA-LITERACY.SL.6-8.6: Adapt speech to a variety of contexts and tasks, demonstrating command of formal English when indicated or appropriate. • CCSS.ELA-LITERACY.SL.9-10.4: Present information, findings, and supporting evidence clearly, concisely, and logically such that listeners can follow the line of reasoning and the organization, development, substance, and style are appropriate to purpose, audience, and task. • CCSS.ELA-LITERACY.SL.9-10.5: Make strategic use of digital media (e.g., textual, graphical, audio, visual, and interactive elements) in presentations to enhance understanding of findings, reasoning, and evidence and to add interest. • CCSS.ELA-LITERACY.SL.9-10.6: Adapt speech to a variety of contexts and tasks, demonstrating command of formal English when indicated or appropriate.

* Only the standard for grade 6 is provided because the standards for grades 7 and 8 are similar. Please see *www. corestandards.org/ELA-Literacy/SL* for the exact wording of the standards for grades 7 and 8.

Gravity and Orbits

How Does Changing the Mass and Velocity of a Satellite and the Mass of the Object
That It Revolves Around Affect the Nature of the Satellite's Orbit?

Lab Handout

Lab 3. Gravity and Orbits: How Does Changing the Mass and Velocity of a Satellite and the Mass of the Object That It Revolves Around Affect the Nature of the Satellite's Orbit?

Introduction

The motion of an object is the result of all the different *forces* that act on it. If you pull on a door, the door will move in the direction that you pulled it. If you push on a marble that is resting on a table, the marble will move in the direction you pushed it. Pulling on a door and pushing on a marble are examples of a *contact force*, which is a force that is applied to an object through direct contact. There are other types of forces that can push or pull on an object without touching it. A magnet, for example, can pull or push on another magnet without touching it. Static electricity, which is the buildup of electrical charge on an object, can also pull or push on an object. Magnetic and electrical forces are therefore called *non-contact forces* because they act at a distance. Perhaps the most common non-contact force is *gravity*. Gravity is a force of attraction between two objects; the force due to gravity always works to bring objects closer together.

Any two objects, as long as they have some mass, will have a gravitational force of attraction between them. The strength or magnitude of the gravitational force that exists between any two objects is influenced by the masses of those two objects and the distance between them. The magnitude of gravitational attraction increases with greater mass. This means that the gravitational force that exists between Earth and a car is greater than the gravitational force that exists between Earth and a marble. The magnitude of gravitational attraction, however, decreases as the distance between any two objects increases. The magnitude of the gravitational force that exists between Earth and an object that is moving away from it will therefore get weaker and weaker as the objects moves farther and farther away from Earth.

The force of gravity keeps planets orbiting a star and moons orbiting planets. An *orbit* is a regular, repeating path that one object in space takes around another one. An object in an orbit is called a *satellite*. A satellite can be natural, like planets, moons, and comets, or it can be something that was created by engineers and scientists, such as the International Space Station or the Hubble Space Telescope.

All orbits are *elliptical,* which means that the satellite follows a path that is round but can range in shape from a perfect circle to a long, thin oval. The shape of the orbit that most of the inner planets of our solar system follow, for example, is nearly circular. Figure L3.1 (p. 90) shows the orbits of Venus, Earth, and Mars. Notice that these orbits look almost like perfect circles. The orbits of comets and some of the outer dwarf plants have a very different shape. They are highly *eccentric.* In other words, their orbits look like a squashed circle.

LAB 3

FIGURE L3.1 _____

The orbits of Venus, Earth, and Mars as they would appear to an observer located above our solar system (the diagram is not to scale)

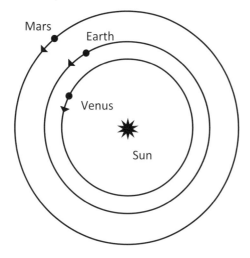

One way to describe the shape of an orbit is to calculate its *eccentricity*. Eccentricity is a way to quantify how much an orbit differs from a perfect circle. It is a value that ranges from 0 to 1. An orbit with an eccentricity of 0 is a perfect circle. Figure L3.2 illustrates orbits with eccentricity values of 0., 0.5, 0.75, and 0.9. The formula used to calculate eccentricity is

$$e = (\sqrt{a^2 - b^2})/a$$

where e is the eccentricity of the ellipse, $a =$ is the major axis of the ellipse, and $b =$ is the minor axis of the ellipse. Figure L3.3 shows how to calculate the eccentricity of an orbit. In this example, the major axis of the ellipse (a) is 7 units long and the minor axis (b) is 5 units long. Substituting these values into the formula gives a value of 0.7. This elliptical orbit would be considered highly eccentric.

FIGURE L3.2 _____

Orbits with different eccentricities

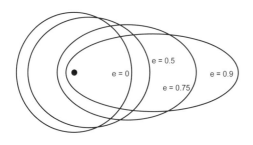

FIGURE L3.3 _____

How to calculate the eccentricity of an orbit

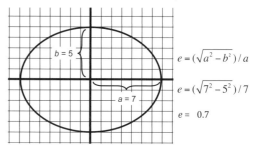

Another way to describe the orbit of a satellite is to measure its orbital distance. Satellites, however, do not always stay the same distance from the star or planet that they orbiting because their orbits are elliptical. For planets, like Earth, the point in their orbit when they are closest to the Sun is called the perihelion (see Figure L3.4). The point where a planet is farthest from the Sun is called the aphelion. The closest point the Moon or a manufactured satellite comes to Earth is called its perigee, and the farthest point is the apogee. Earth reaches its aphelion during July and its perihelion in January. The third, and final, way to describe the orbit of a satellite is to measure the time it takes to make one full orbit. The amount of time required to complete an orbit is called the orbital period. Earth, for example, has an orbital period of one year.

Gravity and Orbits

How Does Changing the Mass and Velocity of a Satellite and the Mass of the Object
That It Revolves Around Affect the Nature of the Satellite's Orbit?

In this investigation, you will have an opportunity to use an online simulation to explore how three different factors affect the shape, distance, and period of a satellite's orbit. The first factor is the mass of the satellite. The second factor is its initial velocity (speed in a given direction). The third factor is the mass of the object that it is orbiting. This type of investigation can be difficult because identifying the exact nature of the relationship that exists between multiple factors is challenging. Take mass as an example. There are many potential ways that the mass of a satellite or the mass of the object it is orbiting could influence the satellite's orbit. The shape, distance, and period of the orbit may depend on the mass of the larger object and/or the mass of the smaller object. The mass of the satellite and the mass of the object it is orbiting could also change these three aspects of a satellite's orbit in different ways.

FIGURE L3.4

Illustration of the orbit of Earth around the Sun and of the Moon around Earth showing the aphelion, perihelion, perigee, and apogee. (The orbits of the Earth and the Moon are not as eccentric as they appear in this image.)

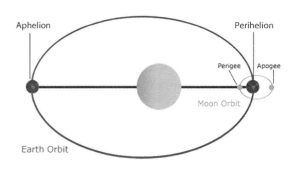

In addition to mass, there are many different ways that an orbit might change due to a change in the initial velocity of a satellite. The eccentricity of an orbit may either increase or decrease as the initial velocity increases. The initial velocity may also affect the orbital period but may not change the distance of its perigee and apogee (or perihelion and aphelion if the satellite is a planet). All of these different relationships are possible, as well as many others. Your goal in this investigation is to determine how all three factors are related to each other so you can better understand and predict the shape, distance, and period of a satellite's orbit.

Your Task

Use what you know about gravity; scale, proportion, and quantity; and the role of models in science to design and carry out an investigation that will allow you to determine how three different factors affect the shape, distance, and period of a satellite's orbit. The three factors you will explore are the mass of the satellite, the initial velocity of the satellite, and the mass of the object that the satellite is orbiting.

The guiding question of this investigation is, *How does changing the mass and velocity of a satellite and the mass of the object that it revolves around affect the nature of the satellite's orbit?*

Materials

You will use an online simulation called *My Solar System* to conduct your investigation; the simulation is available at *https://phet.colorado.edu/en/simulation/legacy/my-solar-system.*

LAB 3

Safety Precautions

Follow all normal lab safety rules.

Investigation Proposal Required?   ☐ Yes   ☐ No

Getting Started

The *My Solar System* simulation (see Figure L3.5) enables you to observe the orbit of a planet as it orbits around a star. It also allows you to change the mass of the planet and the star to see how changes in mass affects the shape, distance, and period of the planet's orbit. You can also add additional bodies to the solar system and change the initial velocity of any object that is orbiting the star.

FIGURE L3.5 _____

A screenshot from the *My Solar System* simulation

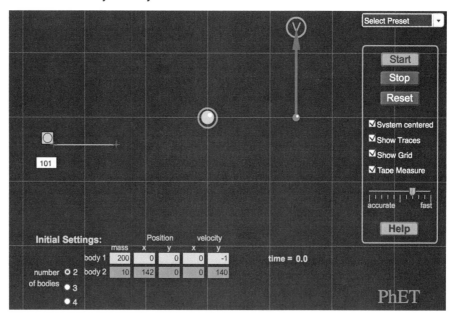

To use this simulation, start by making sure that the boxes next to System Centered, Show Traces, Show Grid, and Tape Measure in the control panel on the right side of the screen are all checked. This will make it easier for you to take the measurements. You can add or remove bodies from the solar system by clicking on the radio buttons in the lower left corner. The mass, initial position, and initial velocity of each body in the solar system can also be changed by typing in new values for each factor using the text boxes at the bottom of the simulation. This simulation is useful because it allows you see the path a planet takes as it orbits a star, and perhaps more important, it provides a way for you to design and carry out controlled experiments. This is important because you must be able

Gravity and Orbits
*How Does Changing the Mass and Velocity of a Satellite and the Mass of the Object
That It Revolves Around Affect the Nature of the Satellite's Orbit?*

to manipulate variables during a controlled experiment, and many of the variables that we are interested in here, such as the mass of a star, the mass of a planet, or the initial velocity of a planet, cannot be changed in the real world.

You will need to design and carry out at least three different experiments using the *My Solar System* simulation to determine the relationship between the three factors and the nature of a satellite's orbit. Remember, any object in an orbit is called a satellite. A satellite can be natural, like planets and moons, or a satellite can something that is manufactured and sent into space. You will need to conduct three different experiments because you will need to be able to answer three specific questions before you will be able to develop an answer to the guiding question for this lab:

- How does changing the mass of the star affect the way a planet orbits around it?
- How does changing the mass of a planet affect the way it orbits around a star?
- How does changing the velocity of a planet affect the way it orbits around a star?

It will be important for you to determine what type of data you need to collect, how you will collect the data, and how you will analyze the data for each experiment because each experiment is slightly different. To determine *what type of data you need to collect,* think about the following questions:

- What are the components of this system and how do they interact?
- How can you describe the components of the system quantitatively?
- What information will you need to determine the perihelion and aphelion (or perigee and apogee) of an orbit during each experiment?
- What information will you need to calculate the eccentricity of an orbit during each experiment?
- What information will you need to determine an orbital period during each experiment?

To determine *how you will collect the data,* think about the following questions:

- What will serve as your independent and dependent variable for each experiment?
- How will you vary the independent variable during each experiment?
- What will you do to hold the other variables constant during each experiment?
- When will you need to take measurements or observations during each experiment?
- What scale or scales should you use when you take your measurements?
- What types of comparisons will you need to make using the simulation?
- How will you keep track of the data you collect and how will you organize it?

To determine *how you will analyze the data,* think about the following questions:

- How will you compare the perihelion and aphelion (or perigee and apogee) of an orbit?
- How will you calculate the eccentricity of an orbit?
- How will you compare the eccentricities of several different orbits?
- How will you determine an orbital period?
- How will you compare the periods of several different orbits?
- What potential proportional relationships can you find in the data?

Once you have carried out all your different experiments, your group will need to develop an answer to the guiding question for this investigation. To be sufficient, your answer must explain how the mass of the satellite, the initial velocity of the satellite, and the mass of the object that the satellite is orbiting affect the eccentricity, the perihelion and aphelion, and the period of an orbit. For it to be valid and acceptable, your answer will also need to be consistent with your findings from all three experiments.

Connections to the Nature of Scientific Knowledge and Scientific Inquiry

As you work through your investigation, be sure to think about

- the difference between laws and theories in science, and
- the assumptions made by scientists about order and consistency in nature.

Initial Argument

Once your group has finished collecting and analyzing your data, your group will need to develop an initial argument. Your initial argument needs to include a claim, evidence to support your claim, and a justification of the evidence. The claim is your group's answer to the guiding question. The evidence is an analysis and interpretation of your data. Finally, the justification of the evidence is why your group thinks the evidence matters. The justification of the evidence is important because scientists can use different kinds of evidence to support their claims. Your group will create your initial argument on a whiteboard. Your whiteboard should include all the information shown in Figure L3.6.

FIGURE L3.6

Argument presentation on a whiteboard

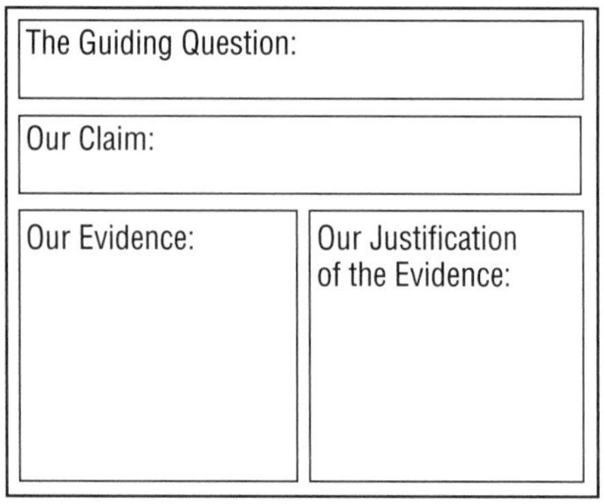

The Guiding Question:	
Our Claim:	
Our Evidence:	Our Justification of the Evidence:

Argumentation Session

The argumentation session allows all of the groups to share their arguments. One or two members of each group will stay at the lab station to share that group's

Gravity and Orbits

*How Does Changing the Mass and Velocity of a Satellite and the Mass of the Object
That It Revolves Around Affect the Nature of the Satellite's Orbit?*

argument, while the other members of the group go to the other lab stations to listen to and critique the other arguments. This is similar to what scientists do when they propose, support, evaluate, and refine new ideas during a poster session at a conference. If you are presenting your group's argument, your goal is to share your ideas and answer questions. You should also keep a record of the critiques and suggestions made by your classmates so you can use this feedback to make your initial argument stronger. You can keep track of specific critiques and suggestions for improvement that your classmates mention in the space below.

Critiques of our initial argument and suggestions for improvement:

If you are critiquing your classmates' arguments, your goal is to look for mistakes in their arguments and offer suggestions for improvement so these mistakes can be fixed. You should look for ways to make your initial argument stronger by looking for things that the other groups did well. You can keep track of interesting ideas that you see and hear during the argumentation in the space below. You can also use this space to keep track of any questions that you will need to discuss with your team.

Interesting ideas from other groups or questions to take back to my group:

LAB 3

Once the argumentation session is complete, you will have a chance to meet with your group and revise your initial argument. Your group might need to gather more data or design a way to test one or more alternative claims as part of this process. Remember, your goal at this stage of the investigation is to develop the best argument possible.

Report

Once you have completed your research, you will need to prepare an investigation report that consists of three sections. Each section should provide an answer for the following questions:

1. What question were you trying to answer and why?

2. What did you do to answer your question and why?

3. What is your argument?

Your report should answer these questions in two pages or less. You should write your report using a word processing application (such as Word, Pages, or Google Docs), if possible, to make it easier for you to edit and revise it later. You should embed any diagrams, figures, or tables into the document. Be sure to write in a persuasive style; you are trying to convince others that your claim is acceptable or valid.

Gravity and Orbits
How Does Changing the Mass and Velocity of a Satellite and the Mass of the Object
That It Revolves Around Affect the Nature of the Satellite's Orbit?

Checkout Questions

Lab 3. Gravity and Orbits: How Does Changing the Mass and Velocity of a Satellite and the Mass of the Object That It Revolves Around Affect the Nature of the Satellite's Orbit?

1. How is an object's mass related to the force of gravity it will exert on other objects?

2. How does the gravitational force of an object change as distance from the object increases?

3. A scientist has drawn a sketch of four planet-satellite pairs. The sizes of the circles represent the relative masses of the objects, and the size and direction of the arrows represent the initial velocity and direction of the satellites.

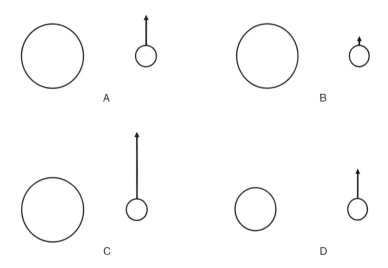

a. Which pair will most likely result in the satellite not entering a stable orbit?

A   B   C   D

b. How do you know?

c. Which pair will most likely result in the satellite crashing into the planet?

A   B   C   D

d. How do you know?

4. Theories are guesses and laws are facts.

  a. I agree with this statement.

  b. I disagree with this statement.

  Explain your answer, using an example from your investigation about gravity and orbits.

Gravity and Orbits

How Does Changing the Mass and Velocity of a Satellite and the Mass of the Object
That It Revolves Around Affect the Nature of the Satellite's Orbit?

5. Scientists assume that the universe is a vast single system in which basic laws are consistent.

    a. I agree with this statement.

    b. I disagree with this statement.

    Explain your answer, using an example from your investigation about gravity and orbits.

6. Scientists often use models to represent the components of a system and how these components interact with each other. Explain why models of systems are useful, using an example from your investigation about gravity and orbits.

7. It is critical for scientists to be able to keep track of changes in a system quantitatively during an investigation. Explain why this is so important, using an example from your investigation about gravity and orbits.

Application Lab

LAB 4

Teacher Notes

Lab 4. Habitable Worlds: Where Should NASA Send a Probe to Look for Life?

Purpose

The purpose of this lab is for students to *apply* what they know about the disciplinary core idea (DCI) of The Universe and Its Stars to identify exoplanets that may support life. In addition, students have an opportunity to learn about the crosscutting concepts (CCs) of (a) Patterns and (b) Scale, Proportion, and Quantity. During the explicit and reflective discussion, students will also learn about (a) the difference between data and evidence in science and (b) the assumptions made by scientists about order and consistency in nature.

Important Earth and Space Science Content

There are several factors that make Earth a suitable place for life to exist. These factors include, but are not limited to, its size, the composition of its atmosphere, and its temperature. These factors make Earth a place where liquid water and carbon-based organic molecules, the building blocks of living organisms, are able to persist.

Since 2009, the Kepler spacecraft has been monitoring the stars around the constellation Cygnus to find *exoplanets,* planets that orbit a star other than the Sun. The brightness of each star is recorded, and if the star dims with a repeated pattern, there is a possibility that there is an orbiting exoplanet passing between the star and the Kepler spacecraft. Based on these dimming events, Kepler objects of interest (KOIs) are identified and analyzed, and scientists determine whether the KOIs are false positives or planetary candidates. Once the planetary candidates are identified, other data are gathered about the exoplanets, their stars, and the relationship between them. These data are compared to data about Earth, the Sun, and their relationship to determine whether the exoplanet is likely to be able to support life.

One important factor in a planet's ability to support life is its size. If a planet is massive enough, its gravitational pull will hold on to gas molecules, creating an atmosphere. An atmosphere protects a planet from harmful radiation and traps heat inside the atmosphere, which causes a warm, relatively stable temperature on the surface of the planet. When a planet is too small, gas molecules escape to space and a planet is left without protection or insulation. Planetary size is also important in determining whether the planet will be terrestrial or gaseous. A planet with a diameter more than twice the size of Earth is unlikely to have a solid surface. To support life, a planet should be more than half the size of Earth to hold an atmosphere, but less than twice the size of Earth to have a rocky surface.

Scientists also use the relationship between Earth and the Sun to make assumptions about requirements for life sustainability. Earth is in the *habitable zone* of the Sun, meaning it is close enough to the Sun to absorb enough heat to sustain life, but not so close that the Sun would heat the planet beyond a livable temperature. The typical livable temperature range is one where liquid water can exist on the planet. The Kepler project reports the temperature of stars and the equilibrium temperature of the planets, which are useful data; however, it is important to understand that the equilibrium temperature reported is an average of temperature readings on the night and day side of a planet. Since it is an average, equilibrium temperature data leave out important information about the range and fluctuation of a planet's temperature. However, if the planet has a sufficient atmosphere, we can assume the surface temperature is at least somewhat regulated.

The Kepler project has identified 12 exoplanets that are less than twice the size of Earth and are within the habitable zone of their star, but it has recently identified one standout exoplanet. Discovered in July 2015, Kepler-452b is the most Earth-like exoplanet discovered so far. Kepler 452b has a diameter 60% larger than Earth, making it a good possibility that it is a rocky planet with an atmosphere (for more information, see *www.nasa.gov/press-release/nasa-kepler-mission-discovers-bigger-older-cousin-to-earth*). It is within the habitable zone of a G2 star (the Sun is also a G2 star), while the other exoplanets in habitable zones orbit K and M stars, which range from "somewhat" to "significantly" smaller and cooler than the Sun. That Kepler 452-b orbits the same type of star as the Sun and that it orbits its star at a similar distance Earth does from the Sun suggests that the planet could be a very similar temperature to Earth. Kepler-452b is the most Earth-like exoplanet and thus the most suitable choice to send a probe looking for life.

Timeline

The instructional time needed to complete this lab investigation is 170–230 minutes. Appendix 3 (p. 573) provides options for implementing this lab investigation over several class periods. Option C (230 minutes) should be used if students are unfamiliar with scientific writing, because this option provides extra instructional time for scaffolding the writing process. You can scaffold the writing process by modeling, providing examples, and providing hints as students write each section of the report. Option D (170 minutes) should be used if students are familiar with scientific writing and have developed the skills needed to write an investigation report on their own. In option D, students complete stage 6 (writing the investigation report) and stage 8 (revising the investigation report) as homework.

Materials and Preparation

The materials needed to implement this investigation are listed in Table 4.1 (p. 104).

LAB 4

TABLE 4.1

Materials list for Lab 4

Item	Quantity
Lab Handout	1 per student
Lab 4 Reference Sheet: Kepler Project Information Packet	1 per student
Whiteboard, 2' × 3'*	1 per group
Peer-review guide and teacher scoring rubric	1 per student
Checkout Questions	1 per student

* As an alternative, students can use computer and presentation software such as Microsoft PowerPoint or Apple Keynote to create their arguments.

Safety Precautions

Remind students to follow all normal lab safety rules.

Topics for the Explicit and Reflective Discussion

Reflecting on the Use of Core Ideas and Crosscutting Concepts During the Investigation

Teachers should begin the explicit and reflective discussion by asking students to discuss what they know about the DCI they used during the investigation. The following are some important concepts related to the DCI of The Universe and Its Stars that students need to be able to identify exoplanets that may support life:

- Life, as we know it, requires liquid water.
- Planets need to be terrestrial to support life.
- Planets need to be large enough to maintain an atmosphere.
- Planets must be in the habitable zone around a star to support life.
- The habitable zone is a range of orbits around a star where a planetary surface can support liquid water given sufficient atmospheric pressure.
- Stars differ in size and temperature.

To help students reflect on what they know about these concepts, we recommend showing them two or three images using presentation software that help illustrate these important ideas. You can then ask the students the following questions to encourage students to share how they are thinking about these important concepts:

1. What do we see going on in this image?
2. Does anyone have anything else to add?
3. What might be going on that we can't see?

4. What are some things that we are not sure about here?

You can then encourage students to think about how CCs played a role in their investigation. There are at least two CCs that students need to be able to identify exoplanets that may support life: (a) Patterns and (b) Scale, Proportion, and Quantity (see Appendix 2 [p. 569] for a brief description of these CCs). To help students reflect on what they know about these CCs, we recommend asking them the following questions:

1. Why is it important to look for, and attempt to explain, patterns in nature during an investigation?

2. What patterns did you identify during your investigation? What did the identification of these patterns allow you to do?

3. Why is it important to be able to describe the components of a system quantitatively during an investigation?

4. What components of the solar system did you need to describe quantitatively during your investigation? What did that allow you to do?

You can then encourage the students to think about how they used all these different concepts to help answer the guiding question and why it is important to use these ideas to help justify their evidence for their final arguments. Be sure to remind your students to explain why they included the evidence in their arguments and make the assumptions underlying their analysis and interpretation of the data explicit in order to provide an adequate justification of their evidence.

Reflecting on Ways to Design Better Investigations

It is important for students to reflect on the strengths and weaknesses of the investigation they designed during the explicit and reflective discussion. Students should therefore be encouraged to discuss ways to eliminate potential flaws, measurement errors, or sources of uncertainty in their investigations. To help students be more reflective about the design of their investigation and what they can do to make their investigations more rigorous in the future, you can ask the following questions:

1. What were some of the strengths of the way you planned and carried out your investigation? In other words, what made it scientific?

2. What were some of the weaknesses of the way you planned and carried out your investigation? In other words, what made it less scientific?

3. What rules can we make, as a class, to ensure that our next investigation is more scientific?

Reflecting on the Nature of Scientific Knowledge and Scientific Inquiry

This investigation can be used to illustrate two important concepts related to the nature of scientific knowledge and the nature of scientific inquiry: (a) the difference between data and evidence in science and (b) the assumptions made by scientists about order and consistency in nature (see Appendix 2 [p. 569] for a brief description of these two concepts). Be sure to review these concepts during and at the end of the explicit and reflective discussion. To help students think about these concepts in relation to what they did during the lab, you can ask the following questions:

1. You had to talk about both data and evidence during your investigation. Can you give me some examples of data and evidence from your investigation?

2. Can you work with your group to come up with a rule that you can use to decide if a piece of information is data or evidence? Be ready to share in a few minutes.

3. Scientists assume that the universe is a vast single system in which basic laws are consistent. Why do you think this assumption is important?

4. Think about what you were trying to do during this investigation. What would you have had to do differently if you could not assume that the universe is a single system in which basic laws are consistent?

You can also use presentation software or other techniques to encourage the students to think about these concepts. You can show examples of information from the investigation that are either data or evidence and ask students to classify each example and explain their thinking. You can also show images of different scientific laws and ask students if these laws would be the same everywhere if the universe was not a single system. Then ask them to think about what scientists would need to do to be able to study the universe if it was made up of many different systems.

Remind your students that, to be proficient in science, it is important that they understand what counts as scientific knowledge and how that knowledge develops over time.

Hints for Implementing the Lab

- Allowing students to determine how they will analyze the supplied data gives students an opportunity to try, to fail, and to learn from their mistakes.

- This lab also provides an excellent opportunity to discuss how scientists identify a signal (a pattern or trend) from the noise (measurement error) in their data. Be sure to use this activity as a concrete example during the explicit and reflective discussion.

Connections to Standards

Table 4.2 highlights how the investigation can be used to address specific (a) performance expectations from the *NGSS* and (b) *Common Core State Standards* in English language arts (*CCSS ELA*).

TABLE 4.2 _____

Lab 4 alignment with standards

***NGSS* performance expectation**	Space systems • MS-ESS1-3: Analyze and interpret data to determine scale properties of objects in the solar system.
***CCSS ELA*—Reading in Science and Technical Subjects**	Key ideas and details • CCSS.ELA-LITERACY.RST.6-8.1: Cite specific textual evidence to support analysis of science and technical texts. • CCSS.ELA-LITERACY.RST.6-8.2: Determine the central ideas or conclusions of a text; provide an accurate summary of the text distinct from prior knowledge or opinions. Craft and structure • CCSS.ELA-LITERACY.RST.6-8.4: Determine the meaning of symbols, key terms, and other domain-specific words and phrases as they are used in a specific scientific or technical context relevant to *grade 6–8 texts and topics*. • CCSS.ELA-LITERACY.RST.6-8.5: Analyze the structure an author uses to organize a text, including how the major sections contribute to the whole and to an understanding of the topic. • CCSS.ELA-LITERACY.RST.6-8.6: Analyze the author's purpose in providing an explanation, describing a procedure, or discussing an experiment in a text. Integration of knowledge and ideas • CCSS.ELA-LITERACY.RST.6-8.7: Integrate quantitative or technical information expressed in words in a text with a version of that information expressed visually (e.g., in a flowchart, diagram, model, graph, or table). • CCSS.ELA-LITERACY.RST.6-8.8: Distinguish among facts, reasoned judgment based on research findings, and speculation in a text. • CCSS.ELA-LITERACY.RST.6-8.9: Compare and contrast the information gained from experiments, simulations, video, or multimedia sources with that gained from reading a text on the same topic.

Continued

TABLE 4.2 (*continued*)

CCSS ELA—**Writing in Science and Technical Subjects**	Text types and purposes • CCSS.ELA-LITERACY.WHST.6-8.1: Write arguments focused on *discipline-specific content.* • CCSS.ELA-LITERACY.WHST.6-8.2: Write informative or explanatory texts, including the narration of historical events, scientific procedures/experiments, or technical processes. Production and distribution of writing • CCSS.ELA-LITERACY.WHST.6-8.4: Produce clear and coherent writing in which the development, organization, and style are appropriate to task, purpose, and audience. • CCSS.ELA-LITERACY.WHST.6-8.5: With some guidance and support from peers and adults, develop and strengthen writing as needed by planning, revising, editing, rewriting, or trying a new approach, focusing on how well purpose and audience have been addressed. • CCSS.ELA-LITERACY.WHST.6-8.6: Use technology, including the internet, to produce and publish writing and present the relationships between information and ideas clearly and efficiently. Range of writing • CCSS.ELA-LITERACY.WHST.6-8.10: Write routinely over extended time frames (time for reflection and revision) and shorter time frames (a single sitting or a day or two) for a range of discipline-specific tasks, purposes, and audiences.
CCSS ELA—**Speaking and Listening**	Comprehension and collaboration • CCSS.ELA-LITERACY.SL.6-8.1: Engage effectively in a range of collaborative discussions (one-on-one, in groups, and teacher-led) with diverse partners on grade 6–8 topics, texts, and issues, building on others' ideas and expressing their own clearly. • CCSS.ELA-LITERACY.SL.6-8.2:* Interpret information presented in diverse media and formats (e.g., visually, quantitatively, orally) and explain how it contributes to a topic, text, or issue under study. • CCSS.ELA-LITERACY.SL.6-8.3:* Delineate a speaker's argument and specific claims, distinguishing claims that are supported by reasons and evidence from claims that are not. Presentation of knowledge and ideas • CCSS.ELA-LITERACY.SL.6-8.4:* Present claims and findings, sequencing ideas logically and using pertinent descriptions, facts, and details to accentuate main ideas or themes; use appropriate eye contact, adequate volume, and clear pronunciation. • CCSS.ELA-LITERACY.SL.6-8.5:* Include multimedia components (e.g., graphics, images, music, sound) and visual displays in presentations to clarify information. • CCSS.ELA-LITERACY.SL.6-8.6: Adapt speech to a variety of contexts and tasks, demonstrating command of formal English when indicated or appropriate.

* Only the standard for grade 6 is provided because the standards for grades 7 and 8 are similar. Please see *www.corestandards.org/ELA-Literacy/SL* for the exact wording of the standards for grades 7 and 8.

Lab Handout

Lab 4. Habitable Worlds: Where Should NASA Send a Probe to Look for Life?

Introduction

Our solar system consists of the star we call the Sun, the planets and dwarf plants that orbit it, and the moons that orbit the planets or dwarf planets. It also contains much smaller objects such as comets, asteroids, and dust. The Sun is one of many stars that have several objects orbiting around it in our Milky Way galaxy. As far as we know, Earth is the only planet that supports life. Earth has several unique properties that allow life to exist; it is solid, it is warm enough to allow liquid water, and it is not too cold to freeze its inhabitants. However, we have only extensively studied the planets and dwarf planets within our own solar system. We know some information about planets outside of our solar system, called *exoplanets*, but exoplanets are difficult to study because they are so far away.

Space scientists with the National Aeronautics and Space Administration (NASA) are currently looking for exoplanets as part of a long-term project called Kepler. To find exoplanets, space scientists aim powerful telescopes at stars outside of our solar system and then measure the brightness of the light that the star emits over long periods of time. They then look for the brightness of the star to dim in a pattern because it indicates that an object is orbiting that star. Figure L4.1 shows what happens to the brightness of a star over time as a planet moves around it. The top row shows what a star would look like to us from Earth as a planet orbits around it. The second row shows how the brightness of the star remains constant except when the planet crosses in front of it—at those moments, the brightness of the star decreases. Kepler space scientists describe this instance as a transit event, and they use computers to identify potential transit events.

FIGURE L4.1

An example of how Kepler space scientists identify exoplanets. The top panels show how the position of a hypothetical exoplanet changes over time as it orbits a hypothetical star when viewed from Earth. The bottom graph shows how the brightness of the star will decrease only when the exoplanet passes between the star and the Earth.

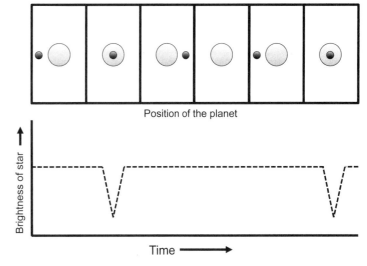

Position of the planet

Time

These potential transit events are classified as *Kepler objects of interest* (KOIs). Kepler space scientists then investigate each KOI to determine if the change in brightness of a star was caused by an exoplanet or something else. So far, space scientists working on the Kepler project have identified over 4,000 exoplanets using this method.

Once the Kepler space scientists have identified an exoplanet, they attempt to determine if it can support life. To accomplish this task, the space scientists look to see if the exoplanet shares three important characteristics with Earth:

1. It must be terrestrial (composed primarily of rocks and metals).

2. It must have an atmosphere.

3. It must orbit within the habitable zone around a star, which is the minimum and maximum distance from a star where the temperature of a planetary surface can support liquid water given sufficient atmospheric pressure.

Space scientists can determine if an exoplanet is terrestrial or not based on its size, because any exoplanet that is more than twice the size of Earth is likely to be a gas giant. Space scientists can also use the size of an exoplanet to determine if it has an atmosphere or not. Any exoplanet that is less than half the size of Earth most likely does not have enough gravitational pull to keep an atmosphere around it. Finally, space scientists can determine if an exoplanet orbits a star within the habitable zone by measuring its orbital period. The location of the habitable zone around a star, however, is not the same for every exoplanet because some stars emit more energy than others.

These three characteristics are important because they are needed to be able to sustain liquid water on the surface of the exoplanet, and liquid water must be present to support life as we know it. Currently, the Kepler space scientists have identified 12 exoplanets that have all three of these characteristics, so it is possible that these 12 planets could support life.

Scientists are interested in finding exoplanets with life on them because learning more about the life found on these exoplanets will help us better understand the evolution of life on Earth. Information about the characteristics of life on exoplanets could either corroborate existing theories about how life began and changes over time or result in new theories. However, it would cost too much money and waste too much time to send exploratory probes to every exoplanet they find to look for life. So scientists must be able to identify the exoplanets with the highest chance of supporting life. In this investigation, you will have an opportunity to use data about several different exoplanets to determine which ones, if any, have the potential to support life.

Your Task

Use what you know about the Earth-Sun system, patterns, and how scientists need to consider different scales, proportional relationships, and quantities during an investigation to examine the characteristics of several exoplanets orbiting around distant stars. Your goal is to determine which exoplanet, if any, is most likely to contain life based on its physical properties, the properties of the star it orbits, and the size and shape of its orbit.

The guiding question of this investigation is, ***Where should NASA send a probe to look for life?***

Materials

You will use the Lab 4 Reference Sheet: Kepler Project Information Packet during your investigation.

Safety Precautions

Follow all normal lab safety rules.

Investigation Proposal Required?   ☐ Yes   ☐ No

Getting Started

To answer the guiding question, you will need to analyze an existing data set. To determine *how you will analyze the data,* think about the following questions:

- Which data are relevant based on the guiding question?
- What type of calculations will you need to make?
- What types of patterns might you look for as you analyze your data?
- Are there any proportional relationships that you can identify?
- How will you determine if the physical properties of the stars and their planet candidates are the same or different?
- How could you use mathematics to determine if there is or is not a difference?
- What type of table or graph could you create to help make sense of your data?

Connections to the Nature of Scientific Knowledge or Scientific Inquiry

As you work through your investigation, be sure to think about

- the difference between data and evidence in science, and
- the assumptions made by scientists about order and consistency in nature.

Initial Argument

Once your group has finished collecting and analyzing your data, your group will need to develop an initial argument. Your initial argument needs to include a claim, evidence to support your claim, and a justification of the evidence. The *claim* is your group's answer to the guiding question. The *evidence* is an analysis and interpretation of your data. Finally, the *justification* of the evidence is why your group thinks the evidence matters. The justification of the evidence is important because scientists can use different kinds of evidence

LAB 4

FIGURE L4.2 _____

Argument presentation on a whiteboard

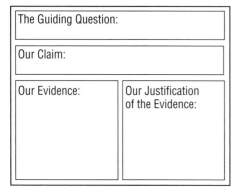

The Guiding Question:	
Our Claim:	
Our Evidence:	Our Justification of the Evidence:

to support their claims. Your group will create your initial argument on a whiteboard. Your whiteboard should include all the information shown in Figure L4.2.

Argumentation Session

The argumentation session allows all of the groups to share their arguments. One or two members of each group will stay at the lab station to share that group's argument, while the other members of the group go to the other lab stations to listen to and critique the other arguments. This is similar to what scientists do when they propose, support, evaluate, and refine new ideas during a poster session at a conference. If you are presenting your group's argument, your goal is to share your ideas and answer questions. You should also keep a record of the critiques and suggestions made by your classmates so you can use this feedback to make your initial argument stronger. You can keep track of specific critiques and suggestions for improvement that your classmates mention in the space below.

Critiques about our initial argument and suggestions for improvement:

If you are critiquing your classmates' arguments, your goal is to look for mistakes in their arguments and offer suggestions for improvement so these mistakes can be fixed. You should look for ways to make your initial argument stronger by looking for things that the other groups did well. You can keep track of interesting ideas that you see and hear during the argumentation in the space below. You can also use this space to keep track of any questions that you will need to discuss with your team.

Interesting ideas from other groups or questions to take back to my group:

Once the argumentation session is complete, you will have a chance to meet with your group and revise your initial argument. Your group might need to gather more data or design a way to test one or more alternative claims as part of this process. Remember, your goal at this stage of the investigation is to develop the best argument possible.

Report

Once you have completed your research, you will need to prepare an *investigation report* that consists of three sections. Each section should provide an answer for the following questions:

1. What question were you trying to answer and why?

2. What did you do to answer your question and why?

3. What is your argument?

Your report should answer these questions in two pages or less. You should write your report using a word processing application (such as Word, Pages, or Google Docs), if possible, to make it easier for you to edit and revise it later. You should embed any diagrams, figures, or tables into the document. Be sure to write in a persuasive style; you are trying to convince others that your claim is acceptable or valid.

LAB 4

Lab 4. Habitable Worlds: Where Should NASA Send a Probe to Look for Life?

1. Draw a model showing the relationship between Earth, the Sun, an exoplanet, its star, the Milky Way galaxy, and the universe.

2. What are three characteristics that make life possible on Earth?

3. An exoplanet is discovered that has the following characteristics:

   - A diameter 1.7 times that of Earth
   - An orbital period of 42.6 days
   - An orbit semi-major axis of 0.8823 AU

   a. Based on the information provided, how confident can you be that this exoplanet is able to support life as we know it? Mark your answer on the line below

   Not at all confident • _____ • Very confident

   b. Explain your answer.

   c. What additional information would make you more confident and why?

4. A list of planet-star radius ratios is an example of evidence.

    a. I agree with this statement.

    b. I disagree with this statement.

Explain your answer, using an example from your investigation about habitable worlds.

5. Scientists assume the universe is a vast single system in which basic laws are consistent.

    a. I agree with this statement.

    b. I disagree with this statement.

Explain your answer, using an example from your investigation about habitable worlds.

6. Scientists often need to look for patterns that occur in the data they collect and analyze. Explain why identifying patterns is important, using an example from your investigation about habitable worlds.

7. It is critical for scientists to be able to describe components of a system quantitatively. Explain why it is important to be able to describe a system quantitatively, using an example from your investigation about habitable worlds.

SECTION 3
History of Earth

Introduction Labs

LAB 5

Teacher Notes

Lab 5. Geologic Time and the Fossil Record: Which Time Intervals in the Past 650 Million Years of Earth's History Are Associated With the Most Extinctions and Which Are Associated With the Most Diversification of Life?

Purpose

The purpose of this lab is to *introduce* students to the disciplinary core idea (DCI) of The History of Planet Earth by having them identify major extinction and diversification events that happened during the last 650 million years of Earth's history. In addition, students have an opportunity to learn about the crosscutting concepts (CCs) of (a) Patterns and (b) Scale, Proportion, and Quantity. During the explicit and reflective discussion, students will also learn about (a) how scientific knowledge changes over time and (b) how scientists use different methods to answer different types of questions.

Important Earth and Space Science Content

Earth scientists reconstruct events that have taken place in past by observing and comparing the structure, sequence, and properties of rocks, sediments, and fossils. They also use the location of current and past rivers, lakes, and ocean basins to reconstruct the history of Earth. An important part of this process is the analysis of rock strata and stratification. A *rock stratum* is a layer of sedimentary rock or soil with internally consistent characteristics that distinguish it from other layers. *Stratification* is the layering of rock strata (see Figure 5.1; *strata* is the plural of *stratum*).

FIGURE 5.1

Sedimentary rock layers (strata) at St Audrie's Bay, West Somerset, England

Earth scientists can also determine when these events happened by determining the *absolute age* or *relative age* of different layers. Earth scientists can determine the absolute age of a layer of rock by measuring the amount of different radioactive elements found in a layer, and they can determine the relative ages of different rock layers using some fundamental ideas about the ways layers of rock form over time that are based on our understanding of geologic processes. There are four fundamental ideas that Earth scientists use to study and determine the relative age of rock layers:

Geologic Time and the Fossil Record

Which Time Intervals in the Past 650 Million Years of Earth's History Are Associated With the Most Extinctions and Which Are Associated With the Most Diversification of Life?

- *The principle of original horizontality,* which was introduced by Nicholas Steno in 1669 (see *www.ucmp.berkeley.edu/history/steno.html*), states that sedimentary rocks are originally laid down in horizontal layers. Figure 5.1 provides an example of this principle.

- *The law of superposition,* which was also introduced by Steno in 1669, states that in an undisturbed column of rock, the youngest rocks are at the top and the oldest rocks are at the bottom.

- *The principle of uniformitarianism,* which was introduced by James Hutton in 1785 and expanded by Charles Lyell in the early 1800s (see *www.uniformitarianism.net*), means geologic processes are consistent throughout time.

- *The principle of faunal succession,* which was introduced by William Smith in 1816 (see *https://earthobservatory.nasa.gov/Features/WilliamSmith/page2.php*), states that fossils are found in rocks in a specific order. This principle led Earth scientists to use fossils as a way to define increments of time within the geologic time scale. Because many individual plant and animal species existed during known time periods, the location of certain types of fossils in a rock layer can reveal the age of the rocks and help Earth scientists decipher the history of landforms.

The geologic history of the Earth, or the geologic time scale, is broken up into hierarchical chunks of time based on the rock strata. Rock strata provide a record of major events that occurred in the past. These major events can be catastrophic, occurring over hours to years, or gradual, occurring over thousands to millions of years. Records of fossils and other rocks also show past periods of massive extinctions and extensive volcanic activity. From largest to smallest, this hierarchical organization of the geologic time scale, which is based on the major events in Earth's history, includes eons, eras, periods, epochs, and stages. All of these are displayed in the portion of the geologic time scale shown in Table 5.1 (p. 124).

The majority of macroscopic organisms, which include algae, fungi, plants, and animals, have lived during the Phanerozoic eon. When Earth scientists first proposed the Phanerozoic eon as a division of geologic time, they believed that the beginning of this eon (542 million years ago [mya]) marked the beginning of life on Earth. We now know that this eon only marks the appearance of macroscopic organisms and that life on Earth actually began about 3.8 billion years ago as single-celled organisms. The Phanerozoic is subdivided into three major divisions, called eras: the Cenozoic, the Mesozoic, and the Paleozoic (see the "Geologic Time Scale" web page at *www.ucmp.berkeley.edu/help/timeform.php*). The suffix *zoic* means animal. The root *Ceno* means recent, the root *Meso* means middle, and the root *Paleo* means ancient. These divisions mark major changes in the nature or composition of life found on Earth. For example, the Cenozoic (65.5 mya–present) is sometimes called the age of mammals because the largest animals on Earth during this era have been mammals, whereas the Mesozoic (251–65.5 mya) is sometimes called the age of dinosaurs because these animals were found on Earth during this time period. The Paleozoic (542–251 mya), in contrast, is divided into six different periods that mark the

LAB 5

TABLE 5.1 _____

Portion of the geologic time scale showing eon, era, periods, epochs, stages, and time span (mya = millions of years ago), based on data from Gradstein, Ogg, and Hilgen (2012)

Eon	Era	Period	Epoch	Stage	Time (mya)
Phanerozoic	Cenozoic	Neogene	Pliocene	Piacenzian	2.6-3.6
				Zanclean	3.6-5.3
			Miocene	Messinian	5.3-7.3
				Tortonian	7.3-11.6
				Serravalian	11.6-13.8
				Langhian	13.8-15.9
				Burdigalian	15.9-20.4
				Aquitanian	20.4-23.0
		Paleogene	Oligocene	Chattian	23.0-28.1
				Rupelian	28.1-33.9
			Eocene	Priabonian	33.9-37.8
				Bartonian	37.8-41.2
				Lutetian	41.2-47.8
				Ypresian	47.8-56.0

appearance of different kinds of invertebrate and vertebrate animals. These descriptions of these eras and periods can be somewhat misleading, however, because many different groups of animals lived during each of them. There were also many kinds of plants living during these different eras and periods.

In this investigation, students will have an opportunity to examine the fossil record to identify major extinction and diversification events in the geologic time scale. The data set they will use during the investigation is included in an Excel file called Fossil Record 2 Database Summary Counts. This file includes counts of the number of different families, orders, classes, or phyla found at specific time intervals in the geologic time scale. It also contains information about the number of families, orders, classes, or phyla that first appeared and were last observed in each time interval. These counts were made from a much larger database called The Fossil Record 2, which is a near-complete listing of the diversity of life through geologic time, compiled at the level of the family (Benton 1993, 1995). The entire database can be downloaded for free from "The Fossil Record" web page (see *http://palaeo.gly.bris.ac.uk/fossilrecord2/fossilrecord/index.html*).

There have been five mass extinction events over the history of life on Earth (Gradstein, Ogg, and Hilgen 2012):

1. The Ordovician-Silurian mass extinction occurred about 444 mya and marked the end of the Ordovician period and the start of the Silurian period. It was the third

Geologic Time and the Fossil Record

Which Time Intervals in the Past 650 Million Years of Earth's History Are Associated With the Most Extinctions and Which Are Associated With the Most Diversification of Life?

largest extinction in Earth's history. During the Ordovician, most life was in the sea, so it was sea creatures such as trilobites, brachiopods, and graptolites that went extinct. In all, some 85% of sea life was wiped out.

2. The late Devonian mass extinction occurred about 359 mya and marked the end of the Devonian period and the start of the Carboniferous period. Three-quarters of all species on Earth died out during this event.

3. The Permian mass extinction was the largest one in Earth's history. It occurred about 252 mya and marked the end of the Permian period and the start of the Triassic period. Ninety-six percent of all species on Earth died during this mass extinction. Marine creatures were affected the most, and insects suffered the only mass extinction of their history.

4. The Triassic-Jurassic mass extinction occurred about 201 mya and marked the end of the Triassic period and the start of the Jurassic period. Many types of animal died out, including many different marine reptiles, some large amphibians, and large numbers of cephalopod mollusks. Roughly half of all the species alive at the time became extinct.

5. The Cretaceous-Paleogene mass extinction occurred about 66 mya and marks the end of the Cretaceous period and the start of the Paleocene epoch. Dinosaurs and many different flowering plants went extinct during this event.

Each of the mass extinctions, however, was followed by a period of great diversification of life. The other great diversification event occurred during the Cambrian period. This event, which began about 541 million years ago and lasted for about 25 million years, marks the first appearance of all major animal phyla in the fossil record.

Timeline

The instructional time needed to complete this lab investigation is 170–230 minutes. Appendix 3 (p. 573) provides options for implementing this lab investigation over several class periods. Option C (230 minutes) should be used if students are unfamiliar with scientific writing, because this option provides extra instructional time for scaffolding the writing process. You can scaffold the writing process by modeling, providing examples, and providing hints as students write each section of the report. Option D (170 minutes) should be used if students are familiar with scientific writing and have developed the skills needed to write an investigation report on their own. In option D, students complete stage 6 (writing the investigation report) and stage 8 (revising the investigation report) as homework.

Materials and Preparation

The materials needed to implement this investigation are listed in Table 5.2 (p. 126). The Fossil Record 2 Database Summary Counts Excel file can be downloaded from the book's

Extras page at *www.nsta.org/adi-ess*. This file can be loaded onto student computers before the investigation, e-mailed to students, or uploaded to a class website that students can access. It is important that Excel be available on the computers that students will use so they can analyze the data set using the tools built into the spreadsheet application. It is also important for you to look over the file before the investigation begins so you can learn how the data in the file are organized. This will enable you to give students suggestions on how to analyze the data.

The Geologic Time Scale information sheet, which is based on an article by Gradstein, Ogg, and Hilgen (2012), can also be downloaded from the book's Extras page at *www.nsta. org/adi-ess* and then printed, photocopied, and distributed to each group. This information sheet provides the time span of each eon, era, period, epoch, and stage in the Geologic Time Scale.

TABLE 5.2

Materials list for Lab 5

Item	Quantity
Computer with Excel or other spreadsheet application	1 per group
Fossil Record 2 Database Summary Counts Excel file	1 per group
Geologic Time Scale information sheet	1 per group
Investigation Proposal A (optional)	1 per group
Whiteboard, 2' × 3'*	1 per group
Lab Handout	1 per student
Peer-review guide and teacher scoring rubric	1 per student
Checkout Questions	1 per student

* As an alternative, students can use computer and presentation software such as Microsoft PowerPoint or Apple Keynote to create their arguments.

Safety Precautions

Remind students to follow all normal lab safety rules.

Topics for the Explicit and Reflective Discussion

Reflecting on the Use of Core Ideas and Crosscutting Concepts During the Investigation

Teachers should begin the explicit and reflective discussion by asking students to discuss what they know about the DCI they used during the investigation. The following are some important concepts related to the DCI of The History of Planet Earth that students need to identify major extinction and diversification events that happened during the last 650 million years of Earth's history:

Geologic Time and the Fossil Record

Which Time Intervals in the Past 650 Million Years of Earth's History Are Associated With the Most Extinctions and Which Are Associated With the Most Diversification of Life?

- The geologic time scale, which is interpreted from rock strata, provides a way to organize Earth's history.

- The geologic time scale is broken up into hierarchical chunks of time.

- Divisions in geologic time are based on the major events in Earth's history and often correspond to major changes in the composition of ancient faunas.

- Scientists can identify major changes in the composition of ancient faunas using fossils.

- Earth scientists can determine when major changes in the composition of an ancient fauna occurred by determining the absolute age or relative age of different rock layers.

- The fossil record is incomplete because life forms that are common, widespread, and have hard shells or skeletons are more likely to be preserved as fossils than life forms that are rare, isolated, and have soft bodies.

- The term *family* refers to a taxonomic rank that falls between order and genus. The levels of classification include species, genus, family, order, class, phylum, and kingdom.

To help students reflect on what they know about these concepts, we recommend showing them two or three images using presentation software that help illustrate these important ideas. You can then ask the students the following questions in order to encourage students to share how they are thinking about these important concepts:

1. What do we see going on in this image?

2. Does anyone have anything else to add?

3. What might be going on that we can't see?

4. What are some things that we are not sure about here?

You can then encourage students to think about how CCs played a role in their investigation. There are at least two CCs that students need to identify major extinction and diversification events: (a) Patterns and (b) Scale, Proportion, and Quantity (see Appendix 2 [p. 569] for a brief description of these CCs). To help students reflect on what they know about these CCs, we recommend asking them the following questions:

1. Why do scientists look for and attempt to explain patterns in nature?

2. What patterns did you identify and use during your investigation? Why was that useful?

3. Why is it important to consider what measurement scale or scales to use during an investigation? Why is useful to look for proportional relationships when analyzing data?

4. What measurement scale or scales did you use during your investigation? What did that allow you to do? Did you attempt to look for proportional relationships when you were analyzing your data? Why or why not?

You can then encourage students to think about how they used all these different concepts to help answer the guiding question and why it is important to use these ideas to help justify their evidence for their final arguments. Be sure to remind your students to explain why they included the evidence in their arguments and make the assumptions underlying their analysis and interpretation of the data explicit in order to provide an adequate justification of their evidence.

Reflecting on Ways to Design Better Investigations

It is important for students to reflect on the strengths and weaknesses of the investigation they designed during the explicit and reflective discussion. Students should therefore be encouraged to discuss ways to eliminate potential flaws, measurement errors, or sources of uncertainty in their investigations. To help students be more reflective about the design of their investigation and what they can do to make their investigations more rigorous in the future, you can ask the following questions:

1. What were some of the strengths of the way you planned and carried out your investigation? In other words, what made it scientific?

2. What were some of the weaknesses of the way you planned and carried out your investigation? In other words, what made it less scientific?

3. What rules can we make, as a class, to ensure that our next investigation is more scientific?

Reflecting on the Nature of Scientific Knowledge and Scientific Inquiry

This investigation can be used to illustrate two important concepts related to the nature of scientific knowledge and the nature of scientific inquiry: (a) how scientific knowledge changes over time and (b) how scientists use different methods to answer different types of questions (see Appendix 2 [p. 569] for a brief description of these two concepts). Be sure to review these concepts during and at the end of the explicit and reflective discussion. To help students think about these concepts in relation to what they did during the lab, you can ask the following questions:

1. Scientific knowledge can and does change over time. Can you tell me why it changes?

2. Can you work with your group to come up with some examples of how scientific knowledge has changed over time? Be ready to share in a few minutes.

3. There is no universal step-by-step scientific method that all scientists follow. Why do you think there is no universal scientific method?

4. Think about what you did during this investigation. How would you describe the method you used to examine the fossil record? Why would you call it that?

You can also use presentation software or other techniques to encourage your students to think about these concepts. You can show examples of how our thinking about how to study and determine the relative age of rock layers has changed over time and ask students to discuss what they think led to those changes. You can also show one or more images of a "universal scientific method" that misrepresent the nature of scientific inquiry (see, e.g., *https://commons.wikimedia.org/wiki/File:The_Scientific_Method_as_an_Ongoing_Process. svg*) and ask students why each image is *not* a good representation of what scientists do to develop scientific knowledge. You can also ask students to suggest revisions to the image that would make it more consistent with the way scientists develop scientific knowledge.

Remind your students that, to be proficient in science, it is important that they understand what counts as scientific knowledge and how that knowledge develops over time.

Hints for Implementing the Lab

- Learn how to use the Fossil Record 2 Database Summary Counts Excel file before the lab begins. It is important for you to know what is included in the file and how you can analyze these data so you can help students when they get stuck or confused.

- A group of three students per computer tends to work well.

- Allow the students to play with the Fossil Record 2 Database Summary Counts Excel file as part of the tool talk before they begin to design their investigation. This gives students a chance to see what they can and cannot do with the data in Excel (or another spreadsheet application).

- Allowing students to design their own procedures for analyzing the data in the Excel file gives students an opportunity to try, to fail, and to learn from their mistakes. However, you can scaffold students as they develop their procedure by having them fill out an investigation proposal. These proposals provide a way for you to offer students hints and suggestions without telling them how to do it. You can also check the proposals quickly during a class period. We recommend using Investigation Proposal A for this lab.

- Encourage students to tabulate the data and make graphs using Excel (or another spreadsheet application). The best way to help students learn how to use Excel is to provide "just-in-time" instruction. In other words, wait for students to get stuck and then give a brief mini-lesson on how to use a specific tool in Excel based on what students are trying to do. They will be much more interested in learning about how to use the tools in Excel if they know it will help solve a problem they are having or will allow them to accomplish one of their goals.

- This lab provides an excellent opportunity to discuss how scientists must make choices about how to analyze the data they have and how the choice of analysis reflects the nature of the question they are trying to answer. Be sure to use this activity as a concrete example during the explicit and reflective discussion.

- This lab also provides an excellent opportunity to discuss how scientists identify a signal (a pattern or trend) from the noise (measurement error) in their data. Be sure to use this activity as a concrete example during the explicit and reflective discussion.

Connections to Standards

Table 5.3 highlights how the investigation can be used to address specific (a) performance expectations from the *NGSS* and (b) *Common Core State Standards* in English language arts (*CCSS ELA*).

TABLE 5.3

Lab 5 alignment with standards

NGSS performance expectation	History of Earth • MS-ESS1-4: Construct a scientific explanation based on evidence from rock strata for how the geologic time scale is used to organize Earth's 4.6-billion-year-old history.
CCSS ELA—Reading in Science and Technical Subjects	Key ideas and details • CCSS.ELA-LITERACY.RST.6-8.1: Cite specific textual evidence to support analysis of science and technical texts. • CCSS.ELA-LITERACY.RST.6-8.2: Determine the central ideas or conclusions of a text; provide an accurate summary of the text distinct from prior knowledge or opinions. Craft and structure • CCSS.ELA-LITERACY.RST.6-8.4: Determine the meaning of symbols, key terms, and other domain-specific words and phrases as they are used in a specific scientific or technical context relevant to *grade 6–8 texts and topics*. • CCSS.ELA-LITERACY.RST.6-8.5: Analyze the structure an author uses to organize a text, including how the major sections contribute to the whole and to an understanding of the topic. • CCSS.ELA-LITERACY.RST.6-8.6: Analyze the author's purpose in providing an explanation, describing a procedure, or discussing an experiment in a text.

Continued

Geologic Time and the Fossil Record

Which Time Intervals in the Past 650 Million Years of Earth's History Are Associated With the Most Extinctions and Which Are Associated With the Most Diversification of Life?

TABLE 5.3 (*continued*)

CCSS ELA—**Reading in Science and Technical Subjects** (*continued*)	Integration of knowledge and ideas • CCSS.ELA-LITERACY.RST.6-8.7: Integrate quantitative or technical information expressed in words in a text with a version of that information expressed visually (e.g., in a flowchart, diagram, model, graph, or table). • CCSS.ELA-LITERACY.RST.6-8.8: Distinguish among facts, reasoned judgment based on research findings, and speculation in a text. • CCSS.ELA-LITERACY.RST.6-8.9: Compare and contrast the information gained from experiments, simulations, video, or multimedia sources with that gained from reading a text on the same topic.
CCSS ELA—**Writing in Science and Technical Subjects**	Text types and purposes • CCSS.ELA-LITERACY.WHST.6-8.1: Write arguments focused on *discipline-specific content.* • CCSS.ELA-LITERACY.WHST.6-8.2: Write informative or explanatory texts, including the narration of historical events, scientific procedures/experiments, or technical processes. Production and distribution of writing • CCSS.ELA-LITERACY.WHST.6-8.4: Produce clear and coherent writing in which the development, organization, and style are appropriate to task, purpose, and audience. • CCSS.ELA-LITERACY.WHST.6-8.4: With some guidance and support from peers and adults, develop and strengthen writing as needed by planning, revising, editing, rewriting, or trying a new approach, focusing on how well purpose and audience have been addressed. • CCSS.ELA-LITERACY.WHST.6-8.6: Use technology, including the internet, to produce and publish writing and present the relationships between information and ideas clearly and efficiently. Range of writing • CCSS.ELA-LITERACY.WHST.6-8.10: Write routinely over extended time frames (time for reflection and revision) and shorter time frames (a single sitting or a day or two) for a range of discipline-specific tasks, purposes, and audiences.

Continued

TABLE 5.3 (*continued*)

***CCSS ELA*—Speaking and Listening**	Comprehension and collaboration • CCSS.ELA-LITERACY.SL.6-8.1: Engage effectively in a range of collaborative discussions (one-on-one, in groups, and teacher-led) with diverse partners on grade 6–8 topics, texts, and issues, building on others' ideas and expressing their own clearly. • CCSS.ELA-LITERACY.SL.6-8.2:* Interpret information presented in diverse media and formats (e.g., visually, quantitatively, orally) and explain how it contributes to a topic, text, or issue under study. • CCSS.ELA-LITERACY.SL.6-8.3:* Delineate a speaker's argument and specific claims, distinguishing claims that are supported by reasons and evidence from claims that are not. Presentation of Knowledge and Ideas • CCSS.ELA-LITERACY.SL.6-8.4:* Present claims and findings, sequencing ideas logically and using pertinent descriptions, facts, and details to accentuate main ideas or themes; use appropriate eye contact, adequate volume, and clear pronunciation. • CCSS.ELA-LITERACY.SL.6-8.5:* Include multimedia components (e.g., graphics, images, music, sound) and visual displays in presentations to clarify information. • CCSS.ELA-LITERACY.SL.6-8.6: Adapt speech to a variety of contexts and tasks, demonstrating command of formal English when indicated or appropriate.

* Only the standard for grade 6 is provided because the standards for grades 7 and 8 are similar. Please see *www.corestandards.org/ELA-Literacy/SL* for the exact wording of the standards for grades 7 and 8.

References

Benton, M. J. 1993. *The fossil record 2.* London: Chapman & Hall.

Benton, M. J. 1995. Diversification and extinction in the history of life. *Science* 268 (5207): 52–58.

Gradstein, F. M., J. G. Ogg, and F. J. Hilgen. 2012. On the geologic time scale. *Newsletters on Stratigraphy* 45 (2): 171–188.

Geologic Time and the Fossil Record

Which Time Intervals in the Past 650 Million Years of Earth's History Are Associated With the Most Extinctions and Which Are Associated With the Most Diversification of Life?

Lab Handout

Lab 5. Geologic Time and the Fossil Record: Which Time Intervals in the Past 650 Million Years of Earth's History Are Associated With the Most Extinctions and Which Are Associated With the Most Diversification of Life?

Introduction

Earth scientists use the structure, sequence, and properties of rocks, sediments, and fossils, as well as the locations of current and past ocean basins, lakes, and rivers, to learn about the major events in Earth's history. Major historical events include the formation of mountain chains and ocean basins, volcanic eruptions, periods of massive glaciation (when ice glaciers increase in size because of colder than average global temperatures), the development of watersheds and rivers, and the evolution and extinction of different types of organisms. Earth scientists can determine when and where these major events happened because rock layers, such as the ones pictured in Figure L5.1, provide a lot of information about how an area has changed over time. Earth scientists can also determine when these changes happened by determining the *absolute age* or *relative age* of different layers. Earth scientists can determine the absolute age of a layer of rock by measuring the amount of different radioactive elements found in a layer, and they can determine the relative ages of different rock layers using some fundamental ideas about the ways layers of rock form over time that are based on our understanding of geologic processes.

FIGURE L5.1 ———————————————————————

Horizontal rock layers are easy to see (a) at the Grand Canyon in Arizona and (b) near Khasab in Oman (a country in the Middle East)

a b

There are four fundamental ideas that Earth scientists use to study and determine the relative age of rock layers. Nicholas Steno introduced the first two in 1669 (see *www.ucmp. berkeley.edu/history/steno.html*). The first fundamental idea, which is called the *principle of original horizontality,* states that sedimentary rocks are originally laid down in horizontal layers; see Figure L5.1 for an example of this principle. The second fundamental idea is called the *law of superposition.* This law states that in an undisturbed column of rock, the youngest rocks are at the top and the oldest are at the bottom. The third fundamental idea is known as the *principle of uniformitarianism*; it was introduced by James Hutton in 1785, and later expanded by Charles Lyell in the early 1800s (see *www.uniformitarianism. net*). This idea states that geologic processes are consistent throughout time. William Smith introduced the fourth fundamental idea in 1816. This idea is called the *principle of faunal succession,* which states that fossils are found in rocks in a specific order (see *https://earthob- servatory.nasa.gov/Features/WilliamSmith/page2.php*). This principle led Earth scientists to use fossils as a way to define increments of time within the geologic time scale. Because many individual plant and animal species existed during known time periods, the location of certain types of fossils in a rock layer can reveal the age of the rocks and help Earth scientists decipher the history of landforms.

The geologic history of the Earth, or geologic time scale, is broken up into hierarchical chunks of time based on the major events in Earth's history. These major events can be cata- strophic, occurring over hours to years, or gradual, occurring over thousands to millions of years. Records of fossils and other rocks also show past periods of massive extinctions and extensive volcanic activity. From largest to smallest, this hierarchical organization of the geological time based on the major events includes eons, eras, periods, epochs, and stages. All of these are displayed in the portion of the geologic time scale shown in Table L5.1.

The majority of macroscopic organisms (organisms that can be seen by the human eye without a microscope), have lived during the Phanerozoic eon; these organisms include algae, fungi, plants, and animals. When Earth scientists first proposed the Phanerozoic eon as a division of geologic time, they believed that the beginning of this eon (542 million years ago [mya]) marked the beginning of life on Earth. We now know that this eon only marks the appearance of macroscopic organisms and that life on Earth actually began about 3.8 billion years ago as single-celled organisms. The Phanerozoic is subdivided into three major divi- sions, called eras: the Cenozoic, the Mesozoic, and the Paleozoic (see the "Geologic Time Scale" web page at *www.ucmp.berkeley.edu/help/timeform.php*). The suffix *zoic* means animal. The root *Ceno* means recent, the root *Meso* means middle, and the root *Paleo* means ancient. These divisions mark major changes in the nature or composition of life found on Earth. For example, the Cenozoic (65.5 mya–present) is sometimes called the age of mammals because the largest animals on Earth during this era have been mammals, whereas the Mesozoic (251–65.5 mya) is sometimes called the age of dinosaurs because these animals were found on Earth during this time period. The Paleozoic (542–251 mya), in contrast, is divided into six different periods that mark the appearance of different kinds of invertebrate and ver- tebrate animals. The descriptions of these eras and periods can be somewhat misleading,

TABLE L5.1

Some of the eons, eras, periods, epochs, and stages in the geologic time scale, based on data from Gradstein, Ogg, and Hilgen (2012). Notice that different stages are nested within an epoch, different epochs are nested within a period, and different periods are nested within an era.

Eon	Era	Period	Epoch	Stage	Time (mya)
Phanerozoic	Cenozoic	Neogene	Pliocene	Piacenzian	2.6-3.6
				Zanclean	3.6-5.3
			Miocene	Messinian	5.3-7.3
				Tortonian	7.3-11.6
				Serravalian	11.6-13.8
				Langhian	13.8-15.9
				Burdigalian	15.9-20.4
				Aquitanian	20.4-23.0
		Paleogene	Oligocene	Chattian	23.0-28.1
				Rupelian	28.1-33.9
			Eocene	Priabonian	33.9-37.8
				Bartonian	37.8-41.2
				Lutetian	41.2-47.8
				Ypresian	47.8-56.0

however, because many different groups of animals lived during each of them. There were also many kinds of plants living during these different eras and periods.

Earth scientists have collected a lot of data about the history of life on Earth over the last 400 years. This information not only allows scientists to determine what the conditions were like on Earth in the past, but also allows them to track when major groups of animals and plants appeared and disappeared during the past 650 million years of Earth's history. In this investigation, you will have an opportunity to learn about the extinction and diversification of life on Earth. It is important to note, however, that the fossil record only provides a partial picture how life on Earth has changed over time. Although it is substantial, the fossil record is incomplete because life forms that are common, widespread, and have hard shells or skeletons are more likely to be preserved as fossils than life forms that are rare, isolated, and have soft bodies. The fossil record, therefore, can only provide limited information about the history of life on Earth.

Your Task

Use a database called The Fossil Record 2 and what you know about geologic time, patterns, scales, proportions, and quantities to identify any major extinction and diversification events that happened during the last 650 million years. You goal is to determine how many of each of these important events occurred and when they happened in the geologic time scale.

The guiding question of this investigation is, *Which time intervals in the past 650 million years of Earth's history are associated with the most extinctions and which are associated with the most diversification of life?*

Materials

You will use a computer with Excel or other spreadsheet application during your investigation. You will also use the following resources:

- Fossil Record 2 Database Summary Counts Excel file
- Geologic Time Scale information sheet

Your teacher will tell you how to access the Excel file.

Safety Precautions

Follow all normal lab safety rules.

Investigation Proposal Required?
☐ Yes   ☐ No

Getting Started

The Fossil Record 2 database is a near-complete listing of the diversity of life through geologic time, compiled at the level of the family (Benton 1993, 1995). In biology, levels of classification include species, genus, family, order, class, phylum, and kingdom (see Figure L5.2), so *family* refers to a level of classification that falls between order and genus. For example, *Pan paniscus*, a species of ape commonly called a bonobo, is a part of

FIGURE L5.2 _____

Levels of biological classification

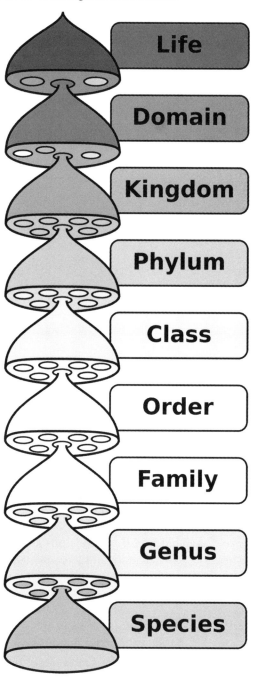

Geologic Time and the Fossil Record

Which Time Intervals in the Past 650 Million Years of Earth's History Are Associated With the Most Extinctions and Which Are Associated With the Most Diversification of Life?

the genus *Pan,* the family Hominidae, the order Primates, the class Mammalia, the phylum Chordata, and the kingdom Animalia.

The Fossil Record 2 database (Benton 1993) is an Excel file. Each row in this file represents a different biological family, and each column represents a different time interval. The geologic time range of each family is entered on the worksheet, with the first appearance of the family labeled as F1, presence in the fossil record labeled as 1, and last recorded appearance labeled as L1. You can download the entire database for free from this website: *http://palaeo.gly.bris.ac.uk/fossilrecord2/fossilrecord/index.html.*

You will not be using the entire database for this investigation; instead, you will be using an Excel file called Fossil Record 2 Database Summary Counts. This Excel file is a simplified version of The Fossil Record 2 database. It includes counts of the number of different families, orders, classes, or phyla found at specific time intervals in the geologic time scale. It also contains information about the number of families, orders, classes, or phyla that first appeared and were last observed in each time interval.

To answer the guiding question, you will need to analyze the data in the Fossil Record 2 Database Summary Counts. Be sure to think about the following questions before you begin analyzing your data:

- Are any of the data irrelevant based on the guiding question?
- How could you use mathematics to describe a change over time?
- What types of patterns might you look for as you analyze your data?
- How does the ratio of new families, orders, classes, or phyla and the total number of families, orders, classes, or phyla at each time interval compare with the other time intervals?
- How does the ratio of extinct families, orders, classes, or phyla and the total number of families, orders, classes, or phyla at each time interval compare with the other time intervals?
- Are there any other proportional relationships that you can look for that will help you answer the guiding question?
- What type of graph could you create to help make sense of your data?

Connections to the Nature of Scientific Knowledge or Scientific Inquiry

As you work through your investigation, be sure to think about

- how scientific knowledge changes over time, and
- how scientists use different methods to answer different types of questions.

Initial Argument

Once your group has finished collecting and analyzing your data, your group will need to develop an initial argument. Your initial argument needs to include a claim, evidence to support your claim, and a justification of the evidence. The *claim* is your group's answer to the guiding question. The *evidence* is an analysis and interpretation of your data. Finally, the *justification* of the evidence is why your group thinks the evidence matters. The justification of the evidence is important because scientists can use different kinds of evidence to support their claims. Your group will create your initial argument on a whiteboard. Your whiteboard should include all the information shown in Figure L5.3.

FIGURE L5.3 ─────────

Argument presentation on a whiteboard

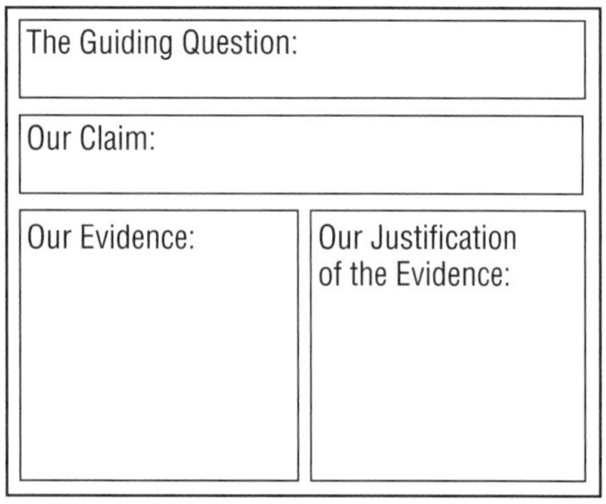

The Guiding Question:	
Our Claim:	
Our Evidence:	Our Justification of the Evidence:

Argumentation Session

The argumentation session allows all of the groups to share their arguments. One or two members of each group will stay at the lab station to share that group's argument, while the other members of the group go to the other lab stations to listen to and critique the other arguments. This is similar to what scientists do when they propose, support, evaluate, and refine new ideas during a poster session at a conference. If you are presenting your group's argument, your goal is to share your ideas and answer questions. You should also keep a record of the critiques and suggestions made by your classmates so you can use this feedback to make your initial argument stronger. You can keep track of specific critiques and suggestions for improvement that your classmates mention in the space below.

Critiques of our initial argument and suggestions for improvement:

Geologic Time and the Fossil Record

Which Time Intervals in the Past 650 Million Years of Earth's History Are Associated With the Most Extinctions and Which Are Associated With the Most Diversification of Life?

If you are critiquing your classmates' arguments, your goal is to look for mistakes in their arguments and offer suggestions for improvement so these mistakes can be fixed. You should look for ways to make your initial argument stronger by looking for things that the other groups did well. You can keep track of interesting ideas that you see and hear during the argumentation in the space below. You can also use this space to keep track of any questions that you will need to discuss with your team.

Interesting ideas from other groups or questions to take back to my group:

Once the argumentation session is complete, you will have a chance to meet with your group and revise your initial argument. Your group might need to gather more data or design a way to test one or more alternative claims as part of this process. Remember, your goal at this stage of the investigation is to develop the best argument possible.

Report

Once you have completed your research, you will need to prepare an *investigation report* that consists of three sections. Each section should provide an answer for the following questions:

1. What question were you trying to answer and why?

2. What did you do to answer your question and why?

3. What is your argument?

Your report should answer these questions in two pages or less. You should write your report using a word processing application (such as Word, Pages, or Google Docs), if possible, to make it easier for you to edit and revise it later. You should embed any diagrams, figures, or tables into the document. Be sure to write in a persuasive style; you are trying to convince others that your claim is acceptable or valid.

References

Benton, M. J. 1993. *The fossil record 2.* London: Chapman & Hall.

Benton, M. J. 1995. Diversification and extinction in the history of life. Science 268 (5207): 52–58.

Gradstein, F. M., J. G. Ogg, and F. J. Hilgen. 2012. On the geologic time scale. *Newsletters on Stratigraphy* 45 (2): 171–188.

Checkout Questions

Lab 5. Geological Time and the Fossil Record: Which Time Intervals in the Past 650 Million Years of Earth's History Are Associated With the Most Extinctions and Which Are Associated With the Most Diversification of Life?

Use the following diagram to answer questions 1–4. The picture shows a cross-section of Earth's strata and the fossils with the associated organisms discovered in those layers.

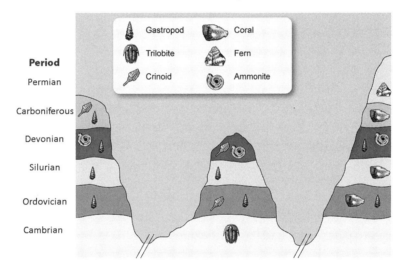

1. Crinoids evolved prior to ammonites.

    a. I agree with this statement

    b. I disagree with this statement

   How do you know?

2. Which geologic period had the greatest diversification of life?

    a. Cambrian

    b. Ordovician

    c. Devonian

How do you know?

3. Which geologic period had the most extinctions?

    a. Cambrian

    b. Ordovician

    c. Devonian

How do you know?

Geologic Time and the Fossil Record

Which Time Intervals in the Past 650 Million Years of Earth's History Are Associated With the Most Extinctions and Which Are Associated With the Most Diversification of Life?

4. The reason that crinoids do not appear in the Silurian period is because they went extinct and then re-evolved in the Devonian Period.

    a. I agree with this statement.

    b. I disagree with this statement.

    How do you know?

5. Scientific knowledge does not change once it has been proven to be a fact.

    a. I agree with this statement.

    b. I disagree with this statement.

    Explain your answer, using an example from your investigation about geologic time and the fossil record.

6. All scientific investigations follow the scientific method.

   a. I agree with this statement.

   b. I disagree with this statement.

   Explain your answer, using an example from your investigation about geologic time and the fossil record.

7. Scientists often need to look for patterns that occur in the data they collect and analyze. Explain why identifying patterns is important, using an example from your investigation about geologic time and the fossil record.

Geologic Time and the Fossil Record

Which Time Intervals in the Past 650 Million Years of Earth's History Are Associated With the Most Extinctions and Which Are Associated With the Most Diversification of Life?

8. Natural phenomena occur at varying scales. Explain why scientists need to consider using different measurement or time scales when deciding how to collect and analyze data, using an example from your investigation about geologic time and the fossil record.

LAB 6

Teacher Notes

Lab 6. Plate Interactions: How Is the Nature of the Geologic Activity That Is Observed Near a Plate Boundary Related to the Type of Plate Interaction That Occurs at That Boundary?

Purpose

The purpose of this lab is to *introduce* students to the disciplinary core idea (DCI) of Plate Tectonics and Large-Scale System Interactions by having them explore plate tectonics and the geologic activity that occurs near different types of plate boundaries. In addition, students have an opportunity to learn about the crosscutting concepts (CCs) of (a) Patterns and (b) Scale, Proportion, and Quantity. During the explicit and reflective discussion, students will also learn about (a) the difference between observations and inferences in science and (b) how the culture of science, societal needs, and current events influence the work of scientists.

Important Earth and Space Science Content

The interior structure of Earth is composed of several layers (see Figure 6.1). At the center of Earth is the inner core. The inner core is a solid sphere with a radius of about 1,120 km, and it consists of mostly iron. The next layer is the outer core, which is liquid and extends beyond the inner core another 2,270 km. The next, and thickest, layer is the mantle. The mantle, which is 2,900 thick, is often divided into three sublayers: the lower mesosphere, the upper mesosphere, and the asthenosphere. The outermost layer of Earth is the lithosphere. The lithosphere includes the crust, which is relatively thin and rocky, and the uppermost mantle.

There are two types of crust found in the lithosphere: continental and oceanic. Continental crust is 35–40 km (22–25 miles) thick on average but may exceed 70 km (43 miles) in some mountainous regions such as the Himalayas and the Rockies. The continental crust is made up of many different types of rock. The average density of the rocks that make up the continental crust is about 2.7 g/cm^3, and these rocks can be as old as 4 billion years. Oceanic crust, in contrast, is only about 7 km (4 miles) thick and is composed of basalt. The rocks that make up the oceanic crust are younger and denser than the rocks that make up the continental crust. Oceanic rocks are 180 million years old and have an average density of 3.0 g/cm^3.

The theory of plate tectonics states that the lithosphere is broken into several plates that move over time (see Figure 6.2). These plates are solid and rigid relative to the asthenosphere below them and thus "float" on the asthenosphere. The plates vary in thickness. Oceanic plates are thinner than continental plates. However, oceanic plates are formed from denser rocks than continental plates. Therefore, oceanic plates are denser than continental plates.

Plate Interactions

How Is the Nature of the Geologic Activity That Is Observed Near a Plate Boundary Related to the Type of Plate Interaction That Occurs at That Boundary?

FIGURE 6.1

The layers of Earth (illustration not drawn to scale)

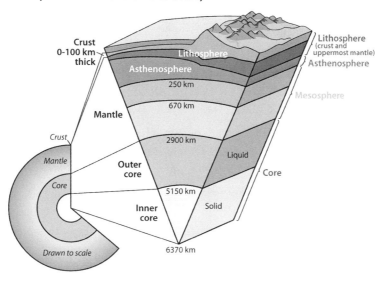

FIGURE 6.2

The major tectonic plates

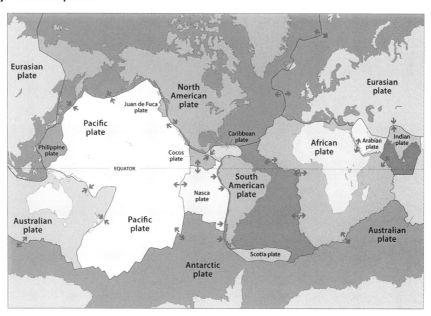

Earth scientists have identified four mechanisms that drive the movement of plates (see Figure 6.3):

- The first mechanism is *the existence of convection currents in the mantle.* These convection currents are driven by an unequal distribution of heat within Earth's interior. The mantle, as a result, is a moving fluid with a high viscosity.

- As this thick and sticky fluid moves under a plate in a given direction, it drags the plate along with it because of the force of friction. This second mechanism is often described as *viscous drag.*

- A third mechanism, called *slab pull,* occurs when one edge of a cold and dense oceanic plate sinks into the asthenosphere. As the leading edge of a plate sinks and melts over time, it "pulls" the rest of plate along. Oceanic plates sink because they are denser then the underlying asthenosphere.

- Another mechanism that contributes to movement of plates is called *ridge push.* The force of gravity drives this mechanism. Oceanic ridges are elevated. As a result, the trailing edge of a plate "slides" down the side of a ridge much like an object placed on a steep ramp slides down that ramp. As this edge of the plate slides down the ridge it pushes the rest of the plate away from the ridge. Ridge push does not contribute to plate motion as much as slab pull does, but it does plays a part.

FIGURE 6.3

Mechanisms that drive the movement of plates

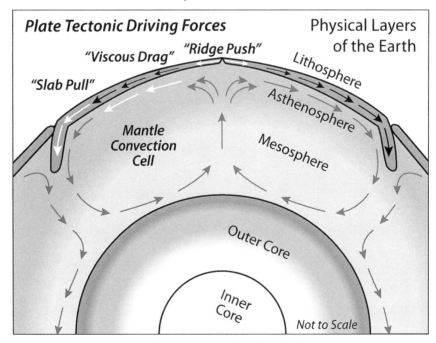

Plate Interactions

How Is the Nature of the Geologic Activity That Is Observed Near a Plate Boundary Related to the Type of Plate Interaction That Occurs at That Boundary?

These different mechanisms help explain the why the plates move in different directions and at different rates over time.

The plates move in different directions and at different speeds in relationship to each other. Plate boundaries are found where one plate interacts with another plate. These boundaries are classified into three different categories (see Figure 6.4): a *convergent boundary* results when two plates collide with each other, a *divergent boundary* results when two plates move away from each other, and a *transform boundary* forms when two plates slide past each other.

FIGURE 6.4

Tectonic plate boundaries

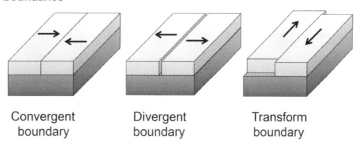

Convergent   Divergent   Transform
boundary   boundary   boundary

Divergent boundaries are constructive boundaries because when two plates move away from each other, molten material from the mantle moves up to fill the gap that is left behind. This is called upwelling. When this rock cools, it becomes part of the existing plates. Earth scientists call this process *seafloor spreading* if it happens in the ocean or *continental rifting* if it happens on land. Divergent plate boundaries tend to be sites of volcanic fissures and shallow earthquakes.

Convergent boundaries, in contrast, are destructive boundaries. Convergent boundaries occur at locations where two plates collide with each other. When an oceanic plate and a continental plate collide, the denser oceanic plate will descend beneath the less dense continental plate, melt, and eventually be reabsorbed into the mantle. This process is called subduction. If two oceanic plates converge, the one capped with denser crust will descend beneath the other one. When two continental plates collide, the crust at the edges of the plates will be pushed up to create a mountain system. The continental crust will push upward and form a mountain system rather than subducting into the mantle, because continental crust is buoyant. A collision between continental plates like this created several mountain ranges, including the Himalayas, the Alps, the Appalachians, and the Urals. Convergent plate boundaries tend to be sites of intense volcanic activity and strong earthquakes.

At transform plate boundaries, no lithosphere is formed or destroyed. These plates slide past each other, often creating earthquakes that are relatively weak compared with those that occur at convergent boundaries.

Timeline

The instructional time needed to complete this lab investigation is 200-280 minutes. Appendix 3 (p. 573) provides options for implementing this lab investigation over several class periods. Option E (280 minutes) should be used if students are unfamiliar with scientific writing, because this option provides extra instructional time for scaffolding the writing process. You can scaffold the writing process by modeling, providing examples, and providing hints as students write each section of the report. Option F (200 minutes) should be used if students are familiar with scientific writing and have developed the skills needed to write an investigation report on their own. In option F, students complete stage 6 (writing the investigation report) and stage 8 (revising the investigation report) as homework.

Materials and Preparation

The materials needed to implement this investigation are listed in Table 6.1. The *Natural Hazards Viewer* interactive map, which was developed by the National Oceanic and Atmospheric Administration's National Geophysical Data Center, is available at *http://maps.ngdc.noaa.gov/viewers/hazards*. It is free to use and can be accessed online using most internet browsers. You should access the website and learn how the interactive map works before beginning the lab investigation. In addition, it is important to check if students can access and use the interactive map from a school computer or tablet, because some schools have set up firewalls and other restrictions on web browsing.

TABLE 6.1

Materials list for Lab 6

Item	Quantity
Computer or tablet with internet access	1 per group
Whiteboard, 2' × 3'*	1 per group
Lab Handout	1 per student
Peer-review guide and instructor scoring rubric	1 per student
Checkout Questions	1 per student

* As an alternative, students can use computer and presentation software such as Microsoft PowerPoint or Apple Keynote to create their arguments.

Safety Precautions

Remind students to follow all normal lab safety rules.

Plate Interactions

How Is the Nature of the Geologic Activity That Is Observed Near a Plate Boundary Related to the Type of Plate Interaction That Occurs at That Boundary?

Topics for the Explicit and Reflective Discussion

Reflecting on the Use of Core Ideas and Crosscutting Concepts During the Investigation

Teachers should begin the explicit and reflective discussion by asking students to discuss what they know about the DCI they used during the investigation. The following are some important concepts related to the DCI of Plate Tectonics and Large-Scale System Interactions that students need to know to determine how the nature of the geologic activity that is observed near a plate boundary is related to the type of plate interaction that occurs at that boundary:

- The lithosphere is broken into several plates that are constantly moving.
- Plate boundaries are found where one plate interacts with another plate.
- Convergent boundaries result when two plates collide with each other.
- Divergent boundaries result when two plates move apart.
- Transform boundaries are formed when two plates slide past each other.

To help students reflect on what they know about these concepts, we recommend showing them two or three images using presentation software that help illustrate these important ideas. You can then ask the students the following questions to encourage students to share how they are thinking about these important concepts:

1. What do we see going on in this image?
2. Does anyone have anything else to add?
3. What might be going on that we can't see?
4. What are some things that we are not sure about here?

You can then encourage students to think about how CCs played a role in their investigation. There are at least two CCs that students need to know to determine how the nature of the geologic activity that is observed near a plate boundary is related to the type of plate interaction that occurs at that boundary: (a) Patterns and (b) Scale, Proportion, and Quantity (see Appendix 2 [p. 569] for a brief description of these CCs). To help students reflect on what they know about these CCs, we recommend asking them the following questions:

1. Why do scientists look for and attempt to explain patterns in nature?
2. What patterns did you identify and use during your investigation? Why was that useful?
3. Why is it important to consider what measurement scale or scales to use during an investigation? Why is useful to look for proportional relationships when analyzing data?

4. What measurement scale or scales did you use during your investigation? What did that allow you to do? Did you attempt to look for proportional relationships when you were analyzing your data? Why or why not?

You can then encourage students to think about how they used all these different concepts to help answer the guiding question and why it is important to use these ideas to help justify their evidence for their final arguments. Be sure to remind your students to explain why they included the evidence in their arguments and make the assumptions underlying their analysis and interpretation of the data explicit in order to provide an adequate justification of their evidence.

Reflecting on Ways to Design Better Investigations

It is important for students to reflect on the strengths and weaknesses of the investigation they designed during the explicit and reflective discussion. Students should therefore be encouraged to discuss ways to eliminate potential flaws, measurement errors, or sources of uncertainty in their investigations. To help students be more reflective about the design of their investigation and what they can do to make their investigations more rigorous in the future, you can ask the following questions:

1. What were some of the strengths of the way you planned and carried out your investigation? In other words, what made it scientific?

2. What were some of the weaknesses of the way you planned and carried out your investigation? In other words, what made it less scientific?

3. What rules can we make, as a class, to ensure that our next investigation is more scientific?

Reflecting on the Nature of Scientific Knowledge and Scientific Inquiry

This investigation can be used to illustrate two important concepts related to the nature of scientific knowledge and the nature of scientific inquiry: (a) the difference between observations and inferences in science and (b) how the culture of science, societal needs, and current events influence the work of scientists (see Appendix 2 [p. 569] for a brief description of these two concepts). Be sure to review these concepts during and at the end of the explicit and reflective discussion. To help students think about these concepts in relation to what they did during the lab, you can ask the following questions:

1. You made observations and inferences during your investigation. Can you give me some examples of these observations and inferences?

Plate Interactions

How Is the Nature of the Geologic Activity That Is Observed Near a Plate Boundary Related to the Type of Plate Interaction That Occurs at That Boundary?

2. Can you work with your group to come up with a rule that you can use to tell the difference between an observation and inference? Be ready to share in a few minutes.

3. People view some types of research as being more important than other types of research because of cultural values and current events. Can you come up with some examples of how cultural values and current events have influenced the work of scientists?

4. Scientists share a set of values, norms, and commitments that shape what counts as knowing, how to represent or communicate information, and how to interact with other scientists. Can you work with your group to come up with a rule that you can use to decide if something is science or not science? Be ready to share in a few minutes.

You can also use presentation software or other techniques to encourage your students to think about these concepts. You can show examples of information from the investigation that are either observations or inferences and ask students to classify each example and explain their thinking. You can also show examples of research projects that were influenced by cultural values and current events and ask students to think about what was going on in society when that research was conducted and why that research was viewed as being important for the greater good.

Remind your students that, to be proficient in science, it is important that they understand what counts as scientific knowledge and how that knowledge develops over time.

Hints for Implementing the Lab

- Learn how to use the interactive map before the lab begins. It is important for you to know how to use the interactive map so you can help students when they get stuck or confused.

- A group of three students per computer or tablet tends to work well.

- Allow the students to play with the interactive map as part of the tool talk before they begin to design their investigation. This gives students a chance to see what they can and cannot do with the interactive map.

- Be sure that students record actual values (e.g., number or severity of geologic events) rather than just attempting to describe what they see on the computer screen (e.g., there are lots of volcanoes, there were more earthquakes). The interactive map contains a number of tools that will allow the students to collect quantitative data during this investigation.

- Students often make mistakes during the data collection stage, but they should quickly realize these mistakes during the argumentation session. It will only take them a short period of time to re-collect data, and they should be allowed to do so. During the explicit and reflective discussion, students will also have the opportunity to reflect on and identify ways to improve the way they design investigations (especially how they attempt to control variables as part of an experiment). This also offers an opportunity to discuss what scientists do when they realize that a mistake is made during a study.

- This lab provides an excellent opportunity to discuss how scientists need to make choices about what data to use when there is too much data to analyze. Be sure to give students advice about how to address this issue as they are working and discuss this issue as part of the explicit and reflective discussion.

- This lab also provides an excellent opportunity to discuss how scientists identify a signal (a pattern or trend) from the noise (measurement error) in their data. Be sure to use this activity as a concrete example during the explicit and reflective discussion.

Connections to Standards

Table 6.2 highlights how the investigation can be used to address specific (a) performance expectations from the *NGSS* and (b) *Common Core State Standards* in English language arts (*CCSS ELA*).

TABLE 6.2

Lab 6 alignment with standards

NGSS performance expectations	History of Earth • MS-ESS2-2: Construct an explanation based on evidence for how geoscience processes have changed Earth's surface at varying time and spatial scales. • MS-ESS2-3: Analyze and interpret data on the distribution of fossils and rocks, continental shapes, and seafloor structures to provide evidence of the past plate motions
CCSS ELA—Reading in Science and Technical Subjects	Key ideas and details • CCSS.ELA-LITERACY.RST.6-8.1: Cite specific textual evidence to support analysis of science and technical texts. • CCSS.ELA-LITERACY.RST.6-8.2: Determine the central ideas or conclusions of a text; provide an accurate summary of the text distinct from prior knowledge or opinions. Craft and structure • CCSS.ELA-LITERACY.RST.6-8.4: Determine the meaning of symbols, key terms, and other domain-specific words and phrases as they are used in a specific scientific or technical context relevant to *grade 6–8 texts and topics*.

Continued

Plate Interactions

How Is the Nature of the Geologic Activity That Is Observed Near a Plate Boundary Related to the Type of Plate Interaction That Occurs at That Boundary?

TABLE 6.2 (*continued*)

***CCSS ELA*—Reading in Science and Technical Subjects** (*continued*)	Craft and structure (*continued*) • CCSS.ELA-LITERACY.RST.6-8.5: Analyze the structure an author uses to organize a text, including how the major sections contribute to the whole and to an understanding of the topic. • CCSS.ELA-LITERACY.RST.6-8.6: Analyze the author's purpose in providing an explanation, describing a procedure, or discussing an experiment in a text. Integration of knowledge and ideas • CCSS.ELA-LITERACY.RST.6-8.7: Integrate quantitative or technical information expressed in words in a text with a version of that information expressed visually (e.g., in a flowchart, diagram, model, graph, or table). • CCSS.ELA-LITERACY.RST.6-8.8: Distinguish among facts, reasoned judgment based on research findings, and speculation in a text. • CCSS.ELA-LITERACY.RST.6-8.9: Compare and contrast the information gained from experiments, simulations, video, or multimedia sources with that gained from reading a text on the same topic.
***CCSS ELA*—Writing in Science and Technical Subjects**	Text types and purposes • CCSS.ELA-LITERACY.WHST.6-8.1: Write arguments focused on *discipline-specific content*. • CCSS.ELA-LITERACY.WHST.6-8.2: Write informative or explanatory texts, including the narration of historical events, scientific procedures/experiments, or technical processes. Production and distribution of writing • CCSS.ELA-LITERACY.WHST.6-8.4: Produce clear and coherent writing in which the development, organization, and style are appropriate to task, purpose, and audience. • CCSS.ELA-LITERACY.WHST.6-8.5: With some guidance and support from peers and adults, develop and strengthen writing as needed by planning, revising, editing, rewriting, or trying a new approach, focusing on how well purpose and audience have been addressed. • CCSS.ELA-LITERACY.WHST.6-8.6: Use technology, including the internet, to produce and publish writing and present the relationships between information and ideas clearly and efficiently. Range of writing • CCSS.ELA-LITERACY.WHST.6-8.10: Write routinely over extended time frames (time for reflection and revision) and shorter time frames (a single sitting or a day or two) for a range of discipline-specific tasks, purposes, and audiences.

Continued

TABLE 6.2 (*continued*)

CCSS ELA—Speaking and Listening	Comprehension and collaboration • CCSS.ELA-LITERACY.SL.6-8.1: Engage effectively in a range of collaborative discussions (one-on-one, in groups, and teacher-led) with diverse partners on grade 6–8 topics, texts, and issues, building on others' ideas and expressing their own clearly. • CCSS.ELA-LITERACY.SL.6-8.2:* Interpret information presented in diverse media and formats (e.g., visually, quantitatively, orally) and explain how it contributes to a topic, text, or issue under study. • CCSS.ELA-LITERACY.SL.6-8.3:* Delineate a speaker's argument and specific claims, distinguishing claims that are supported by reasons and evidence from claims that are not. Presentation of knowledge and ideas • CCSS.ELA-LITERACY.SL.6-8.4:* Present claims and findings, sequencing ideas logically and using pertinent descriptions, facts, and details to accentuate main ideas or themes; use appropriate eye contact, adequate volume, and clear pronunciation. • CCSS.ELA-LITERACY.SL.6-8.5:* Include multimedia components (e.g., graphics, images, music, sound) and visual displays in presentations to clarify information. • CCSS.ELA-LITERACY.SL.6-8.6: Adapt speech to a variety of contexts and tasks, demonstrating command of formal English when indicated or appropriate.

* Only the standard for grade 6 is provided because the standards for grades 7 and 8 are similar. Please see *www.corestandards.org/ELA-Literacy/SL* for the exact wording of the standards for grades 7 and 8.

Plate Interactions

How Is the Nature of the Geologic Activity That Is Observed Near a Plate Boundary Related to the Type of Plate Interaction That Occurs at That Boundary?

Lab Handout

Lab 6. Plate Interactions: How Is the Nature of the Geologic Activity That Is Observed Near a Plate Boundary Related to the Type of Plate Interaction That Occurs at That Boundary?

Introduction

The interior structure of the Earth is composed of several layers (see Figure L6.1). At the center of the Earth is the inner core. The inner core is a solid sphere and consists of mostly iron. It has a radius of about 1,120 km. The next layer is the outer core. The outer core is liquid and extends beyond the inner core another 2,270 km. The next, and thickest, layer is the mantle. The mantle is often divided into three sublayers: the lower mesosphere, the upper mesosphere, and the asthenosphere. The outermost layer of the Earth is the lithosphere. The lithosphere includes the crust and the uppermost mantle.

The theory of plate tectonics states that the lithosphere is broken into several plates that move over time (see Figure L6.2). The plates move in different directions and at different speeds in relationship to each other. Plate boundaries are found where one plate interacts with another plate. These boundaries are classified into three different categories: (a) *convergent boundaries* result when two plates collide with each other, (b) *divergent boundaries* result when two plates move away from each other, and (c) *transform boundaries* form when two

FIGURE L6.1

Earth's layers

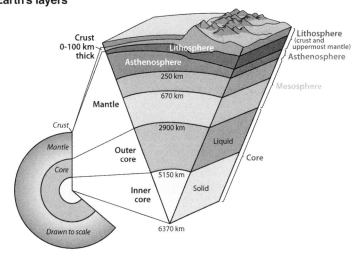

FIGURE L6.2

The major tectonic plates

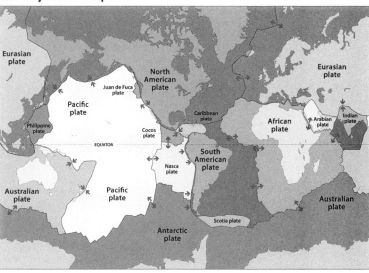

plates slide past each other. Volcanic eruptions and earthquakes often occur along or near plate boundaries.

In this investigation, you will explore where volcanic eruptions and earthquakes tend to happen. You goal is to determine if volcanic eruptions and earthquakes happen more often near a specific type of plate boundary. This type of investigation is important because natural processes, such as the gradual movement of tectonic plates over time, can result in natural hazards. Although it is impossible to prevent volcanic eruptions and earthquakes from happening, we can take steps to reduce their impacts. It is therefore useful for us to understand where these types of hazards are likely to occur because we can prepare for them and respond quickly when they happen. We can, for example, build better buildings, develop warning systems, and increase the response capabilities of cities to help reduce the loss of life and economic costs when we know where volcanic eruptions and earthquake tend to happen.

Your Task

Use an online interactive map to collect data about how often volcanic eruptions and earthquakes happen near the three different types of plate boundaries. Your goal is to use what you know about plate tectonics, patterns, and the use of different scales, proportional relationships, and quantities during an investigation to determine if the way plates interact with each other at a specific location is related to the occurrence of volcanic eruptions and earthquakes at that location.

The guiding question of this investigation is, *How is the nature of the geologic activity that is observed near a plate boundary related to the type of plate interaction that occurs at that boundary?*

Materials

You will use an online interactive map called *Natural Hazards Viewer* to conduct your investigation; the interactive map can be accessed at *http://maps.ngdc.noaa.gov/viewers/hazards*.

Safety Precautions

Be sure to follow all normal lab safety rules.

Investigation Proposal Required?   ☐ Yes   ☐ No

Getting Started

Given the nature of this investigation, you must determine what type of data you need to collect, how you will collect the data, and how will you analyze the data to answer the research question. To determine *what type of data you need to collect*, think about the following questions:

Plate Interactions

How Is the Nature of the Geologic Activity That Is Observed Near a Plate Boundary Related to the Type of Plate Interaction That Occurs at That Boundary?

- How will you identify the location of different types of plate boundary?
- How can you describe an earthquake and a volcanic eruption quantitatively?
- What are the limitations of the available data set?

To determine *how you will collect the data,* think about the following questions:

- What parts of the world will you need to include in your study?
- What scale or scales should you use to quantify the size of an earthquake or a volcanic eruption?
- Will you need to limit the number of samples you include? If so, how will decide what to include?
- What concessions will you need to make to collect the data you need?
- How will you keep track of the data you collect and how will you organize it?

To determine *how you will analyze the data,* think about the following questions:

- What types of comparisons will you need to make?
- What types of patterns might you look for as you analyze the data?
- What potential proportional relationships can you find in the data?
- How could you use mathematics to determine if there are differences between the groups?
- What type of diagram could you create to help make sense of your data?

Connections to the Nature of Scientific Knowledge and Scientific Inquiry

As you work through your investigation, be sure to think about

- the difference between observations and inferences in science, and
- how the culture of science, societal needs, and current events influence the work of scientists.

Initial Argument

Once your group has finished collecting and analyzing your data, your group will need to develop an initial argument. Your initial argument needs to include a claim, evidence to support your claim, and a justification of the evidence. The *claim* is your group's answer to the guiding question. The *evidence* is an analysis and interpretation of your data. Finally, the *justification* of the evidence is why your group thinks the evidence matters. The justification of the evidence is important because scientists can use different kinds of evidence to support their claims. Your group will create your initial argument on a whiteboard. Your whiteboard should include all the information shown in Figure L6.3 (p. 160).

LAB 6

FIGURE L6.3 _____

Argument presentation on a whiteboard

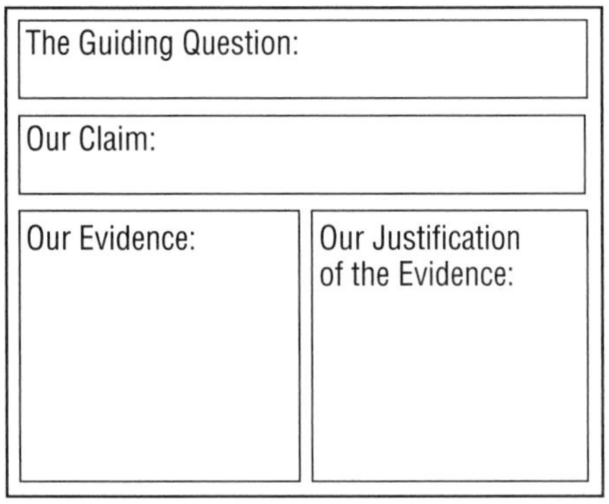

| The Guiding Question: |
| Our Claim: |

| Our Evidence: | Our Justification of the Evidence: |

Argumentation Session

The argumentation session allows all of the groups to share their arguments. One or two members of each group will stay at the lab station to share that group's argument, while the other members of the group go to the other lab stations to listen to and critique the other arguments. This is similar to what scientists do when they propose, support, evaluate, and refine new ideas during a poster session at a conference. If you are presenting your group's argument, your goal is to share your ideas and answer questions. You should also keep a record of the critiques and suggestions made by your classmates so you can use this feedback to make your initial argument stronger. You can keep track of specific critiques and suggestions for improvement that your classmates mention in the space below.

Critiques of our initial argument and suggestions for improvement:

If you are critiquing your classmates' arguments, your goal is to look for mistakes in their arguments and offer suggestions for improvement so these mistakes can be fixed. You should look for ways to make your initial argument stronger by looking for things that the other groups did well. You can keep track of interesting ideas that you see and hear during the argumentation in the space below. You can also use this space to keep track of any questions that you will need to discuss with your team.

Plate Interactions

How Is the Nature of the Geologic Activity That Is Observed Near a Plate Boundary Related to the Type of Plate Interaction That Occurs at That Boundary?

Interesting ideas from other groups or questions to take back to my group:

Once the argumentation session is complete, you will have a chance to meet with your group and revise your initial argument. Your group might need to gather more data or design a way to test one or more alternative claims as part of this process. Remember, your goal at this stage of the investigation is to develop the best argument possible.

Report

Once you have completed your research, you will need to prepare an investigation report that consists of three sections. Each section should provide an answer for the following questions:

1. What question were you trying to answer and why?

2. What did you do to answer your question and why?

3. What is your argument?

Your report should answer these questions in two pages or less. You should write your report using a word processing application (such as Word, Pages, or Google Docs), if possible, to make it easier for you to edit and revise it later. You should embed any diagrams, figures, or tables into the document. Be sure to write in a persuasive style; you are trying to convince others that your claim is acceptable or valid.

Checkout Questions

Lab 6. Plate Interactions: How Is the Nature of the Geologic Activity That Is Observed Near a Plate Boundary Related to the Type of Plate Interaction That Occurs at That Boundary?

Use the map below to answer questions 1 and 2.

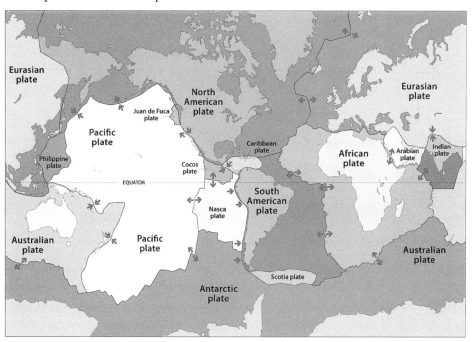

1. On the map above, circle one convergent boundary, one divergent boundary, and one transform boundary. Be sure to label each one. How do you know which boundary is which?

2. Earthquakes occur much more frequently in California than they do in Florida or New York. Using what you learned from your investigation and the map above, why is this the case?

Plate Interactions

*How Is the Nature of the Geologic Activity That Is Observed Near a Plate Boundary Related to
the Type of Plate Interaction That Occurs at That Boundary?*

3. The map below shows the location of a volcanic arc in Central America. Each
    triangle represents the location of a different volcano.

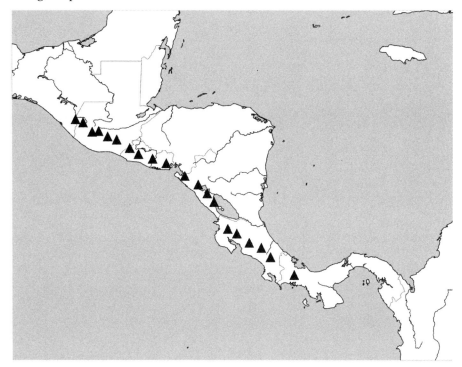

a. What type of boundary is responsible for this volcanic arc and where is it most
    likely located?

b. How do you know?

4. Scientists share a set of values, norms, and commitments that shape what counts as knowing, how to represent or communicate information, and how to interact with other scientists.

   a. I agree with this statement.

   b. I disagree with this statement.

   Explain your answer, using an example from your investigation about plate tectonics.

5. The statement "There were 31 earthquakes at the convergent boundary" is an example of an inference.

   a. I agree with this statement.

   b. I disagree with this statement.

   Explain your answer, using an example from your investigation about plate tectonics.

6. Scientists often need to look for patterns that occur in the data they collect and analyze. Explain why identifying patterns is important for scientists, using an example from your investigation about plate tectonics.

7. Natural phenomena occur at varying scales. Explain why scientists need to consider using different measurement or time scales when deciding how to collect and analyze data, using an example from your investigation about plate tectonics.

Application Lab

Teacher Notes

Lab 7. Formation of Geologic Features: How Can We Explain the Growth of the Hawaiian Archipelago Over the Past 100 Million Years?

Purpose

The purpose of this lab is for students to *apply* what they know about the disciplinary core ideas (DCIs) of (a) The History of Planet Earth and (b) Plate Tectonics and Large-Scale System Interactions by having them develop a conceptual model that explains how the Hawaiian archipelago formed over the last 100 million years. In addition, students have an opportunity to learn about the crosscutting concepts (CCs) of (a) Patterns and (b) Systems and System Models. During the explicit and reflective discussion, students will also learn about (a) the use of models as tools for reasoning about natural phenomena and (b) the assumptions made by scientists about order and consistency in nature.

Important Earth and Space Science Content

Scientists use the theory of plate tectonics to explain the origin of many geologic features on Earth. The theory of plate tectonics indicates that the lithosphere is broken into several plates that are constantly moving (see Figure 7.1). The plates are composed of the

FIGURE 7.1

The major tectonic plates

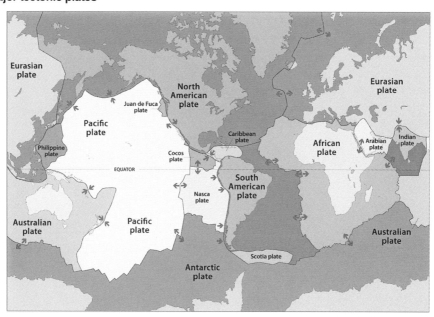

oceanic lithosphere and thicker continental lithosphere. The plates move because they are located on top of giant convection cells in the mantle (see Figure 7.2). These currents bring matter from the hot inner mantle near the outer core up to the cooler surface. The convection cells are driven by the energy that is released when isotopes go through radioactive decay deep within the interior of the Earth. Each of the plates is slowly pushed across Earth's surface in a specific direction by these currents. The plates carry the continents, create or destroy ocean basins, form mountain ranges and plateaus, and produce earthquakes or volcanoes as they move.

Most of the continental and ocean floor features that we see are the result of either constructive or destructive geologic processes that occur along different types of plate boundaries. There are three main categories of plate boundaries (see Figure 7.3): *convergent boundaries* result when two plates collide with each other, *divergent boundaries* result when two plates move away from each other, and *transform boundaries* occur when plates slide past each other. The nature of the geologic features that we see at a particular location

FIGURE 7.2

Mechanisms that drive the movement of plates

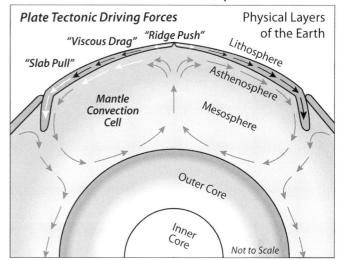

FIGURE 7.3

Tectonic plate boundaries

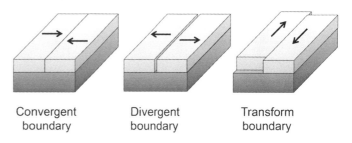

Convergent boundary   Divergent boundary   Transform boundary

depends on whether the plates are being pushed together to create mountains or ocean trenches, being pulled apart to form new ocean floor at mid-oceanic ridges and rift valleys on continents, or sliding past each other along surface faults.

Earth's surface is still being shaped and reshaped because of the movement of plates. One example of this phenomenon is the Hawaiian Islands, an archipelago in the northern part of the Pacific Ocean that consists of eight major islands, several atolls, and numerous smaller islets. It extends from the island of Hawaii over 2,400 km to the Kure Atoll. Each island is made up of one or more volcanoes (see Figure 7.4, p. 170). The island of Hawaii, for instance, is made up of five different volcanoes. Kohala, the oldest volcano on this island, last erupted about 60,000 years ago. It is an extinct volcano because it will never erupt again. Mauna Kea is the next oldest volcano. It is a dormant volcano because the last time it erupted was 3,600 years ago, but it will probably erupt again at some time in the future. The three youngest volcanoes on Hawaii—Hualalai, Mauna Loa, and Kilauea—are active (see *www.nps.gov/havo/faqs.htm* for more information).

LAB 7

FIGURE 7.4

The Hawaiian archipelago and some of its volcanoes (indicated by triangles)

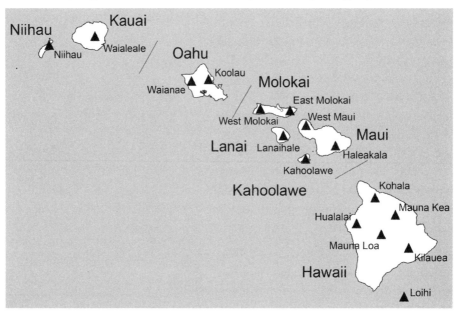

At first glance, these islands look similar to a volcanic island arc, which is a chain of volcanoes that forms above a subducting plate and takes the form of an arc of islands. An example of a volcanic island arc is the Aleutian Islands (see Figure 7.5). The Hawaiian Islands, however, are not located near a plate boundary, so they are not an example of a volcanic island arc. The Hawaiian Islands are a hotspot volcanic chain. A hotspot volcanic chain is created when volcanoes form one after another in the middle of a tectonic plate, as the plate moves over the hotspot, and so the volcanoes increase in age from one end of the chain to the other. In the case of the Hawaiian Islands, the older islands such as Kauai are located in the northwest. These islands are over 4.5 million years old and lush. The big island of Hawaii, in contrast is, about 400,000 years old and much rockier. Volcanic island arcs do not generally exhibit such a simple age pattern like the ones observed with hotspot volcanic chains.

FIGURE 7.5

The Aleutian Islands, an example of a volcanic island arc

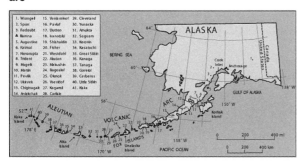

Timeline

The instructional time needed to complete this lab investigation is 220–280 minutes. Appendix 3 (p. 573) provides options for implementing this lab investigation over several class periods. Option A (280 minutes) should be used if students are unfamiliar with scientific writing, because this option provides extra instructional time for scaffolding the writing process. You can scaffold the writing process by modeling, providing examples, and providing hints as students write each section of the report. Option B (220 minutes) should be used if students are familiar with scientific writing and have developed the skills needed to write an investigation report on their own. In option B, students complete stage 6 (writing the investigation report) and stage 8 (revising the investigation report) as homework.

Materials and Preparation

The materials needed to implement this investigation are listed in Table 7.1. The *Natural Hazards Viewer* interactive map, which was developed by the National Oceanic and Atmospheric Administration's National Geophysical Data Center, is available at *http://maps.ngdc.noaa.gov/viewers/hazards*. It is free to use and can be accessed using most internet browsers. You should access the website and learn how the interactive map works before beginning the lab investigation. In addition, it is important to check if students can access and use the interactive map from a school computer or tablet, because some schools have set up firewalls and other restrictions on web browsing.

The Ages of Volcanoes in Hawaiian Islands Excel file can be downloaded from the book's Extras page at *www.nsta.org/adi-ess.* It can be loaded onto student computers before the investigation, e-mailed to students, or uploaded to a class website that students can access. It is important that the computers the students will use during this lab have a spreadsheet application such as Microsoft Excel or Apple Numbers loaded on them, or students must have access to an online spreadsheet application such as Google Sheets. In this way, students can analyze the data set using the computational and graphing tools built into the spreadsheet application. It is also important for you to look over the file before the investigation begins so you can learn how the data in the file are organized. This will enable you to give students suggestions on how to analyze the data.

TABLE 7.1

Materials list for Lab 7

Item	Quantity
Computer or tablet with Excel or other spreadsheet application and internet access	1 per group
Ages of Volcanoes in Hawaiian Island Excel file	1 per group
Investigation Proposal A	1 per group
Whiteboard, 2' × 3'*	1 per group
Lab Handout	1 per student
Peer-review guide and instructor scoring rubric	1 per student
Checkout Questions	1 per student

* As an alternative, students can use computer and presentation software such as Microsoft PowerPoint or Apple Keynote to create their arguments.

LAB 7

Safety Precautions
Remind students to follow all normal lab safety rules.

Topics for the Explicit and Reflective Discussion
Reflecting on the Use of Core Ideas and Crosscutting Concepts During the Investigation
Teachers should begin the explicit and reflective discussion by asking students to discuss what they know about the DCIs they used during the investigation. The following are some important concepts related to the DCIs of (a) The History of Planet Earth and (b) Plate Tectonics and Large-Scale System Interactions that students need to be able to develop a conceptual model that explains the formation of the Hawaiian archipelago:

- The lithosphere is broken into several plates that are constantly moving.
- Plate boundaries are found where one plate interacts with another plate.
- Convergent boundaries result when two plates collide with each other.
- Divergent boundaries result when two plates move apart.
- Transform boundaries are formed when two plates slide past each other.

To help students reflect on what they know about these concepts, we recommend showing them two or three images using presentation software that help illustrate these important ideas. You can then ask the students the following questions in order to encourage students to share how they are thinking about these important concepts:

1. What do we see going on in this image?
2. Does anyone have anything else to add?
3. What might be going on that we can't see?
4. What are some things that we are not sure about here?

You can then encourage students to think about how CCs played a role in their investigation. There are at least two CCs that students need to be able to develop a conceptual model that explains the formation of the Hawaiian archipelago: (a) Patterns and (b) Systems and System Models (see Appendix 2 [p. 569] for a brief description of these two CCs). To help students reflect on what they know about these CCs, we recommend asking them the following questions:

1. Why do scientists look for and attempt to explain patterns in nature?
2. What patterns did you identify and use during your investigation? Why was that useful?

3. Why do scientists often define a system and then develop a model of it as part of an investigation?

4. How did you use a model to understand the formation of the Hawaiian Islands? Why was that useful?

You can then encourage students to think about how they used all these different concepts to help answer the guiding question and why it is important to use these ideas to help justify their evidence for their final arguments. Be sure to remind your students to explain why they included the evidence in their arguments and make the assumptions underlying their analysis and interpretation of the data explicit in order to provide an adequate justification of their evidence.

Reflecting on Ways to Design Better Investigations

It is important for students to reflect on the strengths and weaknesses of the investigation they designed during the explicit and reflective discussion. Students should therefore be encouraged to discuss ways to eliminate potential flaws, measurement errors, or sources of uncertainty in their investigations. To help students be more reflective about the design of their investigation and what they can do to make their investigations more rigorous in the future, you can ask the following questions:

1. What were some of the strengths of the way you planned and carried out your investigation? In other words, what made it scientific?

2. What were some of the weaknesses of the way you planned and carried out your investigation? In other words, what made it less scientific?

3. What rules can we make, as a class, to ensure that our next investigation is more scientific?

Reflecting on the Nature of Scientific Knowledge and Scientific Inquiry

This investigation can be used to illustrate two important concepts related to the nature of scientific knowledge and the nature of scientific inquiry: (a) the use of models as tools for reasoning about natural phenomena and (b) the assumptions made by scientists about order and consistency in nature (see Appendix 2 [p. 569] for a brief description of these two concepts). Be sure to review these concepts during and at the end of the explicit and reflective discussion. To help students think about these concepts in relation to what they did during the lab, you can ask the following questions:

1. I asked you to develop a model to explain the formation of the Hawaiian Islands as part of your investigation. Why is it useful to develop models in science?

2. Can you work with your group to come up with a rule that you can use to decide what a model is and what a model is not in science? Be ready to share in a few minutes.

3. Scientists assume that natural laws operate today as they did in the past and that they will continue to do so in the future. Why do you think this assumption is important?

4. Think about what you were trying to do during this investigation. What would you have had to do differently if you could not assume natural laws operate today as they did in the past?

You can also use presentation software or other techniques to encourage your students to think about these concepts. You can show examples and non-examples of scientific models and then ask students to classify each one and explain their thinking. You can also show images of different scientific laws (such as the law of universal gravitation, the law of conservation of mass, or the law of superposition) and ask students if they think these laws have been the same throughout Earth's history. Then ask them to think about what scientists would need to do to be able to study the past if laws are not consistent through time and space.

Remind your students that, to be proficient in science, it is important that they understand what counts as scientific knowledge and how that knowledge develops over time.

Hints for Implementing the Lab

- Learn how to use the *Natural Hazards Viewer* interactive map and the Ages of Volcanoes in Hawaiian Islands Excel file before the lab begins. It is important for you to know how to use the map and what is included in the Excel file, as well as how to analyze the data, so you can help students when they get stuck or confused.

- A group of three students per computer or tablet tends to work well.

- Allow the students to play with the interactive map and the Excel file as part of the tool talk before they begin to design their investigation. This gives students a chance to see what they can and cannot do with the interactive map and with the data in the file.

- Encourage students to analyze the data in the Age of Volcanoes in Hawaiian Islands Excel file by making graphs. The best way to help students to learn how to use Excel (or another spreadsheet application) is to provide "just-in-time" instruction. In other words, wait for students to get stuck and then give a brief mini-lesson on how to use a specific tool in Excel based on what students are trying to do. They will be much more interested in learning about how to use the tools in Excel if they know it will help solve a problem they are having or will allow them to accomplish one of their goals.

- Students often make mistakes when developing their conceptual models and/ or initial arguments, but they should quickly realize these mistakes during the argumentation session. Be sure to allow students to revise their models and arguments at the end of the argumentation session. The explicit and reflective discussion will also give students an opportunity to reflect on and identify ways to improve how they develop and test models. This also offers an opportunity to discuss what scientists do when they realize a mistake is made.

- Students will likely first infer the existence of a plate boundary that has led to the formation of the Hawaiian Islands. Yet, when they use the *Natural Hazards Viewer* interactive map to locate plate boundaries, they will see that there is no plate boundary near Hawaii. This is a good opportunity to help students think about alternate explanations given that they know no boundary exists and that volcanoes are places where magma comes to the surface. This is also a good opportunity to help students think about ways scientists refine models. The model of plate tectonics as originally conceived could not account for the formation of the Hawaiian archipelago. Thus, scientists used new data to refine their model.

- This lab also provides an excellent opportunity to discuss how scientists must make choices about which data to use and how to analyze the data they have. Be sure to use this activity as a concrete example during the explicit and reflective discussion.

Connections to Standards

Table 7.2 highlights how the investigation can be used to address specific (a) performance expectations from the *NGSS* and (b) *Common Core State Standards* in English language arts (*CCSS ELA*).

TABLE 7.2

Lab 7 alignment with standards

NGSS performance expectations	History of Earth
	- MS-ESS2-2: Construct an explanation based on evidence for how geoscience processes have changed Earth's surface at varying time and spatial scales.
	- MS-ESS2-3: Analyze and interpret data on the distribution of fossils and rocks, continental shapes, and seafloor structures to provide evidence of the past plate motions
	- HS-ESS1-5: Evaluate evidence of the past and current movements of continental and oceanic crust and the theory of plate tectonics to explain the ages of crustal rocks.

Continued

TABLE 7.2 (*continued*)

CCSS ELA—Reading in Science and Technical Subjects	Key ideas and details • CCSS.ELA-LITERACY.RST.6-8.1: Cite specific textual evidence to support analysis of science and technical texts. • CCSS.ELA-LITERACY.RST.6-8.2: Determine the central ideas or conclusions of a text; provide an accurate summary of the text distinct from prior knowledge or opinions. • CCSS.ELA-LITERACY.RST.9-10.1: Cite specific textual evidence to support analysis of science and technical texts, attending to the precise details of explanations or descriptions. • CCSS.ELA-LITERACY.RST.9-10.2: Determine the central ideas or conclusions of a text; trace the text's explanation or depiction of a complex process, phenomenon, or concept; provide an accurate summary of the text. • CCSS.ELA-LITERACY.RST.9-10.3: Follow precisely a complex multistep procedure when carrying out experiments, taking measurements, or performing technical tasks, attending to special cases or exceptions defined in the text. Craft and structure • CCSS.ELA-LITERACY.RST.6-8.4: Determine the meaning of symbols, key terms, and other domain-specific words and phrases as they are used in a specific scientific or technical context relevant to *grade 6–8 texts and topics.* • CCSS.ELA-LITERACY.RST.6-8.5: Analyze the structure an author uses to organize a text, including how the major sections contribute to the whole and to an understanding of the topic. • CCSS.ELA-LITERACY.RST.6-8.6: Analyze the author's purpose in providing an explanation, describing a procedure, or discussing an experiment in a text. • CCSS.ELA-LITERACY.RST.9-10.4: Determine the meaning of symbols, key terms, and other domain-specific words and phrases as they are used in a specific scientific or technical context relevant to *grade 9–10 texts and topics.* • CCSS.ELA-LITERACY.RST.9-10.5: Analyze the structure of the relationships among concepts in a text, including relationships among key terms (e.g., *force, friction, reaction force, energy*). • CCSS.ELA-LITERACY.RST.9-10.6: Analyze the author's purpose in providing an explanation, describing a procedure, or discussing an experiment in a text, defining the question the author seeks to address. Integration of knowledge and ideas • CCSS.ELA-LITERACY.RST.6-8.7: Integrate quantitative or technical information expressed in words in a text with a version of that information expressed visually (e.g., in a flowchart, diagram, model, graph, or table).

Continued

TABLE 7.2 (*continued*)

CCSS ELA—**Reading in Science and Technical Subjects** (*continued*)	Integration of knowledge and ideas (*continued*) • CCSS.ELA-LITERACY.RST.6-8.8: Distinguish among facts, reasoned judgment based on research findings, and speculation in a text. • CCSS.ELA-LITERACY.RST.6-8.9: Compare and contrast the information gained from experiments, simulations, video, or multimedia sources with that gained from reading a text on the same topic. • CCSS.ELA-LITERACY.RST.9-10.7: Translate quantitative or technical information expressed in words in a text into visual form (e.g., a table or chart) and translate information expressed visually or mathematically (e.g., in an equation) into words. • CCSS.ELA-LITERACY.RST.9-10.8: Assess the extent to which the reasoning and evidence in a text support the author's claim or a recommendation for solving a scientific or technical problem. • CCSS.ELA-LITERACY.RST.9-10.9: Compare and contrast findings presented in a text to those from other sources (including their own experiments), noting when the findings support or contradict previous explanations or accounts.
CCSS ELA—**Writing in Science and Technical Subjects**	Text types and purposes • CCSS.ELA-LITERACY.WHST.6-10.1: Write arguments focused on *discipline-specific content*. • CCSS.ELA-LITERACY.WHST.6-10.2: Write informative or explanatory texts, including the narration of historical events, scientific procedures/experiments, or technical processes. Production and distribution of writing • CCSS.ELA-LITERACY.WHST.6-10.4: Produce clear and coherent writing in which the development, organization, and style are appropriate to task, purpose, and audience. • CCSS.ELA-LITERACY.WHST.6-8.5: With some guidance and support from peers and adults, develop and strengthen writing as needed by planning, revising, editing, rewriting, or trying a new approach, focusing on how well purpose and audience have been addressed. • CCSS.ELA-LITERACY.WHST.6-8.6: Use technology, including the internet, to produce and publish writing and present the relationships between information and ideas clearly and efficiently. • CCSS.ELA-LITERACY.WHST.9-10.5: Develop and strengthen writing as needed by planning, revising, editing, rewriting, or trying a new approach, focusing on addressing what is most significant for a specific purpose and audience.

Continued

TABLE 7.2 (*continued*)

CCSS ELA—Writing in Science and Technical Subjects (*continued*)	Production and distribution of writing (*continued*) • CCSS.ELA-LITERACY.WHST.9-10.6: Use technology, including the internet, to produce, publish, and update individual or shared writing products, taking advantage of technology's capacity to link to other information and to display information flexibly and dynamically. Range of writing • CCSS.ELA-LITERACY.WHST.6-10.10: Write routinely over extended time frames (time for reflection and revision) and shorter time frames (a single sitting or a day or two) for a range of discipline-specific tasks, purposes, and audiences.
CCSS ELA—Speaking and Listening	Comprehension and collaboration • CCSS.ELA-LITERACY.SL.6-8.1: Engage effectively in a range of collaborative discussions (one-on-one, in groups, and teacher-led) with diverse partners on grade 6–8 topics, texts, and issues, building on others' ideas and expressing their own clearly. • CCSS.ELA-LITERACY.SL.6-8.2:* Interpret information presented in diverse media and formats (e.g., visually, quantitatively, orally) and explain how it contributes to a topic, text, or issue under study. • CCSS.ELA-LITERACY.SL.6-8.3:* Delineate a speaker's argument and specific claims, distinguishing claims that are supported by reasons and evidence from claims that are not. • CCSS.ELA-LITERACY.SL.9-10.1: Initiate and participate effectively in a range of collaborative discussions (one-on-one, in groups, and teacher-led) with diverse partners on grade 9–10 topics, texts, and issues, building on others' ideas and expressing their own clearly and persuasively. • CCSS.ELA-LITERACY.SL.9-10.2: Integrate multiple sources of information presented in diverse media or formats (e.g., visually, quantitatively, orally) evaluating the credibility and accuracy of each source. • CCSS.ELA-LITERACY.SL.9-10.3: Evaluate a speaker's point of view, reasoning, and use of evidence and rhetoric, identifying any fallacious reasoning or exaggerated or distorted evidence. Presentation of knowledge and ideas • CCSS.ELA-LITERACY.SL.6-8.4:* Present claims and findings, sequencing ideas logically and using pertinent descriptions, facts, and details to accentuate main ideas or themes; use appropriate eye contact, adequate volume, and clear pronunciation. • CCSS.ELA-LITERACY.SL.6-8.5:* Include multimedia components (e.g., graphics, images, music, sound) and visual displays in presentations to clarify information.

Continued

TABLE 7.2 (*continued*)

***CCSS ELA*—Speaking and Listening** (*continued*)	Presentation of knowledge and ideas (*continued*) • CCSS.ELA-LITERACY.SL.6-8.6: Adapt speech to a variety of contexts and tasks, demonstrating command of formal English when indicated or appropriate. • CCSS.ELA-LITERACY.SL.9-10.4: Present information, findings, and supporting evidence clearly, concisely, and logically such that listeners can follow the line of reasoning and the organization, development, substance, and style are appropriate to purpose, audience, and task. • CCSS.ELA-LITERACY.SL.9-10.5: Make strategic use of digital media (e.g., textual, graphical, audio, visual, and interactive elements) in presentations to enhance understanding of findings, reasoning, and evidence and to add interest. • CCSS.ELA-LITERACY.SL.9-10.6: Adapt speech to a variety of contexts and tasks, demonstrating command of formal English when indicated or appropriate.

* Only the standard for grade 6 is provided because the standards for grades 7 and 8 are similar. Please see *www.corestandards.org/ELA-Literacy/SL* for the exact wording of the standards for grades 7 and 8.

LAB 7

Lab Handout

Lab 7. Formation of Geologic Features: How Can We Explain the Growth of the Hawaiian Archipelago Over the Past 100 Million Years?

Introduction

Scientists use the theory of plate tectonics to explain current and past movements of the rocks at Earth's surface and the origin of many geologic features such as those shown in Figure L7.1. The theory of plate tectonics indicates that the lithosphere is broken into several plates that are in constant motion. Multiple lines of evidence support this theory. This evidence includes, but is not limited to, the location of earthquakes, chains of volcanoes (see Figure L7.1A), and non-volcanic mountain ranges (see Figure L7.1b) around the globe; how land under massive loads (such as lakes or ice sheets) can bend and even flow; the existence of mid-oceanic ridges; and the age of rocks near these ridges.

FIGURE L7.1 _____

(a) The Aleutian archipelago, a chain of volcanic islands in Alaska; (b) the Himalayas, a nonvolcanic mountain range in Asia separating the plains of the Indian subcontinent from the Tibetan plateau

 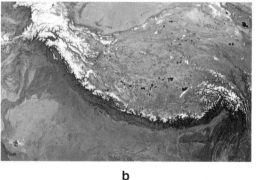

| a | b |

The plates are composed of oceanic and continental lithosphere. The plates move because they are located on top of giant convection cells in the mantle (see Figure L7.2). These currents bring matter from the hot inner mantle near the outer core up to the cooler surface and return cooler matter back to the inner mantle. The convection cells are driven by the energy that is released when isotopes deep within the interior of the Earth go through radioactive decay. The movement of matter in the mantle produces forces, which include viscous drag, slab pull, and ridge push, that together slowly move each of the plates across Earth's surface in a specific direction. The plates carry the continents, create or destroy ocean basins, form mountain ranges and plateaus, and produce earthquakes or volcanoes as they move.

Many interesting Earth surface features, such as the ones shown in Figure L7.1, are the result of either constructive or destructive geologic processes that occur along plate boundaries. There are three main types of plate boundaries (see Figure L7.3): *convergent boundaries* result when two plates collide with each other, *divergent boundaries* result when two plates move away from each other, and *transform boundaries* occur when plates slide past each other. We can explain many of the geologic features we see on Earth's surface when we understand how plates move and interact with each other over time.

Earth's surface is still being shaped and reshaped because of the movement of plates. One example of this phenomenon is the Hawaiian Islands. The Hawaiian Islands is an archipelago in the northern part of the Pacific Ocean that consists of eight major islands, several atolls, and numerous smaller islets. It extends from the island of Hawaii over 2,400 kilometers to the Kure Atoll. Each island is made up of one or more volcanoes (see Figure L7.4). The island of Hawaii, for instance, is made up of five different volcanoes. Two of the volcanoes found on the island of Hawaii are called Mauna Loa and Kilauea. Mauna Loa is the largest active volcano on Earth, and Kilauea is one of the most productive volcanoes in terms of how much lava erupts from it each year.

The number of islands in the Hawaiian archipelago has slowly increased over the last 100 million years. In this investigation, you will attempt to explain why these islands

FIGURE L7.2

Convection cells in the mantle

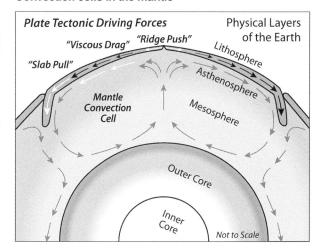

FIGURE L7.3

The three types of plate boundaries

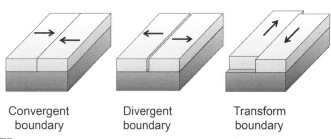

Convergent boundary   Divergent boundary   Transform boundary

FIGURE L7.4

The Hawaiian archipelago and some of its volcanoes

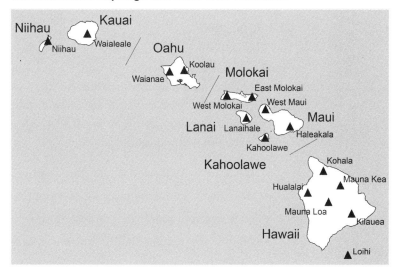

are in the middle of the Pacific Ocean, why they form a chain instead of some other shape, why some of the islands are bigger than other ones, and why the number of islands in the archipelago has slowly increased over time.

Your Task

Develop a conceptual model that you can use to explain how the Hawaiian archipelago formed over the last 100 million years. Your conceptual model must be based on what we know about patterns, systems and system models, and the movement of Earth's plates over time. You should be able to use your conceptual model to predict when and where you will see a new island appear in the Hawaiian archipelago.

The guiding question of this investigation is, **_How can we explain the growth of the Hawaiian archipelago over the past 100 million years?_**

Materials

You will use a computer with Excel or other spreadsheet application during your investigation. You will also use the following resources:

- *Natural Hazards Viewer* online interactive map, available at *http://maps.ngdc.noaa. gov/viewers/hazards*
- Ages of Volcanoes in Hawaiian Islands Excel file; your teacher will tell you how to access the Excel file.

Safety Precautions

Follow all normal lab safety rules.

Investigation Proposal Required?   ☐ Yes   ☐ No

Getting Started

The first step in the development of your conceptual model is to learn as much as you can about the geologic activity around the Hawaiian archipelago. You can use the *Natural Hazards Viewer* interactive map to determine the location of any plate boundaries around the islands, the location of volcanoes on and around each island, and the occurrence and magnitude of earthquakes in the area. As you use the *Natural Hazards Viewer*, be sure to consider the following questions:

- What are the boundaries and the components of the system you are studying?
- How do the components of this system interact with each other?
- How can you quantitatively describe changes within the system over time?
- What scale or scales should you use to when you take your measurements?

- What is going on at the unobservable level that could cause the things that you observe?

The second step in the development of your conceptual model is to learn more about the characteristics of the volcanoes in the Hawaiian archipelago. You can use the Excel file called Ages of Volcanoes in Hawaiian Islands to determine which volcanoes are active and which are dormant, the distances between the volcanoes, and the age of each volcano. As you analyze the data in this Excel file, be sure to consider the following questions:

- What types of patterns could you look for in your data?
- How could you use mathematics to describe a relationship between two variables?
- What could be causing the pattern that you observe?
- What graphs could you create in Excel to help you make sense of the data?

Once you have learned as much as you can about Hawaiian archipelago system, your group can begin to develop your conceptual model. A conceptual model is an idea or set of ideas that explains what causes a particular phenomenon in nature. People often use words, images, and arrows to describe a conceptual model. Your conceptual model needs to be able to explain the origin of the Hawaiian archipelago. It also needs to be able to explain

- why the islands form a chain and not some other shape,
- why the number of islands has increased over the last 100 million years,
- why some islands are bigger than other ones, and
- what will likely happen to the Hawaiian archipelago over the next 100 million years.

The last step in your investigation will be to generate the evidence you to need to convince others that your model is valid and acceptable. To accomplish this goal, you can attempt to show how using a different version of your model or making a specific change to a portion of your model would make your model inconsistent with what we know about the islands in the Hawaiian archipelago. Scientists often make comparisons between different versions of a model in this manner to show that a model they have developed is valid or acceptable. You can also use the *Natural Hazards Viewer* to identify other chains of volcanoes that are similar to ones found in the Hawaiian archipelago. You can then determine if you are able to use your model to explain the formation of other chains of volcanoes. If you are able to show how your conceptual model explains the formation of the Hawaiian archipelago better than other models or that you can use your conceptual model to explain many different phenomena, then you should be able to convince others that it is valid or acceptable.

Connections to the Nature of Scientific Knowledge and Scientific Inquiry

As you work through your investigation, be sure to think about

- the use of models as tools for reasoning about natural phenomena in science, and
- the assumptions made by scientists about order and consistency in nature.

Initial Argument

Once your group has finished collecting and analyzing your data, your group will need to develop an initial argument. Your initial argument needs to include a claim, evidence to support your claim, and a justification of the evidence. The *claim* is your group's answer to the guiding question. The *evidence* is an analysis and interpretation of your data. Finally, the *justification* of the evidence is why your group thinks the evidence matters. The justification of the evidence is important because scientists can use different kinds of evidence to support their claims. Your group will create your initial argument on a whiteboard. Your whiteboard should include all the information shown in Figure L7.5.

FIGURE L7.5 _____

Argument presentation on a whiteboard

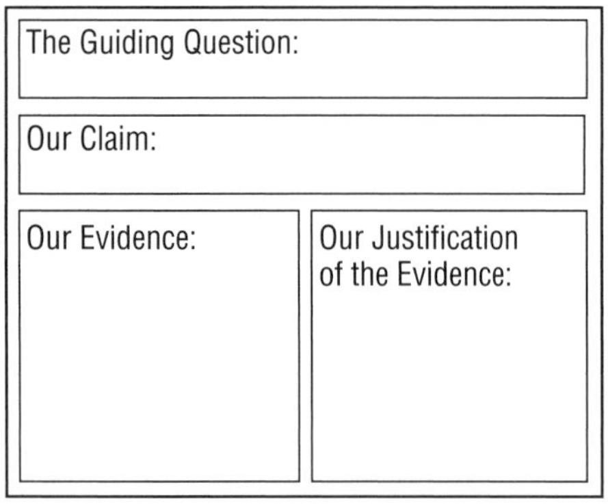

The Guiding Question:	
Our Claim:	
Our Evidence:	Our Justification of the Evidence:

Argumentation Session

The argumentation session allows all of the groups to share their arguments. One or two members of each group will stay at the lab station to share that group's argument, while the other members of the group go to the other lab stations to listen to and critique the other arguments. This is similar to what scientists do when they propose, support, evaluate, and refine new ideas during a poster session at a conference. If you are presenting your group's argument, your goal is to share your ideas and answer questions. You should also keep a record of the critiques and suggestions made by your classmates so you can use this feedback to make your initial argument stronger. You can keep track of specific critiques and suggestions for improvement that your classmates mention in the space below.

Critiques of our initial argument and suggestions for improvement:

If you are critiquing your classmates' arguments, your goal is to look for mistakes in their arguments and offer suggestions for improvement so these mistakes can be fixed. You should look for ways to make your initial argument stronger by looking for things that the other groups did well. You can keep track of interesting ideas that you see and hear during the argumentation in the space below. You can also use this space to keep track of any questions that you will need to discuss with your team.

Interesting ideas from other groups or questions to take back to my group:

Once the argumentation session is complete, you will have a chance to meet with your group and revise your initial argument. Your group might need to gather more data or design a way to test one or more alternative claims as part of this process. Remember, your goal at this stage of the investigation is to develop the best argument possible.

Report

Once you have completed your research, you will need to prepare an *investigation report* that consists of three sections. Each section should provide an answer for the following questions:

1. What question were you trying to answer and why?

2. What did you do to answer your question and why?

3. What is your argument?

Your report should answer these questions in two pages or less. You should write your report using a word processing application (such as Word, Pages, or Google Docs), if possible, to make it easier for you to edit and revise it later. You should embed any diagrams, figures, or tables into the document. Be sure to write in a persuasive style; you are trying to convince others that your claim is acceptable or valid.

LAB 7

Lab 7. Formation of Geologic Features: How Can We Explain the Growth of the Hawaiian Archipelago Over the Past 100 Million Years?

1. Below is a map of the Hawaiian archipelago, shown from above. On the map, draw what you think the archipelago will look like in 100 million years.

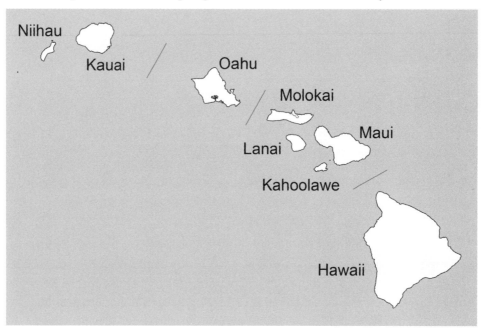

Explain your drawing below.

2. Below is a picture of the Japanese archipelago. What information would you need to determine if the Japanese archipelago formed in the same way the Hawaiian archipelago formed?

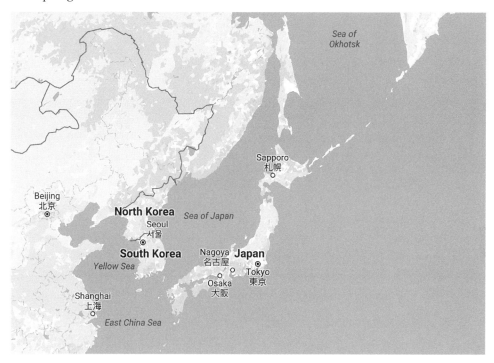

3. Scientists can change or refine a model when presented with new evidence.

   a. I agree with this statement.

   b. I disagree with this statement.

Explain your answer, using an example from your investigation about the formation of the Hawaiian archipelago.

4. When trying to understand events that happened in the past, scientists assume that natural laws operate today in the same way as they did in the past.

   a. I agree with this statement.
   b. I disagree with this statement.

   Explain your answer, using an example from your investigation about the formation of the Hawaiian archipelago.

5. In science, it is important to define a system under study and then develop a model of the system. Explain why this is important to do, using an example from your investigation about the formation of the Hawaiian archipelago.

6. Scientists often look for patterns as part of their work. Explain why it is important to identify patterns during an investigation, using an example from your investigation about the formation of the Hawaiian archipelago.

SECTION 4
Earth's Systems

Introduction Labs

Teacher Notes

Lab 8. Surface Erosion by Wind: Why Do Changes in Wind Speed, Wind Duration, and Soil Moisture Affect the Amount of Soil That Will Be Lost Due to Wind Erosion?

Purpose

The purpose of this lab is to *introduce* students to the disciplinary core idea (DCI) of Earth Materials and Systems by having them develop a conceptual model that can be used to explain why wind speed, wind duration, and soil moisture affect the amount of soil lost due to wind erosion. In addition, students have an opportunity to learn about the crosscutting concepts (CCs) of (a) Scale, Proportion, and Quantity; and (b) Stability and Change. During the explicit and reflective discussion, students will also learn about (a) the difference between data and evidence in science and (b) the nature and role of experiments in science.

Important Earth and Space Science Content

Wind can produce significant erosion effects on Earth's surface through deflation and abrasion. *Deflation* is the removal of soil particles from an area. Soil particles include sand, clay, and silt; the relative size of these three kinds of soil particles is shown in Figure 8.1. *Abrasion* is the wearing down of rock or other objects. Abrasion often results in the creation of soil particles and other, larger, sediments.

FIGURE 8.1

Soil particle types and sizes

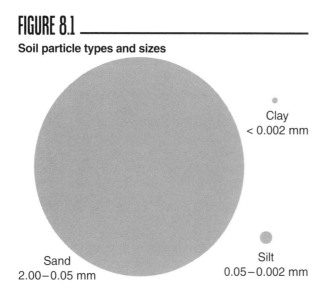

Clay
< 0.002 mm

Sand
2.00–0.05 mm

Silt
0.05–0.002 mm

Wind can transport matter through three different processes. The first process is *suspension*. Wind currents can hold small particles, such as silt and clay, in the atmosphere, and these particles then scatter to cause dust and haze. The second process is *saltation*, which occurs when wind causes soil particles between 0.1 mm and 0.5 mm, such as sand, to skip or bounce along the ground. The third process is *creeping*, a term that Earth scientists use to describe the way that particles that are larger than 0.5 mm roll or slide along the ground because of wind.

Many factors affect the amount of soil that is lost in a given area due to wind erosion. In this investigation, students will only focus on four different factors:

- *Wind speed.* This factor is important because moving air exerts a pushing force on objects. This pushing force, if it is strong enough, will cause soil particles or other larger sediments to move. The amount of force needed to move a soil particle depends on the mass of the particle. Sand particles require more force to move than silt or clay particles because sand particles have more mass. Wind, as a result, needs to be moving at higher speeds to be able to move particles of sand.

- *Wind duration.* In general, the longer wind blows in a given area, the more soil will be lost to wind erosion, because erosion acts on the surface of a material. As the surface layer of soil is removed by wind, the next layer of soil becomes the new surface layer. This interaction at the surface of the material continues until all the soil particles are removed.

- *Soil texture.* This refers to the relative proportion of sand, silt, and clay found in the soil. Scientists classify soils based on the percentage of each type of particle present in the soil (see Figure 8.2). Silt- and sand-rich soils, such as silt loam and sandy loam, are more likely to erode because these types of soils do not easily form aggregates, or groups of particles cemented together.

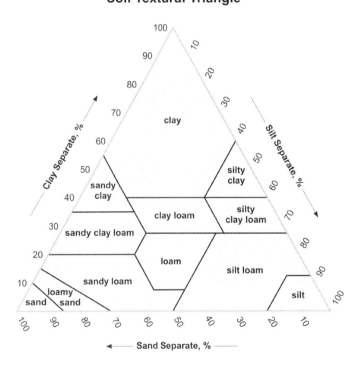

FIGURE 8.2

Soil types by composition

Soil Textural Triangle

- *Soil moisture.* Moist soils do not erode as easily because the water in the soil holds the soil particles together. As a result, the soil particles clump together and become aggregates, which require more force to move.

Timeline

The instructional time needed to complete this lab investigation is 270–330 minutes. Appendix 3 (p. 573) provides options for implementing this lab investigation over several class periods. Option G (330 minutes) should be used if students are unfamiliar with scientific writing, because this option provides extra instructional time for scaffolding the writing process. You can scaffold the writing process by modeling, providing examples, and providing hints as students write each section of the report. This option also gives students time to develop and test their conceptual model. Option H (270 minutes) should be used if students are familiar with scientific writing and have developed the skills needed to write an investigation report on their own. In option H, students complete stage 6 (writing the investigation report) and stage 8 (revising the investigation report) as homework. Both options give students time to test their models on day 3.

Materials and Preparation

The materials needed to implement this investigation are listed in Table 8.1. The consumables and equipment can be purchased from a science supply company such as Carolina, Flinn Scientific, or Ward's Science. The soil samples can also be collected from your local area. To minimize risk of exposure to pesticides or herbicides, we suggest using commercial sterilized potting soil, sand, and other samples instead of samples that you or your students collect. Ensure that there is variety in the texture of soil samples you choose.

Be sure to use a set routine for distributing and collecting the materials during the lab investigation. One option is to set up the materials for each group at each group's lab station before class begins. This option works well when there is a dedicated section of the classroom for lab work and the materials are large and difficult to move (such as a stream table). A second option is to have all the materials on a table or cart at a central location. You can then assign a member of each group to be the "materials manager." This individual is responsible for collecting all the materials his or her group needs from the table or cart during class and for returning all the materials at the end of the class. This option works well when the materials are small and easy to move (such as stopwatches, meter sticks, or thermometers). It also makes it easy to inventory the materials at the end of the class before students leave for the day.

Safety Precautions and Laboratory Waste Disposal

Remind students to follow all normal lab safety rules. In addition, tell the students to take the following safety precautions:

- Wear sanitized indirectly vented chemical-splash goggles and chemical-resistant, nonlatex aprons and gloves throughout the entire investigation (which includes setup and cleanup).

- Handle all glassware with care.

- Plug the fan only into a GFCI-protected circuit. Keep the fan away from water to avoid risk of shock.

- Use caution when working with the fan. Blades can be sharp and cut skin. Do not place fingers into the fan, and never place any object into the turning fan blade.

Surface Erosion by Wind

Why Do Changes in Wind Speed, Wind Duration, and Soil Moisture Affect the Amount of Soil That Will Be Lost Due to Wind Erosion?

TABLE 8.1

Materials list for Lab 8

Item	Quantity
Consumables	
Soil sample A	1 kg per group
Soil sample B	1 kg per group
Soil sample C	1 kg per group
Soil sample D	1 kg per group
Water	As needed
Equipment and other materials	
Safety glasses or goggles	1 per student
Chemical-resistant apron	1 per student
Gloves	1 pair per student
Cookie sheet	1 per group
Trash bags	1–2 per group
Beaker, 250 ml	2 per group
Graduated cylinder, 100 ml	1 per group
Handheld or desk fan with multiple speeds	1 per group
Electronic or triple beam balance	1 per group
Flowchart for soil texture by feel	1 per group
Investigation Proposal C (optional)	3 per group
Whiteboard, 2' × 3'*	1 per group
Lab Handout	1 per student
Peer-review guide and teacher scoring rubric	1 per student
Checkout Questions	1 per student

* As an alternative, students can use computer and presentation software such as Microsoft PowerPoint or Apple Keynote to create their arguments.

- Report and clean up any spills immediately, and avoid walking in areas where water has been spilled.
- Wash hands with soap and water when done collecting the data and after completing the lab.

Waste for this lab will be minimal. You may choose to dry the soil samples after use for later reuse, or the samples can also be disposed of directly into the trash.

LAB 8

Topics for the Explicit and Reflective Discussion

Reflecting on the Use of Core Ideas and Crosscutting Concepts During the Investigation

Teachers should begin the explicit and reflective discussion by asking students to discuss what they know about the DCI they used during the investigation. The following are some important concepts related to the DCI of Earth Materials and Systems that students need to be able to develop a conceptual model that can be used to explain why wind speed, wind duration, and soil moisture affects the amount of soil lost due to wind erosion:

- Wind erosion is caused by abrasion (the wearing down of rock or other objects) or deflation (the removal of soil particles from an area).
- Winds apply a pushing force to soil particles that can cause them to move.
- Large particles have greater mass and require greater force to move.
- Winds of higher velocities apply a greater force on soil particles.
- Moist soils form aggregates that are more massive than individual soil particles.
- Matter cannot be created or destroyed, so it is possible to track the loss of soil by measuring a change in the mass of a sample.

To help students reflect on what they know about these concepts, we recommend showing them two or three images using presentation software that help illustrate these important ideas. You can then ask the students the following questions to encourage students to share how they are thinking about these important concepts:

1. What do we see going on in this image?
2. Does anyone have anything else to add?
3. What might be going on that we can't see?
4. What are some things that we are not sure about here?

You can then encourage students to think about how CCs played a role in their investigation. There are at least two CCs that students need to develop a conceptual model that can be used to explain why wind speed, wind duration, and soil moisture affect the amount of soil lost due to wind erosion: (a) Scale, Proportion, and Quantity; and (b) Stability and Change (see Appendix 2 [p. 569] for a brief description of these CCs. To help students reflect on what they know about these two CCs, we recommend asking them the following questions:

1. Why is it important to think about what controls or affects the rate of change in a system? How can we measure a rate of change?
2. Which factor(s) might have controlled the rate of change in wind erosion in your investigation? What did exploring these factors allow you to do?

Surface Erosion by Wind

Why Do Changes in Wind Speed, Wind Duration, and Soil Moisture Affect the Amount of Soil That Will Be Lost Due to Wind Erosion?

3. Why is it important to consider what measurement scale or scales to use during an investigation? Why is useful to look for proportional relationships when analyzing data?

4. What measurement scale or scales did you use during your investigation? What did that allow you to do? Did you attempt to look for proportional relationships when you were analyzing your data? Why or why not?

You can then encourage the students to think about how they used all these different concepts to help answer the guiding question and why it is important to use these ideas to help justify their evidence for their final arguments. Be sure to remind your students to explain why they included the evidence in their arguments and make the assumptions underlying their analysis and interpretation of the data explicit in order to provide an adequate justification of their evidence.

Reflecting on Ways to Design Better Investigations

It is important for students to reflect on the strengths and weaknesses of the investigation they designed during the explicit and reflective discussion. Students should therefore be encouraged to discuss ways to eliminate potential flaws, measurement errors, or sources of uncertainty in their investigations. To help students be more reflective about the design of their investigation and what they can do to make their investigations more rigorous in the future, you can ask the following questions:

1. What were some of the strengths of the way you planned and carried out your investigation? In other words, what made it scientific?

2. What were some of the weaknesses of the way you planned and carried out your investigation? In other words, what made it less scientific?

3. What rules can we make, as a class, to ensure that our next investigation is more scientific?

Reflecting on the Nature of Scientific Knowledge and Scientific Inquiry

This investigation can be used to illustrate two important concepts related to the nature of scientific knowledge and the nature of scientific inquiry: (a) the difference between data and evidence in science and (b) the nature and role of experiments in science (see Appendix 2 [p. 569] for a brief description of these two concepts). Be sure to review these concepts during and at the end of the explicit and reflective discussion. To help students think about these concepts in relation to what they did during the lab, you can ask the following questions:

1. You had to talk about data and evidence during your investigation. Can you give me some examples of data and evidence from your investigation?

2. Can you work with your group to come up with a rule that you can use to decide if a piece of information is data or evidence? Be ready to share in a few minutes.

3. I asked you to design and carry out several experiments as part of your investigation. Are all investigations in science experiments? Why or why not?

4. Can you work with your group to come up with a rule that you can use to decide if an investigation is an experiment or not? Be ready to share in a few minutes.

You can also use presentation software or other techniques to encourage your students to think about these concepts. You can show examples of information from the investigation that are either data or evidence and ask students to classify each example and explain their thinking. You can also show images of different types of investigations (such as an Earth scientist collecting data in the field as part of an observational or descriptive study, a person working in the library doing a literature review, a person working on a computer to analyze an existing data set, and an actual experiment) and ask students to indicate if they think each image represents an experiment and why or why not.

Remind your students that, to be proficient in science, it is important that they understand what counts as scientific knowledge and how that knowledge develops over time.

Hints for Implementing the Lab

- Be sure to test the fans to ensure that each fan can generate "wind" with enough velocity to transport the soil particles in the sample.

- The soil blown by the fans can be captured using the trash bags. The students can then measure the mass of the captured soil or the mass of the sample of soil before and after being exposed to wind to determine the amount of soil lost to wind erosion.

- Allowing students to design their own procedures for collecting data gives students an opportunity to try, to fail, and to learn from their mistakes. However, you can scaffold students as they develop their procedure by having them fill out an investigation proposal. These proposals provide a way for you to offer students hints and suggestions without telling them how to do it. You can also check the proposals quickly during a class period. We recommend using Investigation Proposal C for this lab. We also recommend that a different investigation proposal be filled out for each experiment. Each group will therefore need to fill out at least three different investigation proposals.

- Investigation Proposal C works best for the nature of the experiments that the students will conduct during this investigation because students can describe a different relationship between the independent and dependent variable for each hypothesis. For example, increasing wind speed might (a) increase the amount of soil lost due to wind erosion (hypothesis 1), (b) decrease the amount of soil lost due to wind erosion (hypothesis 2), or (c) have no effect on the amount of soil lost due to wind erosion (hypothesis 3).

- To create their conceptual models, the students will need to combine what they learned from all three experiments into one explanation. Be sure to remind them that their model must explain why, not just how.

- Students often make mistakes when developing their conceptual models and/ or initial arguments, but they should quickly realize these mistakes during the argumentation session. Be sure to allow students to revise their models and arguments at the end of the argumentation session. The explicit and reflective discussion will also give students an opportunity to reflect on and identify ways to improve how they develop and test models. This also offers an opportunity to discuss what scientists do when they realize a mistake is made.

- Be sure to give students time to test their models. To accomplish this goal, they can make predictions about how much soil will be lost from sample D given its texture, the amount of moisture in it, the wind speed they use, and how long they apply the wind. If their observations match their predictions, they will have evidence that they can use to convince others that their model is valid or acceptable.

- Students often make mistakes during the data collection stage, but they should quickly realize these mistakes during the argumentation session. It will only take them a short period of time to re-collect data, and they should be allowed to do so. During the explicit and reflective discussion, students will also have the opportunity to reflect on and identify ways to improve the way they design investigations (especially how they attempt to control variables as part of an experiment). This provides another opportunity to discuss what scientists do when they realize that a mistake is made during a study.

Connections to Standards

Table 8.2 (p. 202) highlights how the investigation can be used to address specific (a) performance expectations from the *NGSS* and (b) *Common Core State Standards* in English language arts (*CCSS ELA*).

LAB 8

TABLE 8.2

Lab 8 alignment with standards

***NGSS* performance expectations**	History of Earth; Earth's systems • MS-ESS2-1: Develop a model to describe the cycling of Earth's materials and the flow of energy that drives this process. • MS-ESS2-2: Construct an explanation based on evidence for how geoscience processes have changed Earth's surface at varying time and spatial scales.
***CCSS ELA*—Reading in Science and Technical Subjects**	Key ideas and details • CCSS.ELA-LITERACY.RST.6-8.1: Cite specific textual evidence to support analysis of science and technical texts. • CCSS.ELA-LITERACY.RST.6-8.2: Determine the central ideas or conclusions of a text; provide an accurate summary of the text distinct from prior knowledge or opinions. Craft and structure • CCSS.ELA-LITERACY.RST.6-8.4: Determine the meaning of symbols, key terms, and other domain-specific words and phrases as they are used in a specific scientific or technical context relevant to *grade 6–8 texts and topics.* • CCSS.ELA-LITERACY.RST.6-8.5: Analyze the structure an author uses to organize a text, including how the major sections contribute to the whole and to an understanding of the topic. • CCSS.ELA-LITERACY.RST.6-8.6: Analyze the author's purpose in providing an explanation, describing a procedure, or discussing an experiment in a text. Integration of knowledge and ideas • CCSS.ELA-LITERACY.RST.6-8.7: Integrate quantitative or technical information expressed in words in a text with a version of that information expressed visually (e.g., in a flowchart, diagram, model, graph, or table). • CCSS.ELA-LITERACY.RST.6-8.8: Distinguish among facts, reasoned judgment based on research findings, and speculation in a text. • CCSS.ELA-LITERACY.RST.6-8.9: Compare and contrast the information gained from experiments, simulations, video, or multimedia sources with that gained from reading a text on the same topic.
***CCSS ELA*—Writing in Science and Technical Subjects**	Text types and purposes • CCSS.ELA-LITERACY.WHST.6-8.1: Write arguments focused on *discipline-specific content*. • CCSS.ELA-LITERACY.WHST.6-8.2: Write informative or explanatory texts, including the narration of historical events, scientific procedures/experiments, or technical processes.

Continued

TABLE 8.2 (*continued*)

CCSS ELA—**Writing in Science and Technical Subjects** (*continued*)	Production and distribution of writing • CCSS.ELA-LITERACY.WHST.6-8.4: Produce clear and coherent writing in which the development, organization, and style are appropriate to task, purpose, and audience. • CCSS.ELA-LITERACY.WHST.6-8.5: With some guidance and support from peers and adults, develop and strengthen writing as needed by planning, revising, editing, rewriting, or trying a new approach, focusing on how well purpose and audience have been addressed. • CCSS.ELA-LITERACY.WHST.6-8.6: Use technology, including the internet, to produce and publish writing and present the relationships between information and ideas clearly and efficiently. Range of writing • CCSS.ELA-LITERACY.WHST.6-8.10: Write routinely over extended time frames (time for reflection and revision) and shorter time frames (a single sitting or a day or two) for a range of discipline-specific tasks, purposes, and audiences.
CCSS ELA—**Speaking and Listening**	Comprehension and collaboration • CCSS.ELA-LITERACY.SL.6-8.1: Engage effectively in a range of collaborative discussions (one-on-one, in groups, and teacher-led) with diverse partners on grade 6–8 topics, texts, and issues, building on others' ideas and expressing their own clearly. • CCSS.ELA-LITERACY.SL.6-8.2:* Interpret information presented in diverse media and formats (e.g., visually, quantitatively, orally) and explain how it contributes to a topic, text, or issue under study. • CCSS.ELA-LITERACY.SL.6-8.3:* Delineate a speaker's argument and specific claims, distinguishing claims that are supported by reasons and evidence from claims that are not. Presentation of knowledge and ideas • CCSS.ELA-LITERACY.SL.6-8.4:* Present claims and findings, sequencing ideas logically and using pertinent descriptions, facts, and details to accentuate main ideas or themes; use appropriate eye contact, adequate volume, and clear pronunciation. • CCSS.ELA-LITERACY.SL.6-8.5:* Include multimedia components (e.g., graphics, images, music, sound) and visual displays in presentations to clarify information. • CCSS.ELA-LITERACY.SL.6-8.6: Adapt speech to a variety of contexts and tasks, demonstrating command of formal English when indicated or appropriate.

* Only the standard for grade 6 is provided because the standards for grades 7 and 8 are similar. Please see *www.corestandards.org/ELA-Literacy/SL* for the exact wording of the standards for grades 7 and 8.

LAB 8

Lab Handout

Lab 8. Surface Erosion by Wind: Why Do Changes in Wind Speed, Wind Duration, and Soil Moisture Affect the Amount of Soil That Will Be Lost Due to Wind Erosion?

Introduction

Earth scientists use the term *erosion* to describe what happens when liquid water, ice, or wind moves rock, soil, or dissolved materials from one location to another. The process of erosion is responsible for the creation of many interesting landforms or natural phenomena that we observe in the world around us. The rock shown in Figure L8.1, for example, was sculpted by wind erosion. Wind erosion can also produce dust storms such as the one that moved through Casa Grande, Arizona, on July 5, 2011. This enormous storm was 101 miles wide and 7,000 feet high, with wind gusts reaching 60 miles per hour (mph). It obscured mountains and covered freeways and homes with massive amounts of soil (see Figure L8.2).

FIGURE L8.1 _____

A rock sculpted by wind erosion in the Altiplano region of Bolivia

FIGURE L8.2 _____

A dust storm in Casa Grande, Arizona

Wind is the movement of air. It is caused by differences in air pressure within the atmosphere. Wind erodes the surface of Earth through abrasion or deflation. *Abrasion* is the wearing down of rock or other objects. Wind abrasion over a long period of time can make significant changes to the shape of a rock. In fact, the rock in Figure L8.1 looks the way that it does because of wind abrasion. *Deflation* is the removal of soil particles, such as sand, silt, and clay, from an area by wind. Wind can remove soil particles from one location and move them to a different location because moving air exerts a pushing force

Surface Erosion by Wind

Why Do Changes in Wind Speed, Wind Duration, and Soil Moisture Affect the Amount of Soil That Will Be Lost Due to Wind Erosion?

on objects. Wind can move soil particles by suspending them in the air or causing them to bounce or slide along the ground. Soil particles will continue to move until the pushing force of the wind acting on them weakens or stops. These soil particles often accumulate and form mounds or dunes around physical barriers. The dust storm in Figure L8.1 is an example of wind deflation. This dust storm moved a massive amount of soil into the city of Casa Grande.

Wind erosion is a big problem in some regions of the United States because wind can remove and transport a lot of soil in these regions. Some regions struggle with the effects of wind erosion and some regions do not because different regions of the United States have different wind patterns, types of soil, and climates. In some regions of the United States, for example, the average wind speed never exceeds 4.0 mph, but in other areas the average is consistently above 10.5 mph. Wind speed is important to consider because fast-moving air applies a greater pushing force than slow-moving air. The soil found in a region differs in terms of the proportion of sand, silt, and clay that is present in it.

Earth scientists use the soil classification triangle shown in Figure L8.3 to classify soils. Figure L8.4 shows the different types of soil particles and their relative sizes. The size of soil particles is important to consider because objects with more mass require more force to move. The climate of a region affects how much rain a region will typically have at different times of the year. The amount of rainfall affects the moisture of the soil.

Wind speed, wind duration, and soil moisture, as well as several other factors (such as amount and type of vegetation in a region), will affect the amount of soil that will be lost from a given area due to wind erosion. Soil loss due to wind erosion

FIGURE L8.3

Soil types by composition

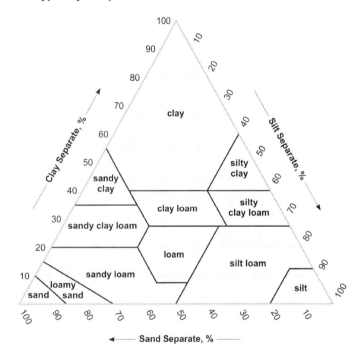

FIGURE L8.4

Soil particle types and sizes

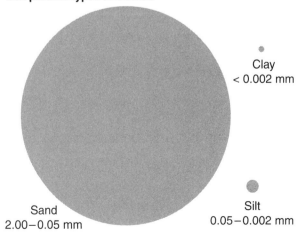

Sand
2.00–0.05 mm

Clay
< 0.002 mm

Silt
0.05–0.002 mm

can destroy crops, put farmers out of business, lead to food shortages, and devastate the economy. During the 1930s, for example, winds removed so much soil on farms in the Midwest that entire crops failed. The winds that blew across the plains suspended so much topsoil in the air that it sometimes blackened the sky. These billows of dust, which were called black blizzards or black rollers at the time, transported soil across the country and even deposited some of it in East Coast cities such as Washington, D.C., and New York. Many farmers went out of business, and the United States, as a result, faced massive food shortages. To prevent a crisis like this from happening again, people must understand how and why soil erodes due to wind. You will therefore explore how wind speed, wind duration, and soil moisture affect the amount of soil that is lost to wind erosion and then develop a conceptual model that you can use to explain your observations and predict how much soil will be lost due to wind erosion under different conditions.

Your Task

Use what you know about wind erosion, soil types, factors that affect rates of change, and the importance of quantity, scale, and proportion in science to determine *how* wind speed, wind duration, and soil moisture affect the amount of soil lost from different types of soil. Then develop a conceptual model that can be used to explain *why* these factors affect the amount of soil lost due to wind erosion. Once you have developed your conceptual model, you will need to test it using a different type of soil to determine if it allows you to predict how much soil will be lost due to wind erosion under a different condition.

The guiding question of this investigation is, *Why do changes in wind speed, wind duration, and soil moisture affect the amount of soil that will lost due to wind erosion?*

Materials

You may use any of the following materials during your investigation:

Consumables	Equipment	
• Soil sample A	• Safety glasses or goggles (required)	• 2 ml Beakers (each 250 ml)
• Soil sample B	• Chemical-resistant apron (required)	• Graduated cylinder (100 ml)
• Soil sample C	• Gloves (required)	• Fan with multiple speeds
• Soil sample D	• Cookie sheet	• Electronic or triple beam balance
• Water	• Trash bag	• Flowchart for soil texture by feel

Safety Precautions

Follow all normal lab safety rules. In addition, take the following safety precautions:

- Wear sanitized indirectly vented chemical-splash goggles and chemical-resistant, nonlatex aprons and gloves throughout the entire investigation (which includes setup and cleanup).

Surface Erosion by Wind

Why Do Changes in Wind Speed, Wind Duration, and Soil Moisture Affect the Amount of Soil That Will Be Lost Due to Wind Erosion?

- Handle all glassware with care.

- Plug the fan only into a GFCI-protected circuit. Keep the fan away from water to avoid risk of shock.

- Use caution when working with the fan. Blades can be sharp and cut skin. Do not place fingers into the fan, and never place any object into the turning fan blade.

- Report and clean up any spills immediately, and avoid walking in areas where water has been spilled.

- Wash hands with soap and water when done collecting the data and after completing the lab.

Investigation Proposal Required? ☐ Yes ☐ No

Getting Started

The first step in developing your model is to plan and carry out three different experiments to determine how wind speed, wind duration, and soil moisture affect the amount of soil lost from different types of soil. You will use soil samples A, B, and C during your experiments. To design each experiment, you will need to determine what type of data you need to collect, how you will collect it, and how you will analyze it.

To determine *what type of data you need to collect,* think about the following questions:

- How will you identify the type of soil?

- How will you determine wind speed?

- How will you determine soil moisture quantitatively?

- How will you determine the amount of soil lost due to wind erosion quantitatively?

- How can you describe the other components of the system?

- How could you keep track of changes over time quantitatively?

To determine *how you will collect the data,* think about the following questions:

- What will be the independent variable and the dependent variable in each experiment?

- What will be the treatment and control conditions in each experiment?

- What other factors will you need to keep constant?

- What scale or scales should you use when you take your measurements?

- What equipment will you need to collect the data you need?

- How will you make sure that your data are of high quality (i.e., how will you reduce error)?

- How will you keep track of and organize the data you collect?

To determine *how you will analyze the data*, think about the following questions:

- How could you use mathematics to describe a change over time?
- How could you use mathematics to determine if there is a difference between the experimental conditions or a relationship between variables?
- What types of proportional relationships might you look for as you analyze your data?
- What type of calculations will you need to make?
- What type of table or graph could you create to help identify a trend in the data?

Once you have carried out your three experiments and understand how changes in wind speed, wind duration, and soil moisture affect the amount of soil lost from different types of soil, your group will need to develop a conceptual model. The model needs to be able to explain why these three factors affect the amount of soil lost to wind erosion in the way that they do. The model also needs to account for the physical properties of the particles that make up soil.

The last step in this investigation is to test your model. To accomplish this goal, you can use soil sample D to determine if your model leads to accurate predictions about the amount of soil that will be lost from a different type of soil under specific conditions (e.g., high wind speed, short duration, and low moisture). If you are able to use your model to make accurate predictions about the effects of wind erosion under different conditions, then you will be able to generate the evidence you need to convince others that the conceptual model you developed is valid or acceptable.

Connections to the Nature of Scientific Knowledge and Scientific Inquiry

As you work through your investigation, be sure to think about

- the difference between data and evidence in science, and
- the nature and role of experiments in science.

Initial Argument

Once your group has finished collecting and analyzing your data, your group will need to develop an initial argument. Your initial argument needs to include a claim, evidence to support your claim, and a justification of the evidence. The *claim* is your group's answer to the guiding question. The *evidence* is an analysis and interpretation of your data. Finally, the *justification* of the evidence is why your group thinks the evidence matters. The justification of the evidence is important because scientists can use different kinds of evidence to support their claims. Your group will create your initial argument on a whiteboard. Your whiteboard should include all the information shown in Figure L8.5.

Argumentation Session

The argumentation session allows all of the groups to share their arguments. One or two members of each group will stay at the lab station to share that group's argument, while the other members of the group go to the other lab stations to listen to and critique the other arguments. This is similar to what scientists do when they propose, support, evaluate, and refine new ideas during a poster session at a conference. If you are presenting your group's argument, your goal is to share your ideas and answer questions. You should also keep a record of the critiques and suggestions made by your classmates so you can use this feedback to make your initial argument stronger. You can keep track of specific critiques and suggestions for improvement that your classmates mention in the space below.

FIGURE L8.5

Argument presentation on a whiteboard

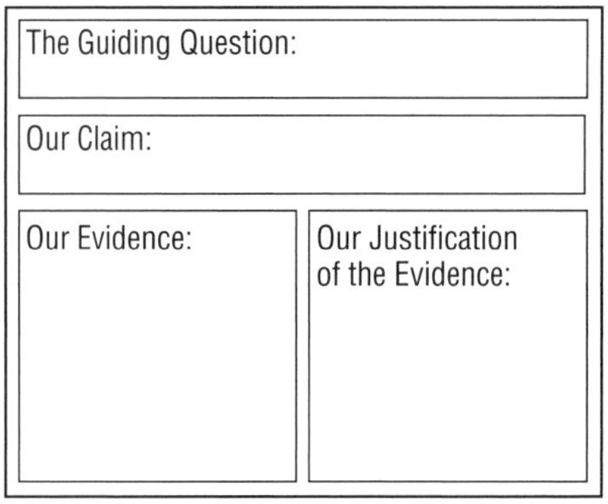

The Guiding Question:	
Our Claim:	
Our Evidence:	Our Justification of the Evidence:

Critiques of our initial argument and suggestions for improvement:

If you are critiquing your classmates' arguments, your goal is to look for mistakes in their arguments and offer suggestions for improvement so these mistakes can be fixed. You should look for ways to make your initial argument stronger by looking for things that the other groups did well. You can keep track of interesting ideas that you see and hear during the argumentation in the space below. You can also use this space to keep track of any questions that you will need to discuss with your team.

Interesting ideas from other groups or questions to take back to my group:

Once the argumentation session is complete, you will have a chance to meet with your group and revise your initial argument. Your group might need to gather more data or design a way to test one or more alternative claims as part of this process. Remember, your goal at this stage of the investigation is to develop the best argument possible.

Report

Once you have completed your research, you will need to prepare an *investigation* report that consists of three sections. Each section should provide an answer for the following questions:

1. What question were you trying to answer and why?

2. What did you do to answer your question and why?

3. What is your argument?

Your report should answer these questions in two pages or less. You should write your report using a word processing application (such as Word, Pages, or Google Docs), if possible, to make it easier for you to edit and revise it later. You should embed any diagrams, figures, or tables into the document. Be sure to write in a persuasive style; you are trying to convince others that your claim is acceptable or valid.

Checkout Questions

Lab 8. Surface Erosion by Wind: Why Do Changes in Wind Speed, Wind Duration, and Soil Moisture Affect the Amount of Soil That Will Be Lost Due to Wind Erosion?

1. A scientist has a collection of soil samples, each classified as shown in the table below.

Sample	Sample classification	Soil moisture (% water by volume)
A	Loamy sand	19
B	Silty clay loam	25
C	Silt loam	10
D	Silty clay loam	15

   a. If exposed to the same wind speed for an equal duration of time, which soil will likely lose the greatest mass to wind erosion?

   b. How do you know? Explain using an example from your investigation about the factors affecting soil loss due to wind erosion.

   c. If exposed to the same wind speed for an equal duration of time, which soil will likely lose the least mass to wind erosion?

   d. How do you know? Explain using an example from your investigation about the factors affecting soil loss due to wind erosion.

2. Given what you know about surface erosion and using examples from your investigation, what are some steps that scientists and engineers can take to help prevent soil erosion in the following areas?

   a. A mountain range

Surface Erosion by Wind

Why Do Changes in Wind Speed, Wind Duration, and Soil Moisture Affect the Amount of Soil That Will Be Lost Due to Wind Erosion?

   b. A park in your local community

   c. Near the ocean or a lake

3. Scientists use experiments to test the validity of a hypothesis (i.e., a tentative explanation) for an observed phenomenon.

   a. I agree with this statement.

   b. I disagree with this statement.

   Explain your answer, using an example from your investigation about the factors affecting soil loss due to wind erosion.

4. In science, there is a difference between data and evidence.

    a. I agree with this statement.

    b. I disagree with this statement.

Explain your answer, using an example from your investigation about the factors affecting soil loss due to wind erosion.

5. In nature, events can occur at varying scales of size and of time. Why is it important to consider the differences in size and scale during an investigation? Give an example of a relatively small effect of wind erosion and a relatively large effect of wind erosion.

6. In science, it is important to understand what factors influence rates of change in system. Explain why this is so important, using an example from your investigation about the factors affecting soil loss due to wind erosion.

LAB 9

Teacher Notes

Lab 9. Sediment Transport by Water: How Do Changes in Stream Flow Affect the Size and Shape of a River Delta?

Purpose

The purpose of this lab is to *introduce* students to the disciplinary core ideas (DCIs) of (a) Earth Materials and Systems and (b) The Roles of Water in Earth's Surface Processes by having them figure out how stream flow volume affects the size and shape of river deltas. In addition, students have an opportunity to learn about the crosscutting concepts (CCs) of (a) Scale, Proportion, and Quantity; and (b) Energy and Matter: Flows, Cycles, and Conservation. During the explicit and reflective discussion, students will also learn about (a) the use of models as tools for reasoning about natural phenomena and (b) the nature and role of experiments in science.

Important Earth and Space Science Content

The land that contributes water to a river system is called a drainage basin. Figure 9.1 is a map of the Mississippi River drainage basin. The drainage basin of one river is separated from the drainage basin of another by a watershed boundary (see Figure 9.1). Watershed boundaries range in size from a ridge separating two small gullies to a continental divide, which splits whole continents into different drainage basins. The Mississippi River has the

FIGURE 9.1 —————————————————————

Mississippi River drainage basin

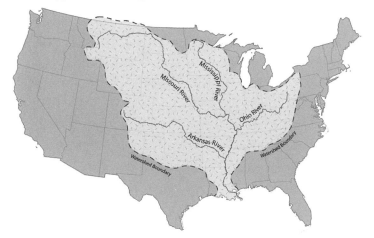

largest drainage basin in North America. It extends between the Rocky Mountains in the west and the Appalachian Mountains in the east. The Mississippi River and its tributaries collect water from more than 3.2 million square kilometers (1.2 million square miles) of the continent.

A *river* can be defined as water flowing in a channel. Water makes its way to the sea under the influence of gravity. The time required for the journey depends on the velocity of the river. *Velocity* is the distance the water travels in a unit of time. The ability of a river to erode and transport materials depends on its velocity. Even slight changes in the velocity of a river can lead to significant changes in the load of sediment that water can transport. Several factors affect the velocity of a river. These factors include the gradient of the channel; the shape, size, and roughness of the channel; and the stream flow.

The *gradient* of a river channel is the vertical drop of channel over a specified distance. The velocity of a river increases as the gradient of the channel increases. The shape, size, and roughness of the channel, however, also affect the velocity of river because water encounters friction as it moves through a channel. Water moves faster through straight, large, and smooth channels than it does through irregular, small, and rough ones. The *stream flow* is the volume of water flowing past a certain point in a given unit of time. The stream flow of most rivers is not constant because it is influenced by the amount of surface water runoff that is currently in the drainage basin. In general, the more surface water runoff that is in a drainage basin, the greater the stream flow of a river. Surface water runoff comes from rain or melting snow.

Rivers are Earth's most important erosional agent. They change the shape of their channels and transport enormous amounts of sediment. Rivers differ in ability to carry a load of sediment. The ability of a river to carry a load is based on two factors: the competence of the river and the capacity of the river. *Competence* is the maximum size of the particles that a river can carry. The velocity of a river determines its competence because faster-moving water is able to carry larger particles than slower-moving water. The *capacity* of a river is the amount of sediment that the river can transport in a certain amount of time. Capacity is measured by combining the *suspended load,* which is the silt and clay particulates carried by the river in suspension, and its *bed load,* which is the amount of larger particles—including rocks, gravel, and sand—carried along the riverbed. The capacity of a river is directly related to the stream flow of the river. The greater the volume of water that flows in a river, the more sediment it is able to transport from one location to another. Much of the sediment that is transported by a river is deposited where the river or stream enters a large body of water. Over time this sediment can accumulate and create a delta.

Deltas are wetlands that form as rivers empty their water and sediment into another body of water, such as a lake, gulf, or ocean. A river moves more slowly as it nears a large body of water. This causes sediment to fall to the river bottom. Deltas are created when the buildup of sediment on the river bottom causes some of the water in the river to break from the channel. Under the right conditions, a river forms a *deltaic lobe.* A mature deltaic lobe includes a series of smaller, shallower channels called distributaries, which branch off from the main channel of the river. In a deltaic lobe, heavier, coarser material settles first. Smaller, finer sediment is carried farther downstream. The finest material, which is called alluvium or silt, is deposited in the larger body of water. As silt builds up, new land is formed. This is the delta. A delta extends a river's mouth into the body of water into which it is emptying.

Like most wetlands, deltas are incredibly diverse and ecologically important ecosystems. Deltas absorb runoff from both floods and storms. Deltas also filter water as it slowly makes its way through the distributaries. This can reduce the amount of pollution entering the larger body of water. Deltas are also important wetland habitats. A wide range of plants (e.g., herbs, lilies, hibiscus trees) grow in deltas. Many animals are also indigenous to deltas. Fish, crustaceans, birds, insects, and even apex predators such as tigers and bears live in or near deltas.

Changes in stream flow can change the size and shape of a delta over time. Changes in stream flow can result from natural causes; for example, decreased precipitation will result in decreased stream flow. Humans can also alter stream flow by controlling or redirecting the flow of water at any point along a river. The Mississippi River delta is an excellent example of how deltas change over time and how human influence can affect the shape, size, and location of deltas. As seen in Figure 9.2, the Mississippi River delta has undergone extensive changes over the last 6,000 years. The land formed by deposited sediment has increased in size, pushing the river channel farther outward into the Gulf of Mexico. The increased stream flow has also resulted in increased sedimentation deposition and increased size of the southern portion of the delta. Although not shown in Figure 9.2, without human intervention, the natural path of the Mississippi would likely gravitate into the Atchafalaya River, which has a much shorter path to the Gulf of Mexico. The Atchafalaya River is located just west of the Mississippi, slightly north of the delta. However, dams have been constructed to prevent the Mississippi from abandoning its current channel. If the Mississippi were to join the Atchafalaya, the delta would certainly change as a result, affecting the cities and towns built on it.

FIGURE 9.2

Evolution of the Mississippi River delta

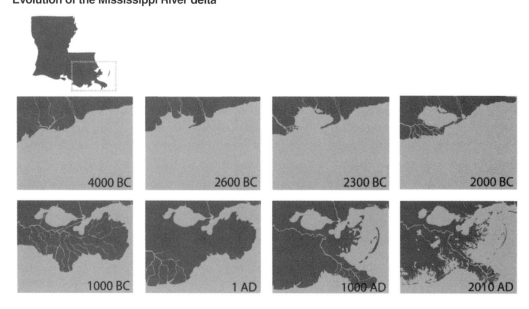

Another change to note is the difference in size of the delta barrier islands from AD 1000 to the present. The most significant reason for the decrease in delta and barrier island size is human interference. Before humans began interfering with the flow of the Mississippi, the river would regularly flood and deposit sediment, ensuring that the land was elevated above sea level. Now, levees for flood control and maintaining a navigable channel to the river's mouth, along with dams far upstream, prevent the river from transporting and depositing a sufficient amount of sediment to maintain the size of the delta.

Timeline

The instructional time needed to complete this lab investigation is 220–280 minutes. Appendix 3 (p. 573) provides options for implementing this lab investigation over several class periods. Option A (280 minutes) should be used if students are unfamiliar with scientific writing, because this option provides extra instructional time for scaffolding the writing process. You can scaffold the writing process by modeling, providing examples, and providing hints as students write each section of the report. Option B (220 minutes) should be used if students are familiar with scientific writing and have developed the skills needed to write an investigation report on their own. In option B, students complete stage 6 (writing the investigation report) and stage 8 (revising the investigation report) as homework.

Materials and Preparation

The materials needed to implement this investigation are listed in Table 9.1 (p. 220). The consumables and equipment can be purchased from a science supply company such as Carolina, Flinn Scientific, or Ward's Science. Carolina and Ward's Science make economy stream table kits that can be purchased for under $100. You can also purchase stream table sand from these companies, or you can use local sediment. If you do not have access to a sufficient number of stream tables, you can make them yourself from inexpensive items. There are many tutorials available online to guide you through constructing stream tables most appropriate for your needs.

LAB 9

TABLE 9.1

Materials list for Lab 9

Item	Quantity
Consumables	
Water (to run through a pump or from a faucet attached to stream table)	As needed
Stream table sand	10 kg per group
Paper towels	As needed
Equipment	
Safety glasses or goggles	1 per student
Chemical-resistant apron	1 per student
Gloves	1 pair per student
Stream table with water volume adjustment	1 per group
Plastic sheet, 12" × 18"	1 per group
Beaker, 1,000 ml	1–2 per group
Graduated cylinder, 250 ml	1 per group
Electronic or triple beam balance	1 per group
Stopwatch	1 per group
Funnel	1 per group
Ruler	1 per group
Protractor	1 per group
Investigation Proposal C (optional)	1 per group
Whiteboard, 2' × 3'*	1 per group
Lab Handout	1 per student
Peer-review guide and instructor scoring rubric	1 per student
Checkout Questions	1 per student

* As an alternative, students can use computer and presentation software such as Microsoft PowerPoint or Apple Keynote to create their arguments.

To create a river delta, students can set up the stream table as shown in Figure 9.3. The sand should be molded in a smooth, gently sloped surface. The channel should be carved into the sand so it is straight and runs the length of the sand. A 12" × 18" sheet of plastic should be placed flush against the foot of the channel. The delta will form on the plastic sheet.

Be sure to use a set routine for distributing and collecting the materials during the lab investigation. One option is to set up the materials for each group at each group's lab station before class begins. This option works well when there is a dedicated section of the classroom for lab work and the materials are large and difficult to move (such as a stream table). A second option is to have all the materials on a table or cart at a central location. You can then assign a member of each group to be the "materials manager." This individual is responsible for collecting all the materials his or her group needs from the table or cart during class and for returning all the materials at the end of the class. This option works well when the materials are small and easy to move (such as stopwatches, metersticks, or thermometers). It also makes it easy to inventory the materials at the end of the class before students leave for the day.

Safety Precautions and Laboratory Waste Disposal

Remind students to follow all normal lab safety rules. In addition, tell the students to take the following safety precautions:

- Wear sanitized indirectly vented chemical-splash goggles and chemical-resistant, nonlatex aprons and gloves throughout the entire investigation (which includes setup and cleanup).

- Keep away from electrical outlets when working with water to prevent or reduce risk of shock.

- Report and clean up spills immediately, and avoid walking in areas where water has been spilled.

- Handle all glassware with care.

- Wash hands with soap and water when done collecting the data and after completing the lab.

No waste disposal is needed in this lab investigation. The sand can be dried and reused.

Topics for the Explicit and Reflective Discussion
Reflecting on the Use of Core Ideas and Crosscutting Concepts During the Investigation

Teachers should begin the explicit and reflective discussion by asking students to discuss what they know about the DCIs they used during the investigation. The following are some important concepts related to the DCIs of (a) Earth Materials and Systems and (b) The Roles of Water in Earth's Surface Processes that students need to be able to determine how stream flow volume affects the size and shape of river deltas:

- production, transportation, and deposition of sediment;

- relationships among stream flow, stream velocity, surface water runoff, channel size, suspended and bed load, capacity and competence; and

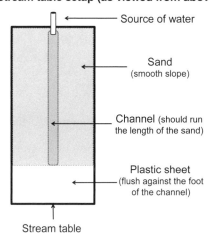

FIGURE 9.3
Stream table setup (as viewed from above)

- Source of water
- Sand (smooth slope)
- Channel (should run the length of the sand)
- Plastic sheet (flush against the foot of the channel)
- Stream table

- the effects of flooding, damming, and levees on stream flow.

To help students reflect on what they know about these concepts, we recommend showing them two or three images using presentation software that help illustrate these important ideas. You can then ask the students the following questions to encourage students to share how they are thinking about these important concepts:

1. What do we see going on in this image?

2. Does anyone have anything else to add?

3. What might be going on that we can't see?

4. What are some things that we are not sure about here?

You can then encourage students to think about how CCs played a role in their investigation. There are at least two CCs that students need to be able to determine how stream flow volume affects the size and shape of river deltas: (a) Scale, Proportion, and Quantity; and (b) Energy and Matter: Flows, Cycles, and Conservation (see Appendix 2 [p. 569] for a brief description of these CCs). To help students reflect on what they know about these CCs, we recommend asking them the following questions:

1. Why is it important to consider what measurement scale or scales to use during an investigation? Why is useful to look for proportional relationships when analyzing data?

2. What measurement scale or scales did you use during your investigation? What did that allow you to do? Did you attempt to look for proportional relationships when you were analyzing your data? Why or why not?

3. Why is it important to track how matter flow into, out of, or within a system during an investigation?

4. How did you track the flow of matter in the system you were studying? What did tracking the flow of matter allow you to do during your investigation?

You can then encourage the students to think about how they used all these different concepts to help answer the guiding question and why it is important to use these ideas to help justify their evidence for their final arguments. Be sure to remind your students to explain why they included the evidence in their arguments and make the assumptions underlying their analysis and interpretation of the data explicit in order to provide an adequate justification of their evidence.

Reflecting on Ways to Design Better Investigations

It is important for students to reflect on the strengths and weaknesses of the investigation they designed during the explicit and reflective discussion. Students should therefore be encouraged to discuss ways to eliminate potential flaws, measurement errors, or sources of uncertainty in their investigations. To help students be more reflective about the design of their investigation and what they can do to make their investigations more rigorous in the future, you can ask the following questions:

1. What were some of the strengths of the way you planned and carried out your investigation? In other words, what made it scientific?

2. What were some of the weaknesses of the way you planned and carried out your investigation? In other words, what made it less scientific?

3. What rules can we make, as a class, to ensure that our next investigation is more scientific?

Reflecting on the Nature of Scientific Knowledge and Scientific Inquiry

This investigation can be used to illustrate two important concepts related to the nature of scientific knowledge and the nature of scientific inquiry: (a) the use of models as tools for reasoning about natural phenomena and (b) the nature and role of experiments in science (see Appendix 2 [p. 569] for a brief description of these two concepts). Be sure to review these concepts during and at the end of the explicit and reflective discussion. To help students think about these concepts in relation to what they did during the lab, you can ask the following questions:

1. Scientists often develop and use physical models and conceptual models during an investigation. What is the difference between the two types of models? Why are models useful in science?

2. What types of model or models did you develop and use during your investigation? How did these models help you answer the guiding question of the lab? What were some limitations of your models?

3. I asked you to design an experiment during your investigation. Are all investigations in science experiments? Why or why not?

4. Can you work with your group to come up with a rule that you can use to decide if an investigation is an experiment or not? Be ready to share in a few minutes.

You can also use presentation software or other techniques to encourage your students to think about these concepts. You can show images of physical and/or conceptual models and ask students to share what they think they are used to do, why they are useful, and some potential limitations of these models. You can also show images of different types of investigations (such as an Earth scientist collecting data in the field as part of an observational or descriptive study, a person working in the library doing a literature review, a person working on a computer to analyze an existing data set, and an actual experiment) and ask them to indicate if they think each image is an example of an experiment and why or why not.

Remind your students that, to be proficient in science, it is important that they understand what counts as scientific knowledge and how that knowledge develops over time.

Hints for Implementing the Lab

- Practice producing deltas using the stream table before the lab begins. It is important for you to know how to use the equipment so you can help students when technical issues arise.

- Allow the students to become familiar with the stream table as part of the tool talk before they begin to design their investigation. This gives students a chance to see what they can and cannot do with the equipment.

- Allowing students to design their own procedures for collecting data gives students an opportunity to try, to fail, and to learn from their mistakes. However, you can scaffold students as they develop their procedure by having them fill out an investigation proposal. These proposals provide a way for you to offer students hints and suggestions without telling them how to do it. You can also check the proposals quickly during a class period. We recommend using Investigation Proposal C for this lab.

- The size of the delta can be quantified by measuring it at its longest and widest points. Students can also remove the plastic sheet, drain the water, and then mass the amount of sand in the delta as another way to quantify its size.

- The shape of the delta can be quantified by measuring the angle of the beds using a protractor.

- Students should run water down the channel for at least five minutes before attempting to measure the size or shape of the delta.

- You may choose to complete data collection for this investigation outside depending on your classroom and stream table setup.

- Students often make mistakes when developing their conceptual models and/ or initial arguments, but they should quickly realize these mistakes during the argumentation session. Be sure to allow students to revise their models and arguments at the end of the argumentation session. The explicit and reflective discussion will also give students an opportunity to reflect on and identify ways to improve how they develop and test models. This also offers an opportunity to discuss what scientists do when they realize a mistake is made.

- Students often make mistakes during the data collection stage, but they should quickly realize these mistakes during the argumentation session. It will only take them a short period of time to re-collect data, and they should be allowed to do so. During the explicit and reflective discussion, students will also have the opportunity to reflect on and identify ways to improve the way they design investigations (especially how they attempt to control variables as part of an experiment). This provides another opportunity to discuss what scientists do when they realize that a mistake is made during a study.

Connections to Standards

Table 9.2 highlights how the investigation can be used to address specific (a) performance expectations from the *NGSS* and (b) *Common Core State Standards* in English language arts (*CCSS ELA*).

TABLE 9.2 ————————————————————————————————

Lab 9 alignment with standards

NGSS performance expectations	History of Earth; Earth's systems • MS-ESS2-1: Develop a model to describe the cycling of Earth's materials and the flow of energy that drives this process. • MS-ESS2-2: Construct an explanation based on evidence for how geoscience processes have changed Earth's surface at varying time and spatial scales. • HS-ESS2-5: Plan and conduct an investigation of the properties of water and its effects on Earth materials and surface processes.
CCSS ELA—Reading in Science and Technical Subjects	Key ideas and details • CCSS.ELA-LITERACY.RST.6-8.1: Cite specific textual evidence to support analysis of science and technical texts. • CCSS.ELA-LITERACY.RST.6-8.2: Determine the central ideas or conclusions of a text; provide an accurate summary of the text distinct from prior knowledge or opinions. • CCSS.ELA-LITERACY.RST.9-10.1: Cite specific textual evidence to support analysis of science and technical texts, attending to the precise details of explanations or descriptions. • CCSS.ELA-LITERACY.RST.9-10.2: Determine the central ideas or conclusions of a text; trace the text's explanation or depiction of a complex process, phenomenon, or concept; provide an accurate summary of the text. • CCSS.ELA-LITERACY.RST.9-10.3: Follow precisely a complex multistep procedure when carrying out experiments, taking measurements, or performing technical tasks, attending to special cases or exceptions defined in the text.

Continued

TABLE 9.2 (*continued*)

CCSS ELA—**Reading in Science and Technical Subjects** (*continued*)	Craft and structure • CCSS.ELA-LITERACY.RST.6-8.4: Determine the meaning of symbols, key terms, and other domain-specific words and phrases as they are used in a specific scientific or technical context relevant to *grade 6–8 texts and topics*. • CCSS.ELA-LITERACY.RST.6-8.5: Analyze the structure an author uses to organize a text, including how the major sections contribute to the whole and to an understanding of the topic. • CCSS.ELA-LITERACY.RST.6-8.6: Analyze the author's purpose in providing an explanation, describing a procedure, or discussing an experiment in a text. • CCSS.ELA-LITERACY.RST.9-10.4: Determine the meaning of symbols, key terms, and other domain-specific words and phrases as they are used in a specific scientific or technical context relevant to *grade 9–10 texts and topics*. • CCSS.ELA-LITERACY.RST.9-10.5: Analyze the structure of the relationships among concepts in a text, including relationships among key terms (e.g., *force, friction, reaction force, energy*). • CCSS.ELA-LITERACY.RST.9-10.6: Analyze the author's purpose in providing an explanation, describing a procedure, or discussing an experiment in a text, defining the question the author seeks to address. Integration of knowledge and ideas • CCSS.ELA-LITERACY.RST.6-8.7: Integrate quantitative or technical information expressed in words in a text with a version of that information expressed visually (e.g., in a flowchart, diagram, model, graph, or table). • CCSS.ELA-LITERACY.RST.6-8.8: Distinguish among facts, reasoned judgment based on research findings, and speculation in a text. • CCSS.ELA-LITERACY.RST.6-8.9: Compare and contrast the information gained from experiments, simulations, video, or multimedia sources with that gained from reading a text on the same topic. • CCSS.ELA-LITERACY.RST.9-10.7: Translate quantitative or technical information expressed in words in a text into visual form (e.g., a table or chart) and translate information expressed visually or mathematically (e.g., in an equation) into words. • CCSS.ELA-LITERACY.RST.9-10.8: Assess the extent to which the reasoning and evidence in a text support the author's claim or a recommendation for solving a scientific or technical problem. • CCSS.ELA-LITERACY.RST.9-10.9: Compare and contrast findings presented in a text to those from other sources (including their own experiments), noting when the findings support or contradict previous explanations or accounts.

Continued

TABLE 9.2 (*continued*)

| *CCSS ELA*—Writing in Science and Technical Subjects | Text types and purposes
• CCSS.ELA-LITERACY.WHST.6-10.1: Write arguments focused on *discipline-specific content*.
• CCSS.ELA-LITERACY.WHST.6-10.2: Write informative or explanatory texts, including the narration of historical events, scientific procedures/experiments, or technical processes.

Production and distribution of writing
• CCSS.ELA-LITERACY.WHST.6-8.4: Produce clear and coherent writing in which the development, organization, and style are appropriate to task, purpose, and audience.
• CCSS.ELA-LITERACY.WHST.6-8.5: With some guidance and support from peers and adults, develop and strengthen writing as needed by planning, revising, editing, rewriting, or trying a new approach, focusing on how well purpose and audience have been addressed.
• CCSS.ELA-LITERACY.WHST.6-8.6: Use technology, including the internet, to produce and publish writing and present the relationships between information and ideas clearly and efficiently.
• CCSS.ELA-LITERACY.WHST.9-10.5: Develop and strengthen writing as needed by planning, revising, editing, rewriting, or trying a new approach, focusing on addressing what is most significant for a specific purpose and audience.
• CCSS.ELA-LITERACY.WHST.9-10.6: Use technology, including the internet, to produce, publish, and update individual or shared writing products, taking advantage of technology's capacity to link to other information and to display information flexibly and dynamically.

Range of writing
• CCSS.ELA-LITERACY.WHST.6-10.10: Write routinely over extended time frames (time for reflection and revision) and shorter time frames (a single sitting or a day or two) for a range of discipline-specific tasks, purposes, and audiences. |
| *CCSS ELA*—Speaking and Listening | Comprehension and collaboration
• CCSS.ELA-LITERACY.SL.6-8.1: Engage effectively in a range of collaborative discussions (one-on-one, in groups, and teacher-led) with diverse partners on grade 6–8 topics, texts, and issues, building on others' ideas and expressing their own clearly.
• CCSS.ELA-LITERACY.SL.6-8.2:* Interpret information presented in diverse media and formats (e.g., visually, quantitatively, orally) and explain how it contributes to a topic, text, or issue under study.
• CCSS.ELA-LITERACY.SL.6-8.3:* Delineate a speaker's argument and specific claims, distinguishing claims that are supported by reasons and evidence from claims that are not. |

Continued

TABLE 9.2 (*continued*)

***CCSS ELA*—Speaking and Listening** (*continued*)	Comprehension and collaboration (*continued*)
	• CCSS.ELA-LITERACY.SL.9-10.1: Initiate and participate effectively in a range of collaborative discussions (one-on-one, in groups, and teacher-led) with diverse partners on grade 9–10 topics, texts, and issues, building on others' ideas and expressing their own clearly and persuasively.
	• CCSS.ELA-LITERACY.SL.9-10.2: Integrate multiple sources of information presented in diverse media or formats (e.g., visually, quantitatively, orally) evaluating the credibility and accuracy of each source.
	• CCSS.ELA-LITERACY.SL.9-10.3: Evaluate a speaker's point of view, reasoning, and use of evidence and rhetoric, identifying any fallacious reasoning or exaggerated or distorted evidence.
	Presentation of knowledge and ideas
	• CCSS.ELA-LITERACY.SL.6-8.4:* Present claims and findings, sequencing ideas logically and using pertinent descriptions, facts, and details to accentuate main ideas or themes; use appropriate eye contact, adequate volume, and clear pronunciation.
	• CCSS.ELA-LITERACY.SL.6-8.5:* Include multimedia components (e.g., graphics, images, music, sound) and visual displays in presentations to clarify information.
	• CCSS.ELA-LITERACY.SL.6-8.6: Adapt speech to a variety of contexts and tasks, demonstrating command of formal English when indicated or appropriate.
	• CCSS.ELA-LITERACY.SL.9-10.4: Present information, findings, and supporting evidence clearly, concisely, and logically such that listeners can follow the line of reasoning and the organization, development, substance, and style are appropriate to purpose, audience, and task.
	• CCSS.ELA-LITERACY.SL.9-10.5: Make strategic use of digital media (e.g., textual, graphical, audio, visual, and interactive elements) in presentations to enhance understanding of findings, reasoning, and evidence and to add interest.
	• CCSS.ELA-LITERACY.SL.9-10.6: Adapt speech to a variety of contexts and tasks, demonstrating command of formal English when indicated or appropriate.

* Only the standard for grade 6 is provided because the standards for grades 7 and 8 are similar. Please see *www.corestandards.org/ELA-Literacy/SL* for the exact wording of the standards for grades 7 and 8.

Lab Handout

Lab 9. Sediment Transport by Water: How Do Changes in Stream Flow Affect the Size and Shape of a River Delta?

Introduction

A *drainage basin* is all the land that contributes water to a river system. The land within a drainage basin is marked on a map by watershed boundaries. A *watershed boundary* is the location where water will either flow into a given drainage basin or flow into a different one. Watershed boundaries follow ridgelines because water always flows from a higher elevation to a lower elevation. Figure L9.1 is a map of the Mississippi River drainage basin, which is the largest drainage basin in North America. It extends between the Rocky Mountains in the west and the Appalachian Mountains in the east. The Mississippi River and its tributaries collect water from more than 3.2 million square kilometers (1.2 million square miles) of the continent. All this water empties into the Gulf of Mexico in Louisiana.

A *river* is water that flows in a channel. Water makes its way to the sea in a channel under the influence of gravity. The time required for the journey depends on the velocity of the river. *Velocity* is the distance the water travels in a unit of time. The ability of a river to erode and transport materials depends on its velocity. Even slight changes in velocity can lead to significant changes in the load of sediment that water can transport. Several factors

FIGURE L9.1

The drainage basin of the Mississippi River

affect the velocity of a river. These factors include the gradient of the channel; the shape, size, and roughness of the channel; and the stream flow.

The *gradient* of a river channel is the vertical drop of a river over a specified distance. The velocity of a river increases as the gradient of a channel increases. The shape, size, and roughness of the channel, however, also affect the velocity of river because water encounters friction as it moves through a channel. Water moves faster through straight, large, and smooth channels than it does through irregular, small, and rough ones. The *stream flow* of a river is the volume of water flowing past a certain point in a given unit of time. The stream flow of most rivers is not constant because it is influenced by the amount of surface water runoff that is currently in the drainage basin. In general, the more surface water runoff that is in a drainage basin, the greater the stream flow of a river. Surface water runoff comes from rain or melting snow.

Rivers create different types of landforms because of the way they are able to produce, transfer, and deposit large amounts of sediment. River deltas are an example of a landform that is produced by a river. River deltas form where a river meets a larger body of water, such as a lake, gulf, or ocean. When the two bodies of water meet, the water from the river cannot flow at the same speed it did in the river channel. As the stream flow slows, the river is no longer ability to carry sediment. The sediment therefore accumulates in large amounts where the two bodies of water meet. Over a long period of time, this sediment builds into a fanlike structure such as the delta of the Mississippi River, seen in Figure L9.2.

FIGURE L9.2

The Mississippi River delta; notice the amount of sediment deposited where the river enters the Gulf of Mexico (marked with a circle)

Native vegetation flourishes around river deltas because the soil is rich in minerals and organic nutrients. The land around a river delta is also excellent for growing crops. River deltas also create wetlands and marshes that protect the shorelines and *estuaries* (transition zones between river and marine environments) because they absorb excess runoff from both floods and storms. River deltas provide unique habitats for many different species that live along the coast and within the estuaries. Many species, as a result, need these habitats in order to grow and thrive.

Humans often change the characteristics of rivers to better suit their needs. For example, humans build dams to generate electricity, create levees to protect buildings or homes from flooding, and dig channels to divert water from rivers to farms in order to produce crops. These are just a few examples of the many ways humans alter the characteristics of rivers. The addition of a dam, levee, or channel anywhere in a drainage basin will change the stream flow of a river. It will also affect the amount and type of sediment in the water that is able to reach the mouth of a river. Human activity that changes the characteristics of a river, as a result, will affect how water moves through a drainage basin and could contribute to a number of environmental, economic, or social problems. In this investigation, you have an opportunity to learn how changing the stream flow of a river affects the size or shape of its delta.

Your Task

Use a stream table to create a physical model of a river that empties into a larger body of water. Then use this physical model and what you know about erosion, sediments, how matter flows within a system, and scales, proportion, and quantity to design and carry out an investigation in order to determine the relationship between the river stream flow and the size or shape of its river delta.

The guiding question of this investigation is, ***How do changes in stream flow affect the size and shape of a river delta?***

Materials

You may use any of the following materials during your investigation:

Consumables
- Water for the stream table
- Sand (for the stream table)
- Paper towels

Equipment
- Safety glasses or goggles (required)
- Chemical-resistant apron (required)
- Gloves (required)
- Stream table with water volume adjustment
- Plastic sheet (12" × 18")
- 1–2 Beakers (each 1,000 ml)

- Graduated cylinder (250 ml)
- Electronic or triple beam balance
- Stopwatch
- Funnel
- Ruler
- Protractor

Safety Precautions

Follow all normal lab safety rules. Be sure to use the stream table as instructed by your teacher. In addition, take the following safety precautions:

- Wear sanitized indirectly vented chemical-splash goggles and chemical-resistant, nonlatex aprons and gloves throughout the entire investigation (which includes setup and cleanup).

- Keep away from electrical outlets when working with water to prevent or reduce risk of shock.

- Report and clean up spills immediately, and avoid walking in areas where water has been spilled.

- Handle all glassware with care.

- Wash hands with soap and water when done collecting the data and after completing the lab.

Investigation Proposal Required? ☐ Yes ☐ No

Getting Started

To answer the guiding question, you will need to carry out an experiment using a stream table. The stream table simulates the behavior of a drainage basin. You can set up the stream table as shown in Figure L9.3. When you add sand, carve a channel into the sand, and then add water to one end of the channel, a river will flow through the channel. The delta will then form on the plastic sheet.

Before you collect your data, spend some time familiarizing yourself with the stream table that you will use. Once you understand how the stream table works, you can plan your experiment. To accomplish this task, you must determine what type of data you need to collect, how you will collect it, and how will you analyze it.

To determine *what type of data you need to collect*, think about the following questions:

- What are the boundaries and components of the system you are studying?
- How can you describe the components of the system quantitatively?
- How can you track how matter flows into, out of, or within this system?
- How will you quantify the size and shape of the delta?

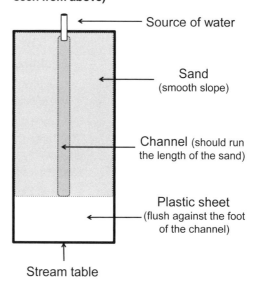

FIGURE L9.3 _____

How to set up a stream table (image is as seen from above)

Source of water

Sand (smooth slope)

Channel (should run the length of the sand)

Plastic sheet (flush against the foot of the channel)

Stream table

To determine *how you will collect the data*, think about the following questions:

- What will be your independent variable and dependent variables in the experiment?
- What will your treatment conditions be and how will you set them up?
- What will serve as a control condition?
- How many trials will be run for each condition?
- How long will each trial run before data are collected?
- When will you need to take measurements?
- What measurement scale or scales should you use to collect data?
- How will you make sure that your data are of high quality (i.e., how will you reduce error)?

- How will you keep track of and organize the data you collect?

To determine *how you will analyze the data,* think about the following questions:

- How could you use mathematics to describe a difference between conditions or a relationship between variables?
- What types of patterns might you look for as you analyze your data?
- What type of calculations will you need to make?
- What type of table or graph could you create to help make sense of your data?

Connections to the Nature of Scientific Knowledge and Scientific Inquiry

As you work through your investigation, be sure to think about

- the use of models as tools for reasoning about natural phenomena, and
- the nature and role of experiments in science.

Initial Argument

Once your group has finished collecting and analyzing your data, your group will need to develop an initial argument. Your initial argument needs to include a claim, evidence to support your claim, and a justification of the evidence. The claim is your group's answer to the guiding question. The evidence is an analysis and interpretation of your data. Finally, the justification of the evidence is why your group thinks the evidence matters. The justification of the evidence is important because scientists can use different kinds of evidence to support their claims. Your group will create your initial argument on a whiteboard. Your whiteboard should include all the information shown in Figure L9.4.

FIGURE L9.4 _____

Argument presentation on a whiteboard

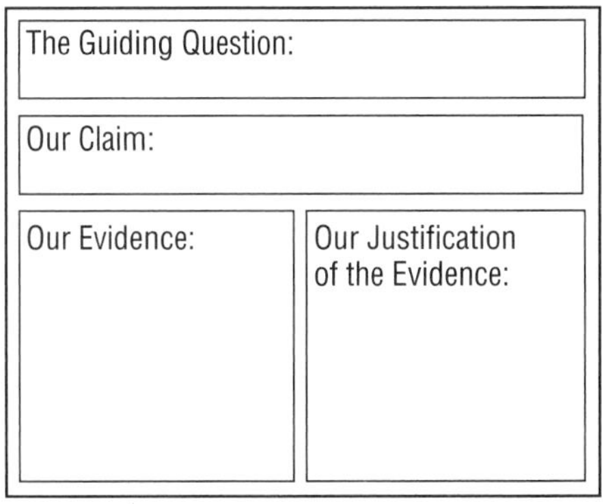

The Guiding Question:	
Our Claim:	
Our Evidence:	Our Justification of the Evidence:

Argumentation Session

The argumentation session allows all of the groups to share their arguments. One or two members of each group will stay at the lab station to share that group's argument, while the other members of the group go to the other lab stations to listen to and critique the other arguments. This is similar to what scientists do when they propose, support, evaluate, and refine new ideas during a poster session at a conference. If you are presenting your group's argument, your goal is to share your ideas and answer questions. You should also keep a record of the critiques and suggestions made by your classmates so you can use this feedback to make your initial argument stronger. You can keep track of specific critiques and suggestions for improvement that your classmates mention in the space below.

Critiques of our initial argument and suggestions for improvement:

If you are critiquing your classmates' arguments, your goal is to look for mistakes in their arguments and offer suggestions for improvement so these mistakes can be fixed. You should look for ways to make your initial argument stronger by looking for things that the other groups did well. You can keep track of interesting ideas that you see and hear during the argumentation in the space below. You can also use this space to keep track of any questions that you will need to discuss with your team.

Interesting ideas from other groups or questions to take back to my group:

Once the argumentation session is complete, you will have a chance to meet with your group and revise your initial argument. Your group might need to gather more data or design a way to test one or more alternative claims as part of this process. Remember, your goal at this stage of the investigation is to develop the best argument possible.

Report

Once you have completed your research, you will need to prepare an investigation report that consists of three sections. Each section should provide an answer for the following questions:

1. What question were you trying to answer and why?

2. What did you do to answer your question and why?

3. What is your argument?

Your report should answer these questions in two pages or less. You should write your report using a word processing application (such as Word, Pages, or Google Docs), if possible, to make it easier for you to edit and revise it later. You should embed any diagrams, figures, or tables into the document. Be sure to write in a persuasive style; you are trying to convince others that your claim is acceptable or valid.

Checkout Questions

Lab 9. Sediment Transport by Water: How Do Changes in Stream Flow Affect the Size and Shape of a River Delta?

Use the figure below to answer questions 1 and 2.

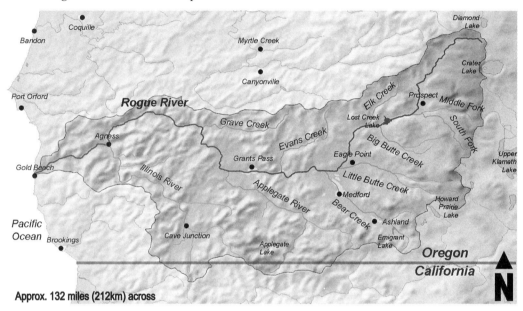

Approx. 132 miles (212km) across

1. A dam is constructed near the town of Agness. What effect will this likely have on the amount of sediment that reaches the town of Gold Beach? Explain why the dam could cause these effects.

2. There is more snowmelt than normal. What effect will this likely have on the amount of sediment that reaches the town of Grants Pass? Explain why an increase in snowmelt could cause these effects.

3. Scientists use experiments to prove a hypothesis is correct.

    a. I agree with this statement.

    b. I disagree with this statement.

    Explain your answer, using an example from your investigation about sediment transport by water.

4. A model is a three-dimensional representation of something on a smaller scale than the original.

   a. I agree with this statement.

   b. I disagree with this statement.

   Explain your answer, using an example from your investigation about sediment transport by water.

5. Scientists often need to be able to track how energy and matter move into, out of, and within systems during an investigation. Explain why it is important track energy and matter, using an example from your investigation about sediment transport by water.

6. Scientists often need to consider what measurement scale or scales to use during an investigation. Explain why it is important for scientist to think about the measurement scales, using an example from your investigation about sediment transport by water.

LAB 10

Teacher Notes

Lab 10. Deposition of Sediments: How Can We Explain the Deposition of Sediments in Water?

Purpose

The purpose of this lab is to *introduce* students to the disciplinary core ideas (DCIs) of (a) Earth Materials and Systems and (b) The Roles of Water in Earth's Surface Processes by having them explore the factors that affect the settling velocity of sediments in water. In addition, students have an opportunity to learn about the crosscutting concepts (CCs) of (a) Cause and Effect: Mechanism and Explanation and (b) Structure and Function. During the explicit and reflective discussion, students will also learn about (a) the use of models as tools for reasoning about natural phenomena and (b) the nature and role of experiments in science.

Important Earth and Space Science Content

Sedimentary rock is formed when sediment particles are produced from other rocks through the process of weathering and those sediments are then transported through erosion, deposited at a new location, and transformed into a new rock through a process called *lithification*. Sediments go through compaction and cementation during the process of lithification. *Compaction* occurs when the individual pieces of sediment in a layer are forced together because of the combined weight of all the sediment in layers above them. *Cementation* happens when the dissolved minerals between the pieces of sediment dry. These minerals then bind the other pieces of sediment together and harden.

Clastic sedimentary rocks are formed when existing rocks are mechanically weathered, producing rock fragment sediments called clasts. These sediments are transported by wind, liquid water, and glacial ice from one location to another. This process inevitably sorts the sediments by size based on the energy required to move them. In this investigation, students will investigate how the density, size, and shape of the sediment particle affects how quickly it will fall through water. The density, size, and shape of the sediment particle affect the order in which deposition occurs, and thus the type of sediment most likely to be found in a given area.

The properties of a stream will also affect the deposition of sediments. This is due to the effects of moving water acting on the particle. Stream systems are complex, and a discussion of all the factors affecting sedimentation in streams would extend beyond the scope of this lab and science courses for grades 6–10. Stream velocity and the turbulence of the stream, however, are two important factors. Stream velocity, which is the speed and direction of the water moving in a stream, and stream turbulence change over the length of a stream. In general, particles that are large and/or dense require greater amounts of energy to remain in suspension (which means the particles are held up and carried along

240

by moving water), so streams that are turbulent and fast moving are able to carry larger and denser particles than a stream that is a calm and slow moving.

There are also properties of the particles themselves that influence sedimentation. The gravitational force acting on the particle is opposed by the buoyant force and is dependent on the particle's mass. The gravitational force acting on a particle is important because of its impact on the particle's *settling velocity,* a measure of how quickly a particle will fall through a still fluid. The larger a particle is, the higher its settling velocity will be. Density and shape also affect a particle's settling velocity in liquid. The higher a particle's density, the higher its settling velocity. Flatter particles have lower settling velocities than round ones because of the increased resistant buoyant force due to their increased surface areas. Particles with a high settling velocity come out of suspension before particles with a low settling velocity

The order of deposition influences the sedimentation process and the types of sedimentary rock likely to be formed in that area. Earth scientists can determine what bodies of water may have once been present in an area by looking at the sedimentary rock formed in that area. They can also gather information about formations upstream and how far particles may have been transported prior to deposition. With their knowledge of current conditions, Earth scientists can study the past and produce maps of previous conditions on Earth.

In this investigation, students will investigate the relationships between settling velocity and density, shape, and mass. They will create a rule or set of rules to help them predict where a river might deposit sediment particles. Using what they know about stream velocity and the rules they have created, they will determine the order in which sediment samples will deposit at the mouth of a river.

Timeline

The instructional time needed to complete this lab investigation is 270–330 minutes. Appendix 3 (p. 573) provides options for implementing this lab investigation over several class periods. Option G (330 minutes) should be used if students are unfamiliar with scientific writing, because this option provides extra instructional time for scaffolding the writing process. You can scaffold the writing process by modeling, providing examples, and providing hints as students write each section of the report. Option H (270 minutes) should be used if students are familiar with scientific writing and have developed the skills needed to write an investigation report on their own. In option H, students complete stage 6 (writing the investigation report) and stage 8 (revising the investigation report) as homework. Both options give students time to test their models on day 3.

LAB 10

Materials and Preparation

The materials needed to implement this investigation are listed in Table 10.1. The sediments and equipment can be purchased from a science supply company such as Carolina, Flinn Scientific, or Ward's Science. The clear plastic tubes can also be purchased from an online retailer such as Amazon.

TABLE 10.1

Materials list for Lab 10

Item	Quantity
Sediments of different sizes	
Gravel	100 g per group
Coarse sand	100 g per group
Medium sand	100 g per group
Fine sand	100 g per group
Sediments of different shapes	
Modeling clay, 2 g pieces	4 per group
Sediments of different densities	
Glass beads, 4 mm	30 per group
Plastic beads, 4 mm	30 per group
Ball bearings, 4 mm	30 per group
Sediment mixture A (gravel, 4 mm plastic beads, and fine sand)	60 ml per group
Sediment mixture B (medium sand, ball bearings, and coarse sand)	60 ml per group
Consumable	
Water	As needed
Equipment and other materials	
Safety glasses or goggles	1 per student
Chemical-resistant apron	1 per student
Gloves	1 pair per student
Clear plastic tube, 120 cm in length and 4 cm inside diameter	1 per group
Rubber stopper, size 6	1 per group
Beaker, 500 ml	1 per group
Beaker, 50 ml	1 per group
Funnel	1 per group
Duct tape	30 cm per group

Continued

TABLE 10.1 *(continued)*

Item	Quantity
Meterstick	1 per group
Bucket	1 per group
Electronic or triple beam balance	1 per group
Stopwatches	2 per group
Investigation Proposal C (optional)	3 per group
Whiteboard, 2' × 3'*	1 per group
Lab Handout	1 per student
Peer-review guide and teacher scoring rubric	1 per student
Checkout Questions	1 per student

* As an alternative, students can use computer and presentation software such as Microsoft PowerPoint or Apple Keynote to create their arguments.

The clear plastic tubes, rubber stoppers, and duct tape are needed to make a water column. Each group will need a water column to collect the data they need to answer the guiding question. We recommend using tubes that are at least 120 cm long with an inside diameter of 4 cm. Tubes that are longer than 120 cm are more expensive, but tubes this long make it easier for students to view and time how long it takes for a sample to fall through the column of water (because some of the sample will settle quite rapidly).

To create a water column, such as the one pictured in Figure 10.1, place a rubber stopper in the end of the tube and wrap tape around the stopper and tube to keep the stopper snug. A size 6 stopper fits snugly in a tube with an inside diameter of 4 cm. The tube can then be filled with water using a beaker or pitcher. The tube should be filled to a level about 10 cm from the top. Once filled, the tube should be kept in an upright position at all times to prevent it from bending. The water can be emptied from the tube by setting the entire tube in a bucket. The tape can then be removed from the stopper, and the stopper can then be loosened so the water can drain out of the tube and into the bucket. Be sure to have the students recover the samples for reuse. The gravel and sand can also be dried out, sieved, and reused year after year.

Be sure to use a set routine for distributing and collecting the materials during the lab investigation. One option is to set up the materials for each group at each group's lab station before class begins. This option works well when there is a dedicated section of the classroom for lab work and the materials are large and difficult to move (such as a stream table). A second option is to have all the materials on a table or cart at a central

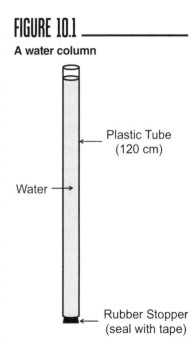

FIGURE 10.1

A water column

Plastic Tube
(120 cm)

Water →

Rubber Stopper
(seal with tape)

location. You can then assign a member of each group to be the "materials manager." This individual is responsible for collecting all the materials his or her group needs from the table or cart during class and for returning all the materials at the end of the class. This option works well when the materials are small and easy to move (such as stopwatches, metersticks, or thermometers). It also makes it easy to inventory the materials at the end of the class before students leave for the day.

Safety Precautions and Laboratory Waste Disposal

Remind students to follow all normal lab safety rules. In addition, tell students to take the following safety precautions:

- Wear sanitized indirectly vented chemical-splash goggles and chemical-resistant aprons and gloves throughout the entire investigation (which includes setup and cleanup).
- Handle all glassware with care.
- Immediately wipe up any spilled water and pick up any spilled beads, ball bearings, or other materials that can cause a slip, trip, or fall hazard.
- Wash hands with soap and water when done collecting the data and after completing the lab.

No special waste disposal is needed for this lab investigation. Tell students to dispose of used water outside (or in a bucket that can be dumped outside later) because some sediments can remain in the water and could clog drains if poured down a sink. The gravel, sand, pieces of modeling clay, glass beads, plastic beads, and ball bearings can be dried and reused.

Topics for the Explicit and Reflective Discussion
Reflecting on the Use of Core Ideas and Crosscutting Concepts During the Investigation

Teachers should begin the explicit and reflective discussion by asking students to discuss what they know about the DCIs they used during the investigation. The following are some important concepts related to the DCIs of (a) Earth Materials and Systems and (b) The Roles of Water in Earth's Surface Processes that students need to be able to explain the deposition of sediments in water:

- The material found in a sedimentary rock comes from other rocks that have weathered over time. As a rock weathers, it is broken into smaller pieces or sediments. These sediments are then carried to other places by wind, liquid water, or glacial ice.
- Sediments eventually settle out of the air or water and accumulate at a specific location. This process is called deposition.

- The rate at which a sediment falls through a fluid is called its settling velocity.

- When a sediment has a settling velocity that is lower than the stream flow velocity of a river, that sediment will be carried downstream. When a sediment has a settling velocity that is higher than the stream flow velocity of a river, that sediment will sink to the bottom of the river and not move farther downstream.

- The physical properties of a sediment particle affect that particle's settling velocity.

To help students reflect on what they know about these concepts, we recommend showing them two or three images using presentation software that help illustrate these important ideas. You can then ask the students the following questions to encourage students to share how they are thinking about these important concepts:

1. What do we see going on in this image?

2. Does anyone have anything else to add?

3. What might be going on that we can't see?

4. What are some things that we are not sure about here?

You can then encourage students to think about how CCs played a role in their investigation. There are at least two CCs that students need to be able to explain the deposition of sediments in water: (a) Cause and Effect: Mechanism and Explanation and (b) Structure and Function (see Appendix 2 [p. 569] for a brief description of these CCs). To help students reflect on what they know about these CCs, we recommend asking them the following questions:

1. Why is it important to identify cause-and-effect relationships in science?

2. What did you have to do during your investigation to determine how a physical property affects the settling velocity of a sediment in water? Why was that useful to do?

3. The way an object is shaped or structured determines many of its properties and how it functions. Why is useful to think about the relationship between structure and function during an investigation?

4. How did the structure of sediment affect how it quickly it falls through water?

You can then encourage the students to think about how they used all these different concepts to help answer the guiding question and why it is important to use these ideas to help justify their evidence for their final arguments. Be sure to remind your students to explain why they included the evidence in their arguments and make the assumptions underlying their analysis and interpretation of the data explicit in order to provide an adequate justification of their evidence.

Reflecting on Ways to Design Better Investigations

It is important for students to reflect on the strengths and weaknesses of the investigation they designed during the explicit and reflective discussion. Students should therefore be encouraged to discuss ways to eliminate potential flaws, measurement errors, or sources of uncertainty in their investigations. To help students be more reflective about the design of their investigation and what they can do to make their investigations more rigorous in the future, you can ask the following questions:

1. What were some of the strengths of the way you planned and carried out your investigation? In other words, what made it scientific?

2. What were some of the weaknesses of the way you planned and carried out your investigation? In other words, what made it less scientific?

3. What rules can we make, as a class, to ensure that our next investigation is more scientific?

Reflecting on the Nature of Scientific Knowledge and Scientific Inquiry

This investigation can be used to illustrate two important concepts related to the nature of scientific knowledge and the nature of scientific inquiry: (a) the use of models as tools for reasoning about natural phenomena and (b) the nature and role of experiments in science (see Appendix 2 [p. 569] for a brief description of these two concepts). Be sure to review these concepts during and at the end of the explicit and reflective discussion. To help students think about these concepts in relation to what they did during the lab, you can ask the following questions:

1. Scientists often develop and use physical models and conceptual models during an investigation. What is the difference between the two types of models? Why are models useful in science?

2. What types of model or models did you develop and use during your investigation? How did these models help you answer the guiding question of the lab? What were some limitations of your models?

3. I asked you to design an experiment during your investigation. Why do scientists conduct experiments?

4. Can you work with your group to come up with a rule that you can use to decide if an investigation is an experiment or not? Be ready to share in a few minutes.

You can also use presentation software or other techniques to encourage your students to think about these concepts. You can show images of different physical and/or conceptual models and ask students to share what they think they are used to do, why they are useful, and some potential limitations of these models. You can also show images of different types of investigations (such as an Earth scientist collecting data in the field as part of an observational or descriptive study, a person working in the library doing a literature review, a person working on a computer to analyze an existing data set, and an actual experiment) and ask students to indicate if they think each image represents an experiment and why or why not.

Remind your students that, to be proficient in science, it is important that they understand what counts as scientific knowledge and how that knowledge develops over time.

Hints for Implementing the Lab

- Allowing students to design their own procedures for collecting data gives students an opportunity to try, to fail, and to learn from their mistakes. However, you can scaffold students as they develop their procedure by having them fill out an investigation proposal. These proposals provide a way for you to offer students hints and suggestions without telling them how to do it. You can also check the proposals quickly during a class period. We recommend using Investigation Proposal C for this lab. Encourage students to fill out an investigation proposal for each experiment they design.

- Allow the students to become familiar with the equipment as part of the tool talk before they begin to design their investigation. This gives students a chance to see what they can and cannot do with the equipment.

- When collecting data, we recommend assigning one student in each group to hold and support the tube while it is filled with water, one student to drop the samples into the open end of the tube, and one or two students to time how long it takes for the samples to settle on the bottom.

- Students often make mistakes when developing their conceptual models and/or initial arguments, but they should quickly realize these mistakes during the argumentation session. Be sure to allow students to revise their models and arguments at the end of the argumentation session. The explicit and reflective discussion will also give students an opportunity to reflect on and identify ways to improve how they develop and test models. This also offers an opportunity to discuss what scientists do when they realize a mistake is made.

- Students often make mistakes during the data collection stage, but they should quickly realize these mistakes during the argumentation session. It will only take them a short period of time to re-collect data, and they should be allowed to do so. During the explicit and reflective discussion, students will also have the opportunity to reflect on and identify ways to improve the way they design investigations (especially how they attempt to control variables as part of an experiment). This provides another opportunity to discuss what scientists do when they realize that a mistake is made during a study.

LAB 10

Connections to Standards

Table 10.2 highlights how the investigation can be used to address specific (a) performance expectations from the *NGSS* and (b) *Common Core State Standards* in English language arts (*CCSS ELA*).

TABLE 10.2 _____

Lab 10 alignment with standards

***NGSS* performance expectations**	History of Earth; Earth's systems • MS-ESS2-1: Develop a model to describe the cycling of Earth's materials and the flow of energy that drives this process. • MS-ESS2-2: Construct an explanation based on evidence for how geoscience processes have changed Earth's surface at varying time and spatial scales. • HS-ESS2-5: Plan and conduct an investigation of the properties of water and its effects on Earth materials and surface processes.
***CCSS ELA*—Reading in Science and Technical Subjects**	Key ideas and details • CCSS.ELA-LITERACY.RST.6-8.1: Cite specific textual evidence to support analysis of science and technical texts. • CCSS.ELA-LITERACY.RST.6-8.2: Determine the central ideas or conclusions of a text; provide an accurate summary of the text distinct from prior knowledge or opinions. • CCSS.ELA-LITERACY.RST.9-10.1: Cite specific textual evidence to support analysis of science and technical texts, attending to the precise details of explanations or descriptions. • CCSS.ELA-LITERACY.RST.9-10.2: Determine the central ideas or conclusions of a text; trace the text's explanation or depiction of a complex process, phenomenon, or concept; provide an accurate summary of the text. • CCSS.ELA-LITERACY.RST.9-10.3: Follow precisely a complex multistep procedure when carrying out experiments, taking measurements, or performing technical tasks, attending to special cases or exceptions defined in the text. Craft and structure • CCSS.ELA-LITERACY.RST.6-8.4: Determine the meaning of symbols, key terms, and other domain-specific words and phrases as they are used in a specific scientific or technical context relevant to *grade 6–8 texts and topics*. • CCSS.ELA-LITERACY.RST.6-8.5: Analyze the structure an author uses to organize a text, including how the major sections contribute to the whole and to an understanding of the topic.

Continued

TABLE 10.2 (*continued*)

***CCSS ELA*—Reading in Science and Technical Subjects** (*continued*)	Craft and structure (*continued*) • CCSS.ELA-LITERACY.RST.6-8.6: Analyze the author's purpose in providing an explanation, describing a procedure, or discussing an experiment in a text. • CCSS.ELA-LITERACY.RST.9-10.4: Determine the meaning of symbols, key terms, and other domain-specific words and phrases as they are used in a specific scientific or technical context relevant to *grade 9–10 texts and topics*. • CCSS.ELA-LITERACY.RST.9-10.5: Analyze the structure of the relationships among concepts in a text, including relationships among key terms (e.g., *force, friction, reaction force, energy*). • CCSS.ELA-LITERACY.RST.9-10.6: Analyze the author's purpose in providing an explanation, describing a procedure, or discussing an experiment in a text, defining the question the author seeks to address. Integration of knowledge and ideas • CCSS.ELA-LITERACY.RST.6-8.7: Integrate quantitative or technical information expressed in words in a text with a version of that information expressed visually (e.g., in a flowchart, diagram, model, graph, or table). • CCSS.ELA-LITERACY.RST.6-8.8: Distinguish among facts, reasoned judgment based on research findings, and speculation in a text. • CCSS.ELA-LITERACY.RST.6-8.9: Compare and contrast the information gained from experiments, simulations, video, or multimedia sources with that gained from reading a text on the same topic. • CCSS.ELA-LITERACY.RST.9-10.7: Translate quantitative or technical information expressed in words in a text into visual form (e.g., a table or chart) and translate information expressed visually or mathematically (e.g., in an equation) into words. • CCSS.ELA-LITERACY.RST.9-10.8: Assess the extent to which the reasoning and evidence in a text support the author's claim or a recommendation for solving a scientific or technical problem. • CCSS.ELA-LITERACY.RST.9-10.9: Compare and contrast findings presented in a text to those from other sources (including their own experiments), noting when the findings support or contradict previous explanations or accounts.
***CCSS ELA*—Writing in Science and Technical Subjects**	Text types and purposes • CCSS.ELA-LITERACY.WHST.6-10.1: Write arguments focused on *discipline-specific content*. • CCSS.ELA-LITERACY.WHST.6-10.2: Write informative or explanatory texts, including the narration of historical events, scientific procedures/experiments, or technical processes.

Continued

TABLE 10.2 (*continued*)

CCSS ELA—**Writing in Science and Technical Subjects** (*continued*)	Production and distribution of writing • CCSS.ELA-LITERACY.WHST.6-10.4: Produce clear and coherent writing in which the development, organization, and style are appropriate to task, purpose, and audience. • CCSS.ELA-LITERACY.WHST.6-8.5: With some guidance and support from peers and adults, develop and strengthen writing as needed by planning, revising, editing, rewriting, or trying a new approach, focusing on how well purpose and audience have been addressed. • CCSS.ELA-LITERACY.WHST.6-8.6: Use technology, including the Internet, to produce and publish writing and present the relationships between information and ideas clearly and efficiently. • CCSS.ELA-LITERACY.WHST.9-10.5: Develop and strengthen writing as needed by planning, revising, editing, rewriting, or trying a new approach, focusing on addressing what is most significant for a specific purpose and audience. • CCSS.ELA-LITERACY.WHST.9-10.6: Use technology, including the internet, to produce, publish, and update individual or shared writing products, taking advantage of technology's capacity to link to other information and to display information flexibly and dynamically. Range of writing • CCSS.ELA-LITERACY.WHST.6-10.10: Write routinely over extended time frames (time for reflection and revision) and shorter time frames (a single sitting or a day or two) for a range of discipline-specific tasks, purposes, and audiences.
CCSS ELA—**Speaking and Listening**	Comprehension and collaboration • CCSS.ELA-LITERACY.SL.6-8.1: Engage effectively in a range of collaborative discussions (one-on-one, in groups, and teacher-led) with diverse partners on grade 6-8 topics, texts, and issues, building on others' ideas and expressing their own clearly. • CCSS.ELA-LITERACY.SL.6-8.2:* Interpret information presented in diverse media and formats (e.g., visually, quantitatively, orally) and explain how it contributes to a topic, text, or issue under study. • CCSS.ELA-LITERACY.SL.6-8.3:* Delineate a speaker's argument and specific claims, distinguishing claims that are supported by reasons and evidence from claims that are not. • CCSS.ELA-LITERACY.SL.9-10.1: Initiate and participate effectively in a range of collaborative discussions (one-on-one, in groups, and teacher-led) with diverse partners on grade 9–10 topics, texts, and issues, building on others' ideas and expressing their own clearly and persuasively.

Continued

TABLE 10.2 (*continued*)

CCSS ELA—**Speaking and Listening** (*continued*)	Comprehension and collaboration (*continued*) • CCSS.ELA-LITERACY.SL.9-10.2: Integrate multiple sources of information presented in diverse media or formats (e.g., visually, quantitatively, orally) evaluating the credibility and accuracy of each source. • CCSS.ELA-LITERACY.SL.9-10.3: Evaluate a speaker's point of view, reasoning, and use of evidence and rhetoric, identifying any fallacious reasoning or exaggerated or distorted evidence. Presentation of knowledge and ideas • CCSS.ELA-LITERACY.SL.6-8.4:* Present claims and findings, sequencing ideas logically and using pertinent descriptions, facts, and details to accentuate main ideas or themes; use appropriate eye contact, adequate volume, and clear pronunciation. • CCSS.ELA-LITERACY.SL.6-8.5:* Include multimedia components (e.g., graphics, images, music, sound) and visual displays in presentations to clarify information. • CCSS.ELA-LITERACY.SL.6-8.6: Adapt speech to a variety of contexts and tasks, demonstrating command of formal English when indicated or appropriate. • CCSS.ELA-LITERACY.SL.9-10.4: Present information, findings, and supporting evidence clearly, concisely, and logically such that listeners can follow the line of reasoning and the organization, development, substance, and style are appropriate to purpose, audience, and task. • CCSS.ELA-LITERACY.SL.9-10.5: Make strategic use of digital media (e.g., textual, graphical, audio, visual, and interactive elements) in presentations to enhance understanding of findings, reasoning, and evidence and to add interest. • CCSS.ELA-LITERACY.SL.9-10.6: Adapt speech to a variety of contexts and tasks, demonstrating command of formal English when indicated or appropriate.

* Only the standard for grade 6 is provided because the standards for grades 7 and 8 are similar. Please see *www.corestandards.org/ELA-Literacy/SL* for the exact wording of the standards for grades 7 and 8.

LAB 10

Lab Handout

Lab 10. Deposition of Sediments: How Can We Explain the Deposition of Sediments in Water?

Introduction

People use sedimentary rocks such as siltstone, shale, and sandstone (see Figure L10.1) for many different purposes. Siltstone, for example, is used to build homes and walls or for decorations. Shale is used to make cement, terra-cotta pots, bricks, and roof tiles. Sandstone is another sedimentary rock that is used as a building material. It is used to create floor tiles and decorative walls in homes or businesses and to create monuments and roads. Sandstone is also used as a sharpening stone for knives.

FIGURE L10.1 _____

Some examples of sedimentary rock

Siltstone

Shale

Sandstone

The material that makes up siltstone, shale, and sandstone comes from other rocks that have weathered over time. These rocks are therefore called *clastic sedimentary rocks*. As a rock weathers, it is broken into smaller pieces or sediments. These sediments are then carried to other places by wind, liquid water, or glacial ice. The sediments eventually settle out of the air or water and accumulate at a specific location. This process is called *deposition,* and it results in layers of different types of sediment. These layers of sediment then turn into a rock through a process called *lithification*.

Sediments go through compaction and cementation during lithification. *Compaction* happens when the individual pieces of sediment in a layer are forced together because of the combined weight of all the sediment in layers above them. *Cementation* happens when the dissolved minerals between the pieces of sediment dry. These minerals then bind the other pieces of sediment together and harden—much like cement mix does after water is added to it. The sediments that are cemented together to create a sedimentary rock often come from all different kinds of rocks, and therefore have different physical properties.

Geologists, as a result, classify clastic sedimentary rocks based on the types of sediment found within them.

Sedimentary rocks, such as those shown in Figure L10.1, consist of layers of different types of sediments because different types of sediments fall through a fluid, such as water, at different rates. The rate at which a sediment falls through a fluid is called its *settling velocity.* The settling velocity of a sediment, like its texture, density, or color, is a unique physical prop-erty of that sediment. When a sediment has a settling velocity that is lower than the stream flow velocity of a river, that sediment will be carried downstream. When a sediment has a settling velocity that is higher than the stream flow velocity of a river, that sediment will sink to the bottom of the river and not move farther downstream. The stream velocity of a river at different locations, as a result, will determine where different types of sediments will accumulate and what types of sedimentary rocks will form at different locations.

It is important for geologists to understand how the different characteristics of a sediment affects its settling velocity because this physical property helps them explain how sediments move from one location to another and allows them to predict where differ-ent types of sediments will accumulate over time. Geologists can also learn more about environmental conditions of the past if they understand the factors that affect the deposition of sediments when they examine the nature and location of different types of sedimentary rock.

A sediment has many different physical properties that may or may not affect its settling velocity; these properties include particle size, shape, and density. Sediment particles can range in size from clay that is less than to 0.002 mm in diameter to large pebbles that can be well over 4 mm in diameter. Geologists often use a specific scale, such as the Wentworth scale (see Table L10.1), to classify or describe the particle size of a sediment. Shape is another physical property of a sediment, and geologists often classify or describe the shape of a sediment particle using the terms shown in Figure L10.2. Finally, the density of a sediment is defined as its mass per unit volume.

In this investigation, you will have an opportunity to figure out how these three physical properties affect the settling velocity of a sediment. You will then use this information to develop a conceptual model that explains how sediments will

TABLE L10.1

Modified Wentworth scale for classifying particles by size

Name	Particle size (mm)
Pebble	> 4
Granule	3.9–2.0
Very coarse sand	1.9–1.0
Coarse sand	0.9–0.5
Medium sand	0.49–0.25
Fine sand	0.24–0.125
Very fine sand	0.124–0.0625
Silt	0.0624–0.002
Clay	< 0.002

FIGURE L10.2

The various shapes of a sediment particle

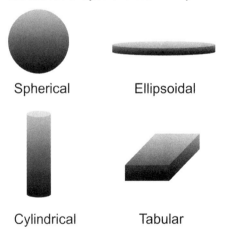

Spherical   Ellipsoidal

Cylindrical   Tabular

settle in water. A conceptual model like this is useful because it allows people to not only understand why sedimentary rocks consist of layers of different materials but also predict the order of the layers that will be found in different kinds of sedimentary rocks.

Your Task

You will be given several different sediments. You will then explore how each type of sediment settles in a column of water over time. Your goal is to use what you know about cause-and-effect relationships and structure and function to design and carry out an investigation that will enable you to develop a conceptual model that explains how particle size, density, and shape affects the settling rate of a sediment in water. Once you have created your conceptual model, you will return these sediments to your teacher. He or she will then give you one or two sediment mixtures. You will use the sediment mixtures to test, and if needed, revise your model. Your model, if valid or acceptable, should allow you to predict how the different types of sediments found in each mixture will accumulate at the bottom of a column of water over time.

The guiding question of this investigation is, *How can we explain the deposition of sediments in water?*

Materials

You may use any of the following materials during your investigation:

Sediments of different sizes
- Gravel
- Coarse sand
- Medium sand
- Fine sand

Sediments of different shapes
- 4 Pieces of modeling clay (each 2 g)

Sediments of different densities
- Glass beads (4 mm)
- Plastic beads (4 mm)
- Ball bearings (4 mm)

Sediment mixtures for testing the model
- Mixture A
- Mixture B

Consumable
- Water

Equipment
- Safety glasses or goggles (required)
- Chemical-resistant apron (required)
- Gloves (required)
- Clear plastic tube
- Rubber stopper
- Beaker (500 ml)
- Beaker (50 ml)
- Funnel
- Duct tape
- Meterstick
- Bucket
- Electronic or triple beam balance
- Stopwatches

Safety Precautions

Follow all normal lab safety rules. In addition, take the following safety precautions:

- Wear sanitized indirectly vented chemical-splash goggles and chemical-resistant aprons and gloves throughout the entire investigation (which includes setup and cleanup).

- Handle all glassware with care.

- Immediately wipe up any spilled water and pick up any spilled beads, ball bearings, or other materials that can cause a slip, trip, or fall hazard.

- Wash hands with soap and water when done collecting the data and after completing the lab.

Investigation Proposal Required?   ☐ Yes   ☐ No

Getting Started

You will need to design and carry out at least three different experiments to determine how the structure of a sediment affects the rate at which it will fall through a column of water. These experiments are necessary because you will need to answer three specific questions before you can develop an answer to the guiding question for this lab:

- How does particle size affect the time it takes a sediment to fall through a column of water?

- How does particle density affect the time it takes a sediment to fall through a column of water?

- How does particle shape affect the time it takes a sediment to fall through a column of water?

You can create a water column, such as the one shown in Figure L10.3, using a large plastic tube and a rubber stopper. Before you create your water column, it will be important for you to determine what type of data you need to collect, how you will collect the data, and how you will analyze the data for each experiment, because each experiment is slightly different.

To determine *what type of data you need to collect*, think about the following questions:

- What conditions need to be satisfied to establish a cause-and-effect relationship?

- How will you determine when a particular sediment type has settled?

- How will you determine how long it takes for a sediment to fall through a column of water?

- What information will you need to calculate the density of a particle?

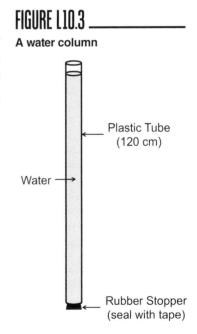

FIGURE L10.3
A water column

Plastic Tube (120 cm)

Water →

Rubber Stopper (seal with tape)

- When will you need to make these measurements or observations?

To determine *how you will collect the data,* think about the following questions:

- What will serve as your independent variable and dependent variable for each experiment?
- How will you vary the independent variable during each experiment?
- What will you do to hold the other variables constant during each experiment?
- What types of comparisons will you need to make?
- What measurement scale or scales should you use to collect data?
- What will you do to reduce error in your measurements?
- How will you keep track of and organize the data you collect?

To determine *how you will analyze the data,* think about the following questions:

- What calculations will you need to make?
- How could you use mathematics to describe a relationship between variables?
- How could you use mathematics to document a difference between groups or conditions?
- What types of patterns might you look for as you analyze your data?
- What type of table or graph could you create to help make sense of your data?

Once you have finished collecting data, your group can develop a conceptual model that explains the deposition of sediments in water. For your conceptual model to be complete, it must be able to explain how the structure of different sediments relates to how different types of sediments will move through water. It must also include information about the stream flow velocity of a body of water such as a river or lake. The stream flow velocity is how fast the water is moving in a specific direction. The stream flow velocity of water in a water column is zero, but in a river it can reach velocities of 25 km/h. Finally, and perhaps most important, you should be able to use your model to predict the order in which different types of sediments will settle at the bottom of a column of water. This type of conceptual model is useful because it enables people to understand where and when different types of sediments will accumulate over time.

The last step in your investigation will be to generate the evidence that you need to convince others that the conceptual model you developed based on your experiments is valid or acceptable. To accomplish this goal, you can use your model to predict how the different sediments in a mixture will settle in a column of water after a set amount of time. If you are able to use your conceptual model to make accurate predictions about how the different sediments in a mixture will move through the water relative to each other, then you should be able to convince others that your model is valid or acceptable.

Connections to the Nature of Scientific Knowledge and Scientific Inquiry

As you work through your investigation, be sure to think about

- the use of models as as tools for reasoning about natural phenomena, and
- the nature and role of experiments in science.

Initial Argument

Once your group has finished collecting and analyzing your data, your group will need to develop an initial argument. Your initial argument needs to include a claim, evidence to support your claim, and a justification of the evidence. The *claim* is your group's answer to the guiding question. The *evidence* is an analysis and interpretation of your data. Finally, the *justification* of the evidence is why your group thinks the evidence matters. The justification of the evidence is important because scientists can use different kinds of evidence to support their claims. Your group will create your initial argument on a whiteboard. Your whiteboard should include all the information shown in Figure L10.4.

FIGURE L10.4 _____

Argument presentation on a whiteboard

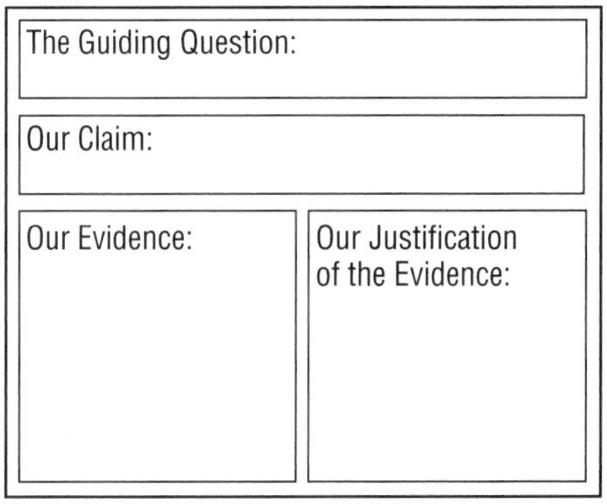

The Guiding Question:	
Our Claim:	
Our Evidence:	Our Justification of the Evidence:

Argumentation Session

The argumentation session allows all of the groups to share their arguments. One or two members of each group will stay at the lab station to share that group's argument, while the other members of the group go to the other lab stations to listen to and critique the other arguments. This is similar to what scientists do when they propose, support, evaluate, and refine new ideas during a poster session at a conference. If you are presenting your group's argument, your goal is to share your ideas and answer questions. You should also keep a record of the critiques and suggestions made by your classmates so you can use this feedback to make your initial argument stronger. You can keep track of specific critiques and suggestions for improvement that your classmates mention in the space below.

Critiques of our initial argument and suggestions for improvement:

If you are critiquing your classmates' arguments, your goal is to look for mistakes in their arguments and offer suggestions for improvement so these mistakes can be fixed. You should look for ways to make your initial argument stronger by looking for things that the other groups did well. You can keep track of interesting ideas that you see and hear during the argumentation in the space below. You can also use this space to keep track of any questions that you will need to discuss with your team.

Interesting ideas from other groups or questions to take back to my group:

Once the argumentation session is complete, you will have a chance to meet with your group and revise your initial argument. Your group might need to gather more data or design a way to test one or more alternative claims as part of this process. Remember, your goal at this stage of the investigation is to develop the best argument possible.

Report

Once you have completed your research, you will need to prepare an *investigation report* that consists of three sections. Each section should provide an answer for the following questions:

1. What question were you trying to answer and why?

2. What did you do to answer your question and why?

3. What is your argument?

Your report should answer these questions in two pages or less. You should write your report using a word processing application (such as Word, Pages, or Google Docs), if possible, to make it easier for you to edit and revise it later. You should embed any diagrams, figures, or tables into the document. Be sure to write in a persuasive style; you are trying to convince others that your claim is acceptable or valid.

Checkout Questions

Lab 10. Deposition of Sediments: How Can We Explain the Deposition of Sediments in Water?

1. Use numbers to rank the following sediments from the greatest to smallest likely settling velocity. Assume that the sediments all have the same density. If you think any two sediments will have the same settling velocity, give them the same number.

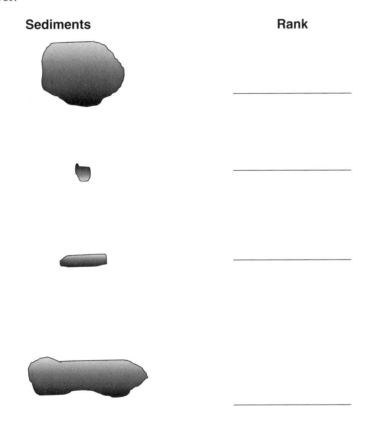

Sediments	Rank
	_____
	_____
	_____
	_____

Explain your answer. Why do you think the order that you chose is correct?

2. Jalen adds a sample of soil to a jar. The soil is made up of fine sand, coarse sand, gravel, and clay. He then fills the remaining space in the jar with water, leaving a little room for air at the top. After putting on the lid, he shakes the jar until all the soil particles are mixed with the water. He leaves the jar on a table overnight. The next morning, he sees four layers of sediment in the jar. Which picture shows how you think the different types of soil particles will settle in the jar: A, B, or C?

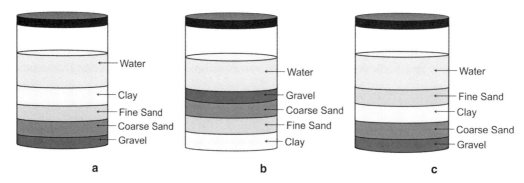

Explain your answer. Why do you think the order that you chose is correct?

3. Scientists often develop models to help them understand complex phenomena.

   a. I agree with this statement.

   b. I disagree with this statement.

   Explain your answer, using an example from your investigation about the deposition of sediments.

4. Scientists use experiments to prove ideas right or wrong.

    a. I agree with this statement.

    a. I disagree with this statement.

    Explain your answer, using an example from your investigation about the deposition of sediments.

5. Scientists are often interested in identifying cause-and-effect relationships. Explain what a cause-and-effect relationship is and why these relationships are important, using an example from your investigation about the deposition of sediments.

6. In nature, the way something is structured often determines its function or places limits on what it can or cannot do. Explain why it is important to keep in mind the relationship between structure and function when attempting to collect or analyze data, using an example from your investigation about the deposition of sediments.

LAB 11

Teacher Notes

Lab 11. Soil Texture and Soil Water Permeability: How Does Soil Texture Affect Soil Water Permeability?

Purpose

The purpose of this lab is to *introduce* students to the disciplinary core ideas (DCIs) of (a) Earth Materials and Systems and (b) The Roles of Water in Earth's Surface Processes by having them determine how soil texture affects soil water permeability. In addition, students have an opportunity to learn about the crosscutting concepts (CCs) of (a) Scale, Proportion, and Quantity; and (b) Energy and Matter: Flows, Cycles, and Conservation. During the explicit and reflective discussion, students will also learn about (a) the difference between observations and inferences in science and (b) how the culture of science, societal needs, and current events influence the work of scientists.

Important Earth and Space Science Content

Soil water permeability is the rate at which water can flow through the pores of a soil. The porosity of a soil is related to the texture of that soil. Scientists define soils by their textures, which are determined by the proportions of sand, silt, and clay particles present in that soil.

FIGURE 11.1

Soil particle types and sizes

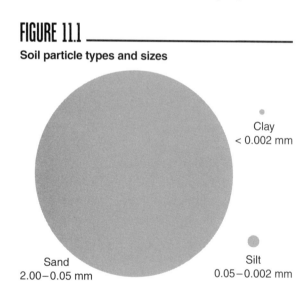

Sand
2.00–0.05 mm

Silt
0.05–0.002 mm

Clay
< 0.002 mm

Sand particles are the largest soil particles, ranging in size from 0.05 mm to 2.0 mm in diameter. Silt particles are the next largest soil particles, ranging in size from 0.002 mm to 0.05 mm in diameter. Clay particles, which are less than 0.002 mm in diameter, are the smallest soil particles. Figure 11.1 shows the particle types and their relative sizes.

Earth scientists use a soil texture triangle, such as the one shown in Figure 11.2, to classify soils based on the proportions of each particle type in the soil sample. *Soil texture* describes the relative proportion of sand, silt, and clay particles by weight in a soil sample. Earth scientists tend to use 12 different categories to classify the texture of a soil. For example, a soil that is composed of 30% clay, 10% silt, and 60% sand is classified as a *sandy clay loam,* whereas a soil that consists of 20% clay, 40% silt, and 40% sand is classified as a *loam.*

The rate of soil water permeability will be greater (more water flowing per unit of time) for soils comprised of larger particles. This is because the pores, or spaces between particles, are larger for larger particles, allowing more water to flow through the larger space. Conversely, soil comprised primarily of smaller particles, such as clay, will have smaller pores, which allows less water to flow through the pores, decreasing the soil water permeability of the soil.

Timeline

The instructional time needed to complete this lab investigation is 220–280 minutes. Appendix 3 (p. 573) provides options for implementing this lab investigation over several class periods. Option A (280 minutes) should be used if students are unfamiliar with scientific writing, because this option provides extra instructional time for scaffolding the writing process. You can scaffold the writing process by modeling, providing examples, and providing hints as students write each section of the report. Option B (220 minutes) should be used if students are familiar with scientific writing and have developed the skills needed to write an investigation report on their own. In option B, students complete stage 6 (writing the investigation report) and stage 8 (revising the investigation report) as homework.

FIGURE 11.2

Soil types by composition

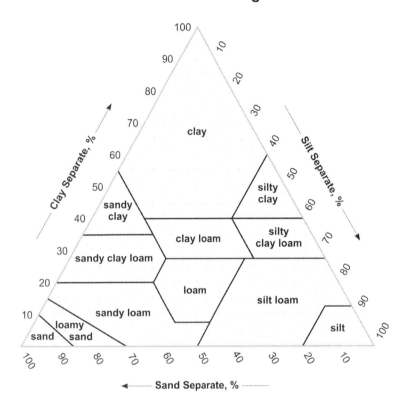

Materials and Preparation

The materials needed to implement this investigation are listed in Table 11.1 (p. 264). The consumables and equipment can be purchased from a science supply company such as Carolina, Flinn Scientific, or Ward's Science. The soil samples can also be collected from your local area. To minimize risk of exposure to pesticides or herbicides, we suggest using commercial sterilized potting soil, sand, and other samples instead of samples that you or your students collect. Be sure to pick soils that represent a range of textures. At a minimum, you will need a soil that is composed of mostly large particles (such as sand), a soil that is composed of mostly small particles (such as clay), and two samples that are composed of a mixture of large and small particles. You can also combine soils such as sand and clay to create a soil with different textures (such as sandy clay or a sandy clay loam).

Soil water permeability can be measured using the equipment shown in Figure 11.3.

TABLE 11.1

Materials list for Lab 11

Item	Quantity
Consumables	
Soil sample A	500 ml per group
Soil sample B	500 ml per group
Soil sample C	500 ml per group
Soil sample D	500 ml per group
Cone coffee filters	12 per group
Paper towels	As needed
Equipment and other materials	
Safety glasses or goggles	1 per student
Chemical-resistant apron	1 per student
Gloves	1 pair per student
Measuring spoon, tablespoon or 15 ml	1 per group
Graduated cylinder, 100 ml	1 per group
Graduated cylinder, 50 ml	1 per group
Funnels	1–4 per group
Beakers, 250 ml	1–4 per group
Support stand	1–4 per group
Ring clamp	1–4 per group
Stopwatch	1–4 per group
Wax marking pencil (or masking tape and marker)	1 per group
Electronic or triple beam balance	1 per group
Flowchart for soil texture by feel	1–2 per group
Investigation Proposal C (optional)	1 per group
Whiteboard, 2' × 3'*	1 per group
Lab Handout	1 per student
Peer-review guide and teacher scoring rubric	1 per student
Checkout Questions	1 per student

* As an alternative, students can use computer and presentation software such as Microsoft PowerPoint or Apple Keynote to create their arguments.

Be sure to use a set routine for distributing and collecting the materials during the lab investigation. One option is to set up the materials for each group at each group's lab station before class begins. This option works well when there is a dedicated section of the classroom for lab work and the materials are large and difficult to move (such as a stream table). A second option is to have all the materials on a table or cart at a central location. You can then assign a member of each group to be the "materials manager." This individual is responsible for collecting all the materials his or her group needs from the table or cart during class and for returning all the materials at the end of the class. This option works well when the materials are small and easy to move (such as stopwatches, metersticks, or thermometers). It also makes it easy to inventory the materials at the end of the class before students leave for the day.

FIGURE 11.3

Equipment needed to measure the rate at which water flows through a sample of soil

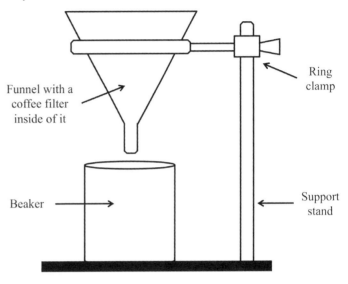

Funnel with a coffee filter inside of it

Ring clamp

Beaker

Support stand

Safety Precautions and Laboratory Waste Disposal

Remind students to follow all normal lab safety rules. In addition, tell the students to take the following safety precautions:

- Wear sanitized indirectly vented chemical-splash goggles and chemical-resistant, nonlatex aprons and gloves throughout the entire investigation (which includes setup and cleanup).
- Handle all glassware with care.
- Report and clean up any spills immediately, and avoid walking in areas where water has been spilled.
- Wash hands with soap and water when done collecting the data and after completing the lab.

Waste for this lab will be minimal. You may choose to dry the soil samples after use for later reuse, or the samples can also be disposed of directly into the trash.

LAB 11

Topics for the Explicit and Reflective Discussion
Reflecting on the Use of Core Ideas and Crosscutting Concepts During the Investigation

Teachers should begin the explicit and reflective discussion by asking students to discuss what they know about the DCIs they used during the investigation. The following are some important concepts related to the DCIs of (a) Earth Materials and Systems and (b) The Roles of Water in Earth's Surface Processes that students need to determine how soil texture affects soil water permeability:

- Soils are comprised of different proportions of sand, silt, and clay particles.
- Sand is the largest particle, and clay is the smallest.
- Soil type is determined by texture, which is based on the proportion of sand, silt, and clay particles in the soil.
- The larger the soil particles, the larger the spaces, or pores, are between particles.
- Matter cannot be created or destroyed, so it is possible to track the loss of soil by measuring a change in the mass of a sample.

To help students reflect on what they know about these concepts, we recommend showing them two or three images using presentation software that help illustrate these important ideas. You can then ask the students the following questions to encourage students to share how they are thinking about these important concepts:

1. What do we see going on in this image?
2. Does anyone have anything else to add?
3. What might be going on that we can't see?
4. What are some things that we are not sure about here?

You can then encourage students to think about how CCs played a role in their investigation. There are at least two CCs that students need to determine how soil texture affects soil water permeability: (a) Scale, Proportion, and Quantity; and (b) Energy and Matter: Flows, Cycles, and Conservation (see Appendix 2 [p. 569] for a brief description of these two CCs). To help students reflect on what they know about these CCs, we recommend asking them the following questions:

1. Why is it important to consider what measurement scale or scales to use during an investigation? Why is useful to look for proportional relationships when analyzing data?
2. What measurement scale or scales did you use during your investigation? What did that allow you to do? Did you attempt to look for proportional relationships when you were analyzing your data? Why or why not?

3. Why is it important to track how matter flows into, out of, or within a system during an investigation?

4. How did you track the flow of matter in the system you were studying? What did tracking the flow of matter allow you to do during your investigation?

You can then encourage the students to think about how they used all these different concepts to help answer the guiding question and why it is important to use these ideas to help justify their evidence for their final arguments. Be sure to remind your students to explain why they included the evidence in their arguments and make the assumptions underlying their analysis and interpretation of the data explicit in order to provide an adequate justification of their evidence.

Reflecting on Ways to Design Better Investigations

It is important for students to reflect on the strengths and weaknesses of the investigation they designed during the explicit and reflective discussion. Students should therefore be encouraged to discuss ways to eliminate potential flaws, measurement errors, or sources of uncertainty in their investigations. To help students be more reflective about the design of their investigation and what they can do to make their investigations more rigorous in the future, you can ask the following questions:

1. What were some of the strengths of the way you planned and carried out your investigation? In other words, what made it scientific?

2. What were some of the weaknesses of the way you planned and carried out your investigation? In other words, what made it less scientific?

3. What rules can we make, as a class, to ensure that our next investigation is more scientific?

Reflecting on the Nature of Scientific Knowledge and Scientific Inquiry

This investigation can be used to illustrate two important concepts related to the nature of scientific knowledge and the nature of scientific inquiry: (a) the difference between observations and inferences in science and (b) how the culture of science, societal needs, and current events influence the work of scientists (see Appendix 2 [p. 569] for a brief description of these two concepts). Be sure to review these concepts during and at the end of the explicit and reflective discussion. To help students think about these concepts in relation to what they did during the lab, you can ask the following questions:

1. You made observations and inferences during your investigation. Can you give me some examples of these observations and inferences?

2. Can you work with your group to come up with a rule that you can use to tell the difference between an observation and inference? Be ready to share in a few minutes.

3. People view some types of research as being more important than other types of research because of cultural values and current events. Can you come up with some examples of how cultural values and current events have influenced the work of scientists?

4. Scientists share a set of values, norms, and commitments that shape what counts as knowing, how to represent or communicate information, and how to interact with other scientists. Can you work with your group to come up with a rule that you can use to decide if something is science or not science? Be ready to share in a few minutes.

You can also use presentation software or other techniques to encourage your students to think about these concepts. You can show examples of information from the investigation that are either observations or inferences and ask students to classify each example and explain their thinking. You can also show images of research projects that were influenced by cultural values and current events and ask students to think about what was going on in society when the research was conducted and why that research was viewed as being important for the greater good.

Remind your students that, to be proficient in science, it is important that they understand what counts as scientific knowledge and how that knowledge develops over time.

Hints for Implementing the Lab

- Allow the students to become familiar with the equipment setup shown in Figure 11.3 as part of the tool talk before they begin to design their investigation. This gives students a chance to see what they can and cannot do with the equipment.

- Allowing students to design their own procedures for collecting data gives students an opportunity to try, to fail, and to learn from their mistakes. However, you can scaffold students as they develop their procedure by having them fill out an investigation proposal. These proposals provide a way for you to offer students hints and suggestions without telling them how to do it. You can also check the proposals quickly during a class period. We recommend using Investigation Proposal C for this lab.

- Investigation Proposal C works best for the experiment that students will conduct during this investigation because they can describe a different relationship between the independent variable and dependent variable for each hypothesis. For example, increasing the percentage of sand in a soil might (a) increase soil water permeability (hypothesis 1), (b) decrease soil water permeability (hypothesis 2), or (c) have no effect on soil water permeability (hypothesis 3).

- Encourage students to use the flowchart for soil texture by feel to determine the compositions of the soils you provide for them. They can then use a soil texture triangle to determine the percentage of sand, silt, and clay in the sample.

- Don't worry about removing sticks or other items from the soil. Students will deal with these inconsistencies in different ways, or may ignore them entirely, which will create valuable variety in student results.

- Provide students with as many ring stands as you have available, up to four per group. The process of filtering water through the soil samples is time-consuming, and some students may recognize that running multiple tests at once may increase their efficiency. Ensure that the funnel provided to students fits into the ring stand without falling through, as shown in Figure 11.3.

- Students often make mistakes during the data collection stage, but they should quickly realize these mistakes during the argumentation session. It will only take them a short period of time to re-collect data, and they should be allowed to do so. During the explicit and reflective discussion, students will also have the opportunity to reflect on and identify ways to improve the way they design investigations (especially how they attempt to control variables as part of an experiment). This also offers an opportunity to discuss what scientists do when they realize that a mistake is made during a study.

Connections to Standards

Table 11.2 highlights how the investigation can be used to address specific (a) performance expectations from the *NGSS* and (b) *Common Core State Standards* in English language arts (*CCSS ELA*).

TABLE 11.2

Lab 11 alignment with standards

NGSS performance expectations	Earth's systems • MS-ESS2-4: Develop a model to describe the cycling of water through Earth's systems driven by energy from the Sun and the force of gravity. • HS-ESS2-5: Plan and conduct an investigation of the properties of water and its effects on Earth materials and surface processes.

Continued

TABLE 11.2 (*continued*)

CCSS ELA—Reading in Science and Technical Subjects	Key ideas and details • CCSS.ELA-LITERACY.RST.6-8.1: Cite specific textual evidence to support analysis of science and technical texts. • CCSS.ELA-LITERACY.RST.6-8.2: Determine the central ideas or conclusions of a text; provide an accurate summary of the text distinct from prior knowledge or opinions. • CCSS.ELA-LITERACY.RST.9-10.1: Cite specific textual evidence to support analysis of science and technical texts, attending to the precise details of explanations or descriptions. • CCSS.ELA-LITERACY.RST.9-10.2: Determine the central ideas or conclusions of a text; trace the text's explanation or depiction of a complex process, phenomenon, or concept; provide an accurate summary of the text. • CCSS.ELA-LITERACY.RST.9-10.3: Follow precisely a complex multistep procedure when carrying out experiments, taking measurements, or performing technical tasks, attending to special cases or exceptions defined in the text. Craft and structure • CCSS.ELA-LITERACY.RST.6-8.4: Determine the meaning of symbols, key terms, and other domain-specific words and phrases as they are used in a specific scientific or technical context relevant to *grade 6–8 texts and topics*. • CCSS.ELA-LITERACY.RST.6-8.5: Analyze the structure an author uses to organize a text, including how the major sections contribute to the whole and to an understanding of the topic. • CCSS.ELA-LITERACY.RST.6-8.6: Analyze the author's purpose in providing an explanation, describing a procedure, or discussing an experiment in a text. • CCSS.ELA-LITERACY.RST.9-10.4: Determine the meaning of symbols, key terms, and other domain-specific words and phrases as they are used in a specific scientific or technical context relevant to *grade 9–10 texts and topics*. • CCSS.ELA-LITERACY.RST.9-10.5: Analyze the structure of the relationships among concepts in a text, including relationships among key terms (e.g., *force, friction, reaction force, energy*). • CCSS.ELA-LITERACY.RST.9-10.6: Analyze the author's purpose in providing an explanation, describing a procedure, or discussing an experiment in a text, defining the question the author seeks to address. Integration of knowledge and ideas • CCSS.ELA-LITERACY.RST.6-8.7: Integrate quantitative or technical information expressed in words in a text with a version of that information expressed visually (e.g., in a flowchart, diagram, model, graph, or table).

Continued

TABLE 11.2 (*continued*)

CCSS ELA—**Reading in Science and Technical Subjects** (*continued*)	Integration of knowledge and ideas (*continued*) • CCSS.ELA-LITERACY.RST.6-8.8: Distinguish among facts, reasoned judgment based on research findings, and speculation in a text. • CCSS.ELA-LITERACY.RST.6-8.9: Compare and contrast the information gained from experiments, simulations, video, or multimedia sources with that gained from reading a text on the same topic. • CCSS.ELA-LITERACY.RST.9-10.7: Translate quantitative or technical information expressed in words in a text into visual form (e.g., a table or chart) and translate information expressed visually or mathematically (e.g., in an equation) into words. • CCSS.ELA-LITERACY.RST.9-10.8: Assess the extent to which the reasoning and evidence in a text support the author's claim or a recommendation for solving a scientific or technical problem. • CCSS.ELA-LITERACY.RST.9-10.9: Compare and contrast findings presented in a text to those from other sources (including their own experiments), noting when the findings support or contradict previous explanations or accounts.
CCSS ELA—**Writing in Science and Technical Subjects**	Text types and purposes • CCSS.ELA-LITERACY.WHST.6-10.1: Write arguments focused on *discipline-specific content.* • CCSS.ELA-LITERACY.WHST.6-10.2: Write informative or explanatory texts, including the narration of historical events, scientific procedures/experiments, or technical processes. Production and distribution of writing • CCSS.ELA-LITERACY.WHST.6-8.4: Produce clear and coherent writing in which the development, organization, and style are appropriate to task, purpose, and audience. • CCSS.ELA-LITERACY.WHST.6-8.5: With some guidance and support from peers and adults, develop and strengthen writing as needed by planning, revising, editing, rewriting, or trying a new approach, focusing on how well purpose and audience have been addressed. • CCSS.ELA-LITERACY.WHST.6-10.6: Use technology, including the internet, to produce and publish writing and present the relationships between information and ideas clearly and efficiently. • CCSS.ELA-LITERACY.WHST.9-10.5: Develop and strengthen writing as needed by planning, revising, editing, rewriting, or trying a new approach, focusing on addressing what is most significant for a specific purpose and audience.

Continued

TABLE 11.2 (*continued*)

CCSS ELA—**Writing in Science and Technical Subjects** (*continued*)	Production and distribution of writing (*continued*) • CCSS.ELA-LITERACY.WHST.9-10.6: Use technology, including the internet, to produce, publish, and update individual or shared writing products, taking advantage of technology's capacity to link to other information and to display information flexibly and dynamically. Range of writing • CCSS.ELA-LITERACY.WHST.6-10.10: Write routinely over extended time frames (time for reflection and revision) and shorter time frames (a single sitting or a day or two) for a range of discipline-specific tasks, purposes, and audiences.
CCSS ELA—**Speaking and Listening**	Comprehension and collaboration • CCSS.ELA-LITERACY.SL.6-8.1: Engage effectively in a range of collaborative discussions (one-on-one, in groups, and teacher-led) with diverse partners on grade 6–8 topics, texts, and issues, building on others' ideas and expressing their own clearly. • CCSS.ELA-LITERACY.SL.6-8.2:* Interpret information presented in diverse media and formats (e.g., visually, quantitatively, orally) and explain how it contributes to a topic, text, or issue under study. • CCSS.ELA-LITERACY.SL.6-8.3:* Delineate a speaker's argument and specific claims, distinguishing claims that are supported by reasons and evidence from claims that are not. • CCSS.ELA-LITERACY.SL.9-10.1: Initiate and participate effectively in a range of collaborative discussions (one-on-one, in groups, and teacher-led) with diverse partners on grade 9–10 topics, texts, and issues, building on others' ideas and expressing their own clearly and persuasively. • CCSS.ELA-LITERACY.SL.9-10.2: Integrate multiple sources of information presented in diverse media or formats (e.g., visually, quantitatively, orally) evaluating the credibility and accuracy of each source. • CCSS.ELA-LITERACY.SL.9-10.3: Evaluate a speaker's point of view, reasoning, and use of evidence and rhetoric, identifying any fallacious reasoning or exaggerated or distorted evidence. Presentation of knowledge and ideas • CCSS.ELA-LITERACY.SL.6-8.4:* Present claims and findings, sequencing ideas logically and using pertinent descriptions, facts, and details to accentuate main ideas or themes; use appropriate eye contact, adequate volume, and clear pronunciation.

Continued

TABLE 11.2 (*continued*)

***CCSS ELA*—Speaking and Listening** (*continued*)	Presentation of knowledge and ideas (*continued*) • CCSS.ELA-LITERACY.SL.6-8.5:* Include multimedia components (e.g., graphics, images, music, sound) and visual displays in presentations to clarify information. • CCSS.ELA-LITERACY.SL.6-8.6: Adapt speech to a variety of contexts and tasks, demonstrating command of formal English when indicated or appropriate. • CCSS.ELA-LITERACY.SL.9-10.4: Present information, findings, and supporting evidence clearly, concisely, and logically such that listeners can follow the line of reasoning and the organization, development, substance, and style are appropriate to purpose, audience, and task. • CCSS.ELA-LITERACY.SL.9-10.5: Make strategic use of digital media (e.g., textual, graphical, audio, visual, and interactive elements) in presentations to enhance understanding of findings, reasoning, and evidence and to add interest. • CCSS.ELA-LITERACY.SL.9-10.6: Adapt speech to a variety of contexts and tasks, demonstrating command of formal English when indicated or appropriate.

* Only the standard for grade 6 is provided because the standards for grades 7 and 8 are similar. Please see *www.corestandards.org/ELA-Literacy/SL* for the exact wording of the standards for grades 7 and 8.

LAB 11

Lab 11. Soil Texture and Soil Water Permeability: How Does Soil Texture Affect Soil Water Permeability?

Soil particle types and sizes

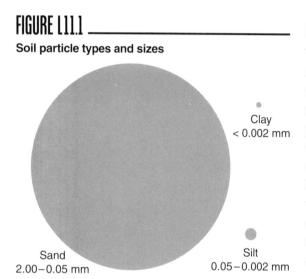

Sand
2.00–0.05 mm

Clay
< 0.002 mm

Silt
0.05–0.002 mm

FIGURE L11.2

Soil types by composition

Introduction

When it rains, water lands on the ground and flows into streams, rivers, gutters, and sewers. It can also soak into the ground. The ground is made up rock and soil. Soil is composed of many small particles. In between these particles are pores. Water can flow into and through these pores. *Soil water permeability* is defined as the rate at which water can flow through the pores of a soil. Rainwater tends to seep into soils that have high soil water permeability and tends to pool on top of or flow over soils with low soil water permeability. There are several different physical properties of soil that can affect the rate at which water flows through it.

One of the physical properties of soil that affects soil water permeability is *soil texture.* Soil is composed of small particles that vary in size, shape, and chemical composition. There are three main types of soil particles (see Figure L11.1): (a) sand particles are between 0.05 mm and 2.0 mm in diameter, (b). silt particles are between 0.002 mm and 0.05 mm in diameter, and (c). clay particles are smaller than 0.002 mm. *Soil texture* describes the relative proportion of sand, silt, and clay particles by weight in soil sample. Earth scientists use a soil texture triangle, such as the one shown in Figure L11.2, to classify soils according to 12 textural categories, based on the proportions of each particle type in the soil sample. For example, a soil that is composed of 30% clay, 10% silt, and 60% sand is classified as a *sandy clay loam;* whereas a soil that consists of 20% clay, 40% silt and 40% sand *is classified as a* loam.

Geotechnical engineers are often concerned about the water permeability of soil because they are responsible for building structures on or in the ground, and the rate at which water flows through soil can have an adverse effect on these structures. Geotechnical engineers must therefore understand how the physical properties of soil affect the rate at which water flows through it. In this investigation, you will have an opportunity to examine how soil texture, which is an important physical property of soil, affects soil water permeability.

Your Task

Use what you know about soil composition; scale, proportion, and quantity; and the importance of tracking how matter moves into, through, and out of a system to plan and carry out an investigation to determine the relationship between soil texture and soil water permeability.

The guiding question of this investigation is, *How does soil texture affect soil water permeability?*

Materials

You may use any of the following materials during your investigation:

Consumables
- Soil sample A
- Soil sample B
- Soil sample C
- Soil sample D
- 12 Coffee filters
- Paper towels

Equipment
- Safety glasses or goggles (required)
- Chemical-resistant apron (required)
- Gloves (required)
- Measuring spoon (15 ml)
- 1 Graduated cylinder (100 ml)
- 1 Graduated cylinder (50 ml)
- 1–4 Funnels
- 1–4 Beakers (each 250 ml)
- 1–4 Support stands
- 1–4 Ring clamps

- 1 Stopwatch
- 1 Wax marking pencil
- Electronic or triple beam balance
- Flowchart for soil texture by feel

Safety Precautions

Follow all normal lab safety rules. In addition, take the following safety precautions:

- Wear sanitized indirectly vented chemical-splash goggles and chemical-resistant, nonlatex aprons and gloves through the entire investigation (which includes setup and cleanup).

- Handle all glassware with care.

- Report and clean up any spills immediately, and avoid walking in areas where water has been spilled.

- Wash hands with soap and water when done collecting the data and after completing the lab.

LAB 11

Getting Started

You can determine the rate at which water flows through a soil sample using the equipment illustrated in Figure L11.3. Once you have your equipment set up, you can place a sample of soil inside the coffee filter and then add water to the soil. The water will flow through the soil and the coffee filter and then land in the beaker. Your teacher will let you know where you can get samples of different types of soil.

Before you begin to design your experiment using this equipment, think about what type of data you need to collect, how you will collect the data, and how you will analyze the data.

FIGURE L11.3

Equipment needed to measure the rate at which water flows through a sample of soil

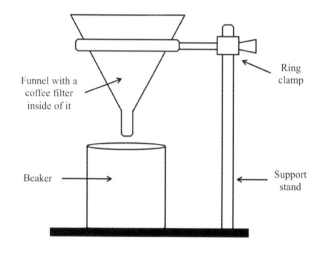

Funnel with a coffee filter inside of it

Ring clamp

Beaker

Support stand

To determine *what type of data you need to collect*, think about the following questions:

- How can you track water flowing through soil?
- What information about a soil sample do you need to determine its texture?
- What information will you need to calculate a rate?
- What type of measurements or observations will you need to record during each experiment?
- When will you need to make these measurements or observations?

To determine *how you will collect the data*, think about the following questions:

- What will serve as your independent variable and dependent variable?
- How will you vary the independent variable during each experiment?
- What will you do to hold the other variables constant during each experiment?
- What types of comparisons will you need to make?
- What measurement scale or scales should you use to collect data?
- What will you do to reduce error in your measurements?
- How will you keep track of and organize the data you collect?

To determine *how you will analyze the data,* think about the following questions:

- What calculations will you need to make?
- How could you use mathematics to describe a relationship between variables?
- How could you use mathematics to document a difference between groups or conditions?
- What types of patterns might you look for as you analyze your data?
- What type of table or graph could you create to help make sense of your data?

Connections to the Nature of Scientific Knowledge and Scientific Inquiry

As you work through your investigation, be sure to think about

- the difference between observations and inferences in science, and
- how the culture of science, societal needs, and current events influence the work of scientists.

Initial Argument

Once your group has finished collecting and analyzing your data, your group will need to develop an initial argument. Your initial argument needs to include a claim, evidence to support your claim, and a justification of the evidence. The *claim* is your group's answer to the guiding question. The *evidence* is an analysis and interpretation of your data. Finally, the *justification* of the evidence is why your group thinks the evidence matters. The justification of the evidence is important because scientists can use different kinds of evidence to support their claims. Your group will create your initial argument on a whiteboard. Your whiteboard should include all the information shown in Figure L11.4.

FIGURE L11.4 _____

Argument presentation on a whiteboard

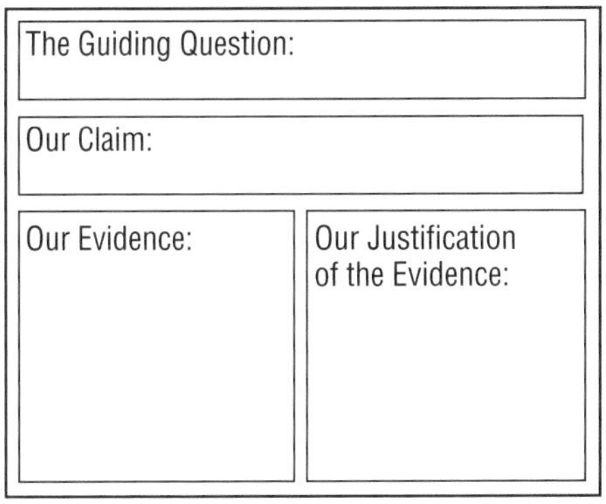

The Guiding Question:	
Our Claim:	
Our Evidence:	Our Justification of the Evidence:

Argumentation Session

The argumentation session allows all of the groups to share their arguments. One or two members of each group will stay at the lab station to share that group's argument, while the other members of the group go to the other lab stations to listen to and critique the other arguments. This is similar to what scientists do when they propose, support, evaluate, and refine new ideas during a poster session at a conference. If you are presenting your group's argument, your goal is to share your ideas and answer questions. You should also keep a record of the critiques and suggestions made by your classmates so you can use this feedback to make your initial argument stronger. You can keep track of specific critiques and suggestions for improvement that your classmates mention in the space provided.

Critiques of our initial argument and suggestions for improvement:

If you are critiquing your classmates' arguments, your goal is to look for mistakes in their arguments and offer suggestions for improvement so these mistakes can be fixed. You should look for ways to make your initial argument stronger by looking for things that the other groups did well. You can keep track of interesting ideas that you see and hear during the argumentation in the space below. You can also use this space to keep track of any questions that you will need to discuss with your team.

Interesting ideas from other groups or questions to take back to my group:

Once the argumentation session is complete, you will have a chance to meet with your group and revise your initial argument. Your group might need to gather more data or design a way to test one or more alternative claims as part of this process. Remember, your goal at this stage of the investigation is to develop the best argument possible.

Report

Once you have completed your research, you will need to prepare an *investigation report* that consists of three sections. Each section should provide an answer for the following questions:

1. What question were you trying to answer and why?

2. What did you do to answer your question and why?

3. What is your argument?

Your report should answer these questions in two pages or less. You should write your report using a word processing application (such as Word, Pages, or Google Docs), if possible, to make it easier for you to edit and revise it later. You should embed any diagrams, figures, or tables into the document. Be sure to write in a persuasive style; you are trying to convince others that your claim is acceptable or valid.

LAB 11

Lab 11. Soil Texture and Soil Water Permeability: How Does Soil Texture Affect Soil Water Permeability?

1. Peyton has collected some soil samples. The figure below shows a microscopic view of his three different soils. Assume each sample is viewed under the same magnification.

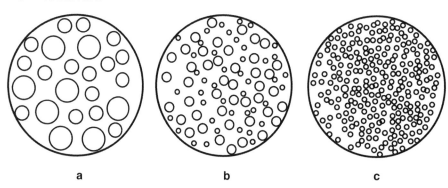

O = .002 mm

a   b   c

   a. Which soil is made mostly of clay?

   b. How do you know?

   c. Which soil has the highest rate of soil water permeability?

   d. How do you know?

2. Jackson has a sample of soil. He sets up a ring stand and a funnel with a filter and places a beaker below. He places 100 ml of water into the soil and waits for the water to flow through it. After two minutes, 40 ml of water is in the beaker. What is the rate of soil water permeability of Jackson's soil sample?

3. Scientists use proportional relationships to classify matter.

   a. I agree with this statement.

   b. I disagree with this statement.

   Explain your answer, using an example from your investigation about soil water permeability.

4. Adrianna identified one of her soil samples as sandy clay loam using the flowchart for soil texture by feel. She tells her group that she has made an inference about the sample's texture.

   a. I agree with this statement.

   b. I disagree with this statement.

   Explain your answer, using an example from your investigation about soil water permeability.

5. Scientists are influenced by many different factors when doing their work. The culture of science represents a shared set of values, norms, and commitments that shape what counts as knowing something in science, how to represent and communicate information, and how to interact with other scientists. How did your group decide that your claim was valid scientific knowledge? Give an example from your investigation about soil porosity.

6. Scientists often track how matter moves into, out of, and within systems during an investigation. Explain why it useful to track the movement of matter into, out of, and within a system using an example from your investigation about soil water permeability.

Teacher Notes

Lab 12. Cycling of Water on Earth: Why Do the Temperature and the Surface Area to Volume Ratio of a Sample of Water Affect Its Rate of Evaporation?

Purpose

The purpose of this lab is to *introduce* students to the disciplinary core idea (DCI) of Earth Materials and Systems by having them determine how temperature and the surface area to volume ratio of water affect the rate of water evaporation. In addition, students have an opportunity to learn about the crosscutting concepts (CCs) of (a) Energy and Matter: Flows, Cycles, and Conservation; and (b) Stability and Change. During the explicit and reflective discussion, students will also learn about (a) the difference between observations and inferences in science and (b) the nature and role of experiments in science.

Important Earth and Space Science Content

The water cycle, shown in Figure 12.1, is a conceptual model that scientists use to explain how water moves into, out of, and within Earth's systems. This process is driven by energy from the Sun and the force of gravity. When energy from the Sun heats liquid water, some of it transforms into gaseous water vapor and enters the atmosphere. This process is called *evaporation*. The water vapor rises into the air, where cooler temperatures cause it to condense into tiny liquid water droplets. A huge concentration of these droplets becomes visible to us as a cloud.

Air currents move clouds around the globe. This process is called *advection*. The water droplets in clouds collide, grow, and eventually fall to the ground as precipitation. Some precipitation falls as snow and can accumulate as ice caps and glaciers, which can store frozen water for thousands of years. Snowpacks in warmer climates often thaw and melt in the spring, and the melted water flows overland as snowmelt. Most precipitation, however, falls back into the oceans or onto land. On land, the water will flow over the ground as surface runoff because of the force of gravity.

A portion of the runoff will enter rivers or streams and move toward an ocean. Some of the runoff accumulates and forms freshwater lakes. Some of the water will also soak into the ground. This water is called groundwater. Plants will uptake some groundwater. This water will move through the plants through a process called *transpiration* and be released to the atmosphere. Transpiration is essentially evaporation of water from plant leaves. The rest of the groundwater will either move deep into the ground and create aquifers or stay close to the surface. The groundwater that remains close to the surface will seep back into lakes or rivers as groundwater discharge or create freshwater springs. Over time,

FIGURE 12.1

The water cycle explains how water molecules move into, out of, and within Earth's systems

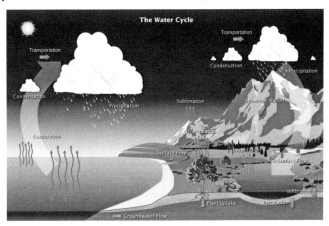

Note: A larger, full-color version of this figure is available on the book's Extras page at *www.nsta.org/adi-ess.*

though, all of this water keeps moving, and eventually reaches an ocean or returns to the atmosphere through the process of evaporation.

The difference between the solid, liquid, and gas states of a substance such as water is due to the amount of kinetic energy the atoms or molecules have and how these particles are moving relative to each other. *Kinetic energy* is the energy of motion. Atoms or molecules that are moving quickly have more kinetic energy than atoms or molecules that move more slowly. For example, the molecules in a sample of gaseous water move around quickly. These molecules therefore have a lot of kinetic energy. The molecules in a sample of solid water (ice), in contrast, move around more slowly and have less kinetic energy compared with the molecules in a sample of gaseous water.

The rate at which water evaporates depends on both the sample's temperature and its surface area to volume ratio. A sample of water at higher temperatures will consist of water molecules that have more kinetic energy and thus are moving more quickly than the water molecules in a sample of cold water and are more likely to dissociate with other water molecules. Thus, the rate of water evaporation increases at higher temperatures. A water sample with a greater surface area to volume ratio will also evaporate more quickly than a water sample that has a low surface area to volume ratio because more water molecules in the sample have contact with the atmosphere and are able to acquire more energy from the surroundings. Once a water molecule in the sample of liquid water gains enough energy to disassociate from the other water molecules in the sample, it can escape the sample. Because this process only happens at the surface of a liquid, or the liquid-air interface, the amount of liquid water that is exposed to the atmosphere affects the evaporation rate of that liquid. Thus, a larger surface area to volume ratio means a quicker rate of evaporation.

Timeline

The instructional time needed to complete this lab investigation is 270–330 minutes. Appendix 3 (p. 573) provides options for implementing this lab investigation over several class periods. Option G (330 minutes) should be used if students are unfamiliar with scientific writing, because this option provides extra instructional time for scaffolding the writing process. You can scaffold the writing process by modeling, providing examples, and providing hints as students write each section of the report. This option also gives students time to develop and test their conceptual model. Option H (270 minutes) should be used if students are familiar with scientific writing and have developed the skills needed to write an investigation report on their own. In option H, students complete stage 6 (writing the investigation report) and stage 8 (revising the investigation report) as homework. This option also gives students time to develop and test their conceptual model.

Materials and Preparation

The materials needed to implement this investigation are listed in Table 12.1. The equipment can be purchased from a science supply company such as Carolina, Flinn Scientific, or Ward's Science. The Pyrex containers can also be purchased from a local grocery or other retail store such as Target or Wal-Mart. It is important that the containers be different shapes and sizes but can hold the same volume of water. Students can use these containers to change the surface area to volume ratio of the water but keep the mass or volume of water constant across conditions during their experiment. The 500 ml beaker and the hot plate can be used to heat water to different temperatures before adding them to the plastic containers.

Be sure to use a set routine for distributing and collecting the materials during the lab investigation. One option is to set up the materials for each group at each group's lab station before class begins. This option works well when there is a dedicated section of the classroom for lab work and the materials are large and difficult to move (such as a stream table). A second option is to have all the materials on a table or cart at a central location. You can then assign a member of each group to be the "materials manager." This individual is responsible for collecting all the materials his or her group needs from the table or cart during class and for returning all the materials at the end of the class. This option works well when the materials are small and easy to move (such as stopwatches, metersticks, or thermometers). It also makes it easy to inventory the materials at the end of the class before students leave for the day.

Safety Precautions

Remind students to follow all normal lab safety rules. In addition, tell students to take the following safety precautions:

- Wear sanitized indirectly vented chemical-splash goggles and chemical-resistant, nonlatex aprons through the entire investigation (which includes setup and cleanup).

TABLE 12.1

Materials list for Lab 12

Item	Quantity
Consumables	
Water	As needed
Ice	As needed
Equipment and other materials	
Safety glasses or goggles	1 per student
Chemical-resistant apron	1 per student
Pyrex containers of different shapes and sizes (all should be able to hold the same volume of water)	6 per group
Electronic or triple beam balance	1 per group
Graduated cylinder, 250 ml	1 per group
Graduated cylinder, 100 ml	1 per group
Beaker, 500 ml	1 per group
Beaker, 250 ml	1 per group
Beaker, 150 ml	1 per group
Thermometer (nonmercury)	1 per group
Heat lamps	2 per group
Hot plate	1 per group
Ruler	1 per group
Investigation Proposal C (optional)	2 per group
Whiteboard, 2' × 3'*	1 per group
Lab Handout	1 per student
Peer-review guide and teacher scoring rubric	1 per student
Checkout Questions	1 per student

* As an alternative, students can use computer and presentation software such as Microsoft PowerPoint or Apple Keynote to create their arguments.

- Report and clean up spills immediately, and avoid walking in areas where water has been spilled

- Use caution when working with heat lamps and hot plates; they can get hot enough to burn skin.

- Never spray water on hot heat lamps—glass will shatter, producing dangerous glass projectiles.

- Use only GFCI-protected electrical receptacles for heat lamps and hot plates to prevent or reduce potential shock hazard.

- Handle all glassware with care.

- Handle glass thermometers with care. They are fragile and can break, causing a sharp hazard that can cut or puncture skin.

- Wash hands with soap and water when done collecting the data and after completing the lab.

Topics for the Explicit and Reflective Discussion

Reflecting on the Use of Core Ideas and Crosscutting Concepts During the Investigation

Teachers should begin the explicit and reflective discussion by asking students to discuss what they know about the DCI they used during the investigation. The following are some important concepts related to the DCI of Earth Materials and Systems that students need to determine how temperature and the surface area to volume ratio of water affect the rate of water evaporation:

- Matter is neither created nor destroyed; it just changes from one form to another (law of conservation of matter).

- Scientists can track how matter moves into, out of, and within a system by tracking changes in mass over time, because all matter has mass.

- Matter consists of tiny particles (atoms or molecules) that are in constant motion, and the average kinetic energy of all these particles depends only on the temperature. The average kinetic energy of the particles increases as the matter becomes warmer (kinetic-molecular theory).

To help students reflect on what they know about these concepts, we recommend showing them two or three images using presentation software that help illustrate these important ideas. You can then ask the students the following questions to encourage students to share how they are thinking about these important concepts:

1. What do we see going on in this image?

2. Does anyone have anything else to add?

3. What might be going on that we can't see?

4. What are some things that we are not sure about here?

You can then encourage students to think about how CCs played a role in their investigation. There are at least two CCs that students need to determine how temperature and the surface area to volume ratio of water affect the rate of water evaporation: (a) Energy and Matter: Flows, Cycles, and Conservation; and (b) Stability and Change (see Appendix 2 [p. 569] for a brief description of these CCs). To help students reflect on what they know about these CCs, we recommend asking them the following questions:

1. Why is it important to track how matter flows into, out of, or within a system during an investigation?

2. How did you track the flow of matter in the system you were studying? What did tracking the flow of matter allow you to do during your investigation

3. Why is it important to think about what controls or affects the rate of change in a system? How can we measure a rate of change?

4. Which factor(s) might have controlled the rate of change water evaporation in your investigation? What did exploring these factors allow you to do?

You can then encourage the students to think about how they used all these different concepts to help answer the guiding question and why it is important to use these ideas to help justify their evidence for their final arguments. Be sure to remind your students to explain why they included the evidence in their arguments and make the assumptions underlying their analysis and interpretation of the data explicit in order to provide an adequate justification of their evidence.

Reflecting on Ways to Design Better Investigations
It is important for students to reflect on the strengths and weaknesses of the investigation they designed during the explicit and reflective discussion. Students should therefore be encouraged to discuss ways to eliminate potential flaws, measurement errors, or sources of uncertainty in their investigations. To help students be more reflective about the design of their investigation and what they can do to make their investigations more rigorous in the future, you can ask the following questions:

1. What were some of the strengths of the way you planned and carried out your investigation? In other words, what made it scientific?

2. What were some of the weaknesses of the way you planned and carried out your investigation? In other words, what made it less scientific?

3. What rules can we make, as a class, to ensure that our next investigation is more scientific?

Reflecting on the Nature of Scientific Knowledge and Scientific Inquiry

This investigation can be used to illustrate two important concepts related to the nature of scientific knowledge and the nature of scientific inquiry: (a) the difference between observations and inferences in science and (b) the nature and role of experiments in science (see Appendix 2 [p. 569] for a brief description of these two concepts). Be sure to review these concepts during and at the end of the explicit and reflective discussion. To help students think about these concepts in relation to what they did during the lab, you can ask the following questions:

1. You made observations and inferences during your investigation. Can you give me some examples of these observations and inferences?

2. Can you work with your group to come up with a rule that you can use to tell the difference between an observation and an inference? Be ready to share in a few minutes.

3. I asked you to design and carry out an experiment as part of your investigation. Are all investigations in science experiments? Why or why not?

4. Can you work with your group to come up with a rule that you can use to decide if an investigation is an experiment or not? Be ready to share in a few minutes.

You can also use presentation software or other techniques to encourage the students to think about these concepts. You can show examples of information from the investigation that are either observations or inferences and ask students to classify each example and explain their thinking. You can also show images of different types of investigations (such as an Earth scientist collecting data in the field as part of an observational or descriptive study, a person working in the library doing a literature review, a person working on a computer to analyze an existing data set, and an actual experiment) and ask students to indicate if they think each image represents an experiment and why or why not.

Remind your students that, to be proficient in science, it is important that they understand what counts as scientific knowledge and how that knowledge develops over time in order to be proficient in science at some point during the discussion.

Hints for Implementing the Lab

- Allow the students to become familiar with the equipment as part of the tool talk before they begin to design their experiments. This gives students a chance to see what they can and cannot do with the equipment.

- Allowing students to design their own procedures for collecting data gives students an opportunity to try, to fail, and to learn from their mistakes. However, you can scaffold students as they develop their procedure by having them fill out an investigation proposal. These proposals provide a way for you to offer students hints and suggestions without telling them how to do it. You can also check the proposals quickly during a class period. We recommend using Investigation Proposal C for this lab. Students should fill out a different investigation proposal for each experiment they plan.

- Investigation Proposal C works best for the two experiments that the students will conduct during this investigation because students can describe a different relationship between the independent variable and dependent variable for each hypothesis. For example, increasing the temperature of the water might (a) increase the rate of water evaporation (hypothesis 1), (b) decrease the rate of water evaporation (hypothesis 2), or (c) have no effect on the rate of water evaporation (hypothesis 3).

- Students do not need to use a large volume of water during each experiment. We recommend limiting the amount of water they use in each sample to 100 ml and 200 ml.

- Students will need to measure the rate of evaporation over at least 24 hours for noticeable results. The recommended timeline options work well because they provide time in the schedule to leave the water samples under a heat lamp or in direct sunlight for longer periods of time. Each group can set up their samples of water and collect initial measurements on day 1. They can then leave them unattended to the next period. Each group can then collect their final measurements at the beginning of the period on day 2. You can set up stations with heat lamps around the room so students can leave their containers of water under the lamps overnight.

- To create their conceptual models, the students will need to combine what they learned from both experiments into one explanation. Be sure to remind them that their model must explain why, not just how.

- Be sure to give students time to test their models. To accomplish this goal, they can make predictions about how much water will be lost from a sample given its temperature and surface area to volume ratio. If their observations match their predictions, they will have evidence that they can use to convince others that their model is valid or acceptable.

- Students often make mistakes when developing their conceptual models and/ or initial arguments, but they should quickly realize these mistakes during the argumentation session. Be sure to allow students to revise their models and arguments at the end of the argumentation session. The explicit and reflective discussion will also give students an opportunity to reflect on and identify ways to improve how they develop and test models. This also offers an opportunity to discuss what scientists do when they realize a mistake is made.

- Students often make mistakes during the data collection stage, but they should quickly realize these mistakes during the argumentation session. It will only take them a short period of time to re-collect data, and they should be allowed to do so. During the explicit and reflective discussion, students will also have the opportunity to reflect on and identify ways to improve the way they design investigations (especially how they attempt to control variables as part of an experiment). This provides another opportunity to discuss what scientists do when they realize that a mistake is made during a study.

Connections to Standards

Table 12.2 highlights how the investigation can be used to address specific (a) performance expectations from the *NGSS* and (b) *Common Core State Standards* in English language arts (*CCSS ELA*).

TABLE 12.2

Lab 12 alignment with standards

NGSS performance expectations	Earth's systems • MS-ESS2-4: Develop a model to describe the cycling of water through Earth's systems driven by energy from the Sun and the force of gravity. • HS-ESS2-5: Plan and conduct an investigation of the properties of water and its effects on Earth materials and surface processes.
CCSS ELA—Reading in Science and Technical Subjects	Key ideas and details • CCSS.ELA-LITERACY.RST.6-8.1: Cite specific textual evidence to support analysis of science and technical texts. • CCSS.ELA-LITERACY.RST.6-8.2: Determine the central ideas or conclusions of a text; provide an accurate summary of the text distinct from prior knowledge or opinions. • CCSS.ELA-LITERACY.RST.9-10.1: Cite specific textual evidence to support analysis of science and technical texts, attending to the precise details of explanations or descriptions. • CCSS.ELA-LITERACY.RST.9-10.2: Determine the central ideas or conclusions of a text; trace the text's explanation or depiction of a complex process, phenomenon, or concept; provide an accurate summary of the text. • CCSS.ELA-LITERACY.RST.9-10.3: Follow precisely a complex multistep procedure when carrying out experiments, taking measurements, or performing technical tasks, attending to special cases or exceptions defined in the text.

Continued

TABLE 12.2 (*continued*)

CCSS ELA—**Reading in Science and Technical Subjects** (*continued*)	Craft and structure • CCSS.ELA-LITERACY.RST.6-8.4: Determine the meaning of symbols, key terms, and other domain-specific words and phrases as they are used in a specific scientific or technical context relevant to *grade 6–8 texts and topics*. • CCSS.ELA-LITERACY.RST.6-8.5: Analyze the structure an author uses to organize a text, including how the major sections contribute to the whole and to an understanding of the topic. • CCSS.ELA-LITERACY.RST.6-8.6: Analyze the author's purpose in providing an explanation, describing a procedure, or discussing an experiment in a text. • CCSS.ELA-LITERACY.RST.9-10.4: Determine the meaning of symbols, key terms, and other domain-specific words and phrases as they are used in a specific scientific or technical context relevant to *grade 9–10 texts and topics*. CCSS.ELA-LITERACY.RST.9-10.5: Analyze the structure of the relationships among concepts in a text, including relationships among key terms (e.g., *force, friction, reaction force, energy*). • CCSS.ELA-LITERACY.RST.9-10.6: Analyze the author's purpose in providing an explanation, describing a procedure, or discussing an experiment in a text, defining the question the author seeks to address. Integration of knowledge and ideas • CCSS.ELA-LITERACY.RST.6-8.7: Integrate quantitative or technical information expressed in words in a text with a version of that information expressed visually (e.g., in a flowchart, diagram, model, graph, or table). • CCSS.ELA-LITERACY.RST.6-8.8: Distinguish among facts, reasoned judgment based on research findings, and speculation in a text. • CCSS.ELA-LITERACY.RST.6-8.9: Compare and contrast the information gained from experiments, simulations, video, or multimedia sources with that gained from reading a text on the same topic. • CCSS.ELA-LITERACY.RST.9-10.7: Translate quantitative or technical information expressed in words in a text into visual form (e.g., a table or chart) and translate information expressed visually or mathematically (e.g., in an equation) into words. • CCSS.ELA-LITERACY.RST.9-10.8: Assess the extent to which the reasoning and evidence in a text support the author's claim or a recommendation for solving a scientific or technical problem. • CCSS.ELA-LITERACY.RST.9-10.9: Compare and contrast findings presented in a text to those from other sources (including their own experiments), noting when the findings support or contradict previous explanations or accounts.

Continued

TABLE 12.2 (*continued*)

CCSS ELA—Writing in Science and Technical Subjects	Text types and purposes • CCSS.ELA-LITERACY.WHST.6-10.1: Write arguments focused on *discipline-specific content.* • CCSS.ELA-LITERACY.WHST.6-10.2: Write informative or explanatory texts, including the narration of historical events, scientific procedures/experiments, or technical processes. Production and distribution of writing • CCSS.ELA-LITERACY.WHST.6-10.4: Produce clear and coherent writing in which the development, organization, and style are appropriate to task, purpose, and audience. • CCSS.ELA-LITERACY.WHST.6-8.5: With some guidance and support from peers and adults, develop and strengthen writing as needed by planning, revising, editing, rewriting, or trying a new approach, focusing on how well purpose and audience have been addressed. • CCSS.ELA-LITERACY.WHST.6-8.6: Use technology, including the internet, to produce and publish writing and present the relationships between information and ideas clearly and efficiently. • CCSS.ELA-LITERACY.WHST.9-10.5: Develop and strengthen writing as needed by planning, revising, editing, rewriting, or trying a new approach, focusing on addressing what is most significant for a specific purpose and audience. • CCSS.ELA-LITERACY.WHST.9-10.6: Use technology, including the internet, to produce, publish, and update individual or shared writing products, taking advantage of technology's capacity to link to other information and to display information flexibly and dynamically. Range of writing • CCSS.ELA-LITERACY.WHST.6-10.10: Write routinely over extended time frames (time for reflection and revision) and shorter time frames (a single sitting or a day or two) for a range of discipline-specific tasks, purposes, and audiences.
CCSS ELA—Speaking and Listening	Comprehension and collaboration • CCSS.ELA-LITERACY.SL.6-8.1: Engage effectively in a range of collaborative discussions (one-on-one, in groups, and teacher-led) with diverse partners on grade 6–8 topics, texts, and issues, building on others' ideas and expressing their own clearly. • CCSS.ELA-LITERACY.SL.6-8.2:*Interpret information presented in diverse media and formats (e.g., visually, quantitatively, orally) and explain how it contributes to a topic, text, or issue under study. • CCSS.ELA-LITERACY.SL.6-8.3:* Delineate a speaker's argument and specific claims, distinguishing claims that are supported by reasons and evidence from claims that are not.

Continued

TABLE 12.2 (*continued*)

***CCSS ELA*—Speaking and Listening** (*continued*)	Comprehension and collaboration (*continued*)
	• CCSS.ELA-LITERACY.SL.9-10.1: Initiate and participate effectively in a range of collaborative discussions (one-on-one, in groups, and teacher-led) with diverse partners on grade 9–10 topics, texts, and issues, building on others' ideas and expressing their own clearly and persuasively.
	• CCSS.ELA-LITERACY.SL.9-10.2: Integrate multiple sources of information presented in diverse media or formats (e.g., visually, quantitatively, orally) evaluating the credibility and accuracy of each source.
	• CCSS.ELA-LITERACY.SL.9-10.3: Evaluate a speaker's point of view, reasoning, and use of evidence and rhetoric, identifying any fallacious reasoning or exaggerated or distorted evidence.
	Presentation of knowledge and ideas
	• CCSS.ELA-LITERACY.SL.6-8.4:* Present claims and findings, sequencing ideas logically and using pertinent descriptions, facts, and details to accentuate main ideas or themes; use appropriate eye contact, adequate volume, and clear pronunciation.
	• CCSS.ELA-LITERACY.SL.6-8.5:* Include multimedia components (e.g., graphics, images, music, sound) and visual displays in presentations to clarify information.
	• CCSS.ELA-LITERACY.SL.6-8.6: Adapt speech to a variety of contexts and tasks, demonstrating command of formal English when indicated or appropriate.
	• CCSS.ELA-LITERACY.SL.9-10.4: Present information, findings, and supporting evidence clearly, concisely, and logically such that listeners can follow the line of reasoning and the organization, development, substance, and style are appropriate to purpose, audience, and task.
	• CCSS.ELA-LITERACY.SL.9-10.5: Make strategic use of digital media (e.g., textual, graphical, audio, visual, and interactive elements) in presentations to enhance understanding of findings, reasoning, and evidence and to add interest.
	• CCSS.ELA-LITERACY.SL.9-10.6: Adapt speech to a variety of contexts and tasks, demonstrating command of formal English when indicated or appropriate.

* Only the standard for grade 6 is provided because the standards for grades 7 and 8 are similar. Please see *www.corestandards.org/ELA-Literacy/SL* for the exact wording of the standards for grades 7 and 8.

LAB 12

Lab Handout

Lab 12. Cycling of Water on Earth: Why Do the Temperature and the Surface Area to Volume Ratio of a Sample of Water Affect Its Rate of Evaporation?

Introduction

Water can be found as a liquid, solid, or gas on Earth. Lakes, rivers, and oceans contain liquid water. For example, Lake Tahoe (shown in Figure L12.1a), which straddles the border between California and Nevada, contains 36.15 cubic miles of liquid water. Polar ice caps and glaciers contain solid water. For example, the Perito Moreno glacier (see Figure L12.1b), located in western Patagonia, Argentina, is a huge ice formation that is 30 km in length, 5 km wide, and has an average height of 74 m. The atmosphere contains gaseous water vapor. Unlike the other two forms of water, gaseous water vapor is invisible. We can see water vapor only when it condenses to form visible clouds of water droplets such as in Figure L12.1c. Water vapor is responsible for humidity.

FIGURE L12.1

Water can be found in all three states on Earth. (a) Lake Tahoe contains liquid water. (b) The Perito Moreno glacier contains solid water. (c) Gaseous water is invisible unless it condenses to form clouds of water droplets, as seen in the Owakudani volcanic valley in Japan.

a   b   c

All water is made up of molecules that are composed of two hydrogen atoms and one oxygen atom. This type of molecule is called a water molecule. All water molecules have the same mass. Water molecules are also constantly in motion. Because water molecules have mass and are constantly in motion, they have *kinetic energy*. Temperature is a way we can measure the average kinetic energy of the molecules in any sample of water. A high temperature means that the molecules in the sample have high average kinetic energy and are moving quickly, while a low temperature means the molecules have low kinetic energy and are moving slowly. The three states of water are determined by temperature. At low temperatures (less than 0°C), water is a solid (ice); at room temperature, water is a liquid; and at high temperatures (more than 100°C), water is a gas.

FIGURE L12.2

The water cycle explains how water molecules move into, out of, and within Earth's systems

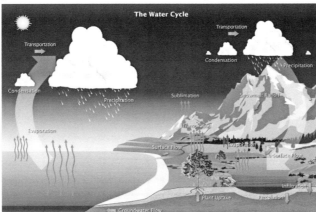

The water cycle, shown in Figure L12.2, is a model that scientists use to explain how water molecules move into, out of, and within Earth's systems. This process is driven by energy from the Sun and the force of gravity. When energy from the Sun heats liquid water, some of it transforms into gaseous water vapor and enters the atmosphere. This process is called evaporation. The water vapor rises into the air, where cooler temperatures cause it to condense into tiny liquid water droplets. A huge concentration of these droplets becomes visible to us as a cloud. Air currents move clouds around the globe. The water droplets in clouds collide, grow, and eventually fall to the ground as precipitation. Some precipitation falls as snow and can accumulate as ice caps and glaciers, which can store frozen water for thousands of years. Snowpacks in warmer climates often thaw and melt in the spring, and the melted water flows overland as snowmelt. Most precipitation, however, falls back into the oceans or onto land. On land, the water will flow over the ground as surface runoff because of the force of gravity.

A portion of the runoff will enter rivers in valleys in the landscape and move toward an ocean. Some of the runoff accumulates and forms freshwater lakes. Some of the water will also soak into the ground. This water is called groundwater. Some of the groundwater will move deep into the ground and create aquifers, which are underground stores of freshwater, and some will stay close to the surface. The groundwater that remains close to the surface will seep back into lakes or rivers as groundwater discharge or will create freshwater springs. Over time, though, all of this water keeps moving, and it eventually reaches an ocean or returns to the atmosphere through the process of evaporation.

Water can evaporate and enter the atmosphere from sources as vast as the ocean and as small as your pet's water dish. At any place where the surface of the water meets with the air, water molecules are able to leave the liquid water and enter the atmosphere. It might seem like evaporation makes liquid water disappear, but recall that the law of conservation of matter states that matter can never be created nor destroyed, but it can change form. When evaporation happens, the molecules are simply changing from a liquid phase, which is visible to us, to a gaseous phase, which we cannot see. Keeping this in mind, we can tell how much evaporation has happened by measuring changes in the mass or volume of the liquid water.

You may have noticed that bodies of water evaporate at different rates. For example, a rain puddle on the street can evaporate in a few hours, but water in a glass may take days or weeks to evaporate. There are many factors that may affect the rate that water evaporates:

- The amount of energy that water absorbs from the Sun

- The temperature of the water

- The *surface area to volume ratio* of the water; or the amount of the water's surface that is exposed to the air compared with its total volume

In this investigation, you will have an opportunity to determine how water temperature and the surface area to volume ratio of a sample of water contribute to the rate that water evaporates. Once you understand how these two factors affect how quickly a sample of water will evaporate, you will then develop a conceptual model that you can use to explain your observations and predict how quickly water will evaporate under different conditions.

Your Task

Use what you know about the properties of water, rates of change, and the importance of tracking the movement of matter into, out of, and within systems during an investigation to plan and carry out an experiment to determine how changes in the temperature and the surface area to volume ratio of a sample of water affect how quickly it will evaporate. Then develop a conceptual model that can be used to explain *why* these factors affect the rate that water evaporates. Once you have developed your conceptual model, you will need to test it using different water samples to determine if it allows you to predict how much liquid water will be lost due to evaporation under different conditions.

The guiding question of this investigation is, *Why do the temperature and the surface area to volume ratio of a sample of water affect its rate of evaporation?*

Materials

You may use any of the following materials during your investigation:

Consumables
- Water
- Ice

Equipment
- Safety glasses or goggles (required)
- Chemical-resistant apron (required)
- Pyrex containers of different shapes and sizes
- Electronic or triple beam balance
- Graduated cylinder (250 ml)
- Graduated cylinder (100 ml)
- Beaker (500 ml)

- Beaker (250 ml)
- Beaker (150 ml)
- Thermometer (nonmercury)
- 2 Heat lamps
- Hot plate
- Ruler

Safety Precautions

Follow all normal lab safety rules. In addition, take the following safety precautions:

- Wear sanitized indirectly vented chemical-splash goggles and chemical-resistant, nonlatex aprons throughout the entire investigation (which includes setup and cleanup).
- Report and clean up spills immediately, and avoid walking in areas where water has been spilled
- Use caution when working with heat lamps and hot plates; they can get hot enough to burn skin.
- Never spray water on hot heat lamps—glass will shatter, producing dangerous glass projectiles.
- Use only GFCI-protected electrical receptacles for heat lamps and hot plates to prevent or reduce potential shock hazard.
- Handle all glassware with care.
- Handle glass thermometers with care. They are fragile and can break, causing a sharp hazard that can cut or puncture skin.
- Wash hands with soap and water when done collecting the data and after completing the lab.

Investigation Proposal Required?   ☐ Yes   ☐ No

Getting Started

The first step in developing your model is to plan and carry out at least two experiments. Figure L12.3 (p. 300) shows how you can use a heat lamp to warm your different samples of water. The heat lamp will serve as the source of energy for each experiment. Your teacher may also allow you to set your water samples outside in direct sunlight depending on the time of year. Figure L12.3 also shows how you can use containers of different sizes and shapes to manipulate the surface area to volume ratio of a sample of water. You can use a hot plate to heat your water samples to different temperatures or to maintain the temperature of a water sample.

Before you begin to design your two experiments using this equipment, be sure to think about what type of data you need to collect, how you will collect the data, and how you will analyze the data. To determine *what type of data you need to collect,* think about the following questions:

- How will you track the flow of energy into each water sample?
- How will you track the amount of water loss from a sample?
- How will you measure the rate of water evaporation (change over time)?

To determine *how you will collect the data,* think about the following questions:

- What will be the independent and dependent variables for each experiment?
- What conditions will you need to set up for each experiment?
- How will you make sure you are only testing one variable at a time?
- How often will you need to take measurements during each experiment?
- What measurement scale or scales should you use to collect data?
- How will you make sure that your data are of high quality (i.e., how will you reduce error)?
- How will you keep track of and organize the data you collect?

To determine *how you will analyze the data,* think about the following questions:

- What type of calculations will you need to make?
- How will you determine if rates of change are the same or different?
- How could you use mathematics to document a difference between conditions?
- What type of table or graph could you create to help make sense of your data?

FIGURE L12.3

How to use a heat lamp to warm different samples of water

Once you have carried out your two experiments, you will need to develop a conceptual model. Your model needs to be able to explain why temperature and the surface area to volume ratio of a sample of water affect the amount of water lost due to evaporation in the way that these two factors do. The model also needs to account for the kinetic energy of water molecules and the conservation of matter.

The last step in this investigation is to test your model. To accomplish this goal, you can set up a third experiment to determine if your model leads to accurate predictions about the amount of water that will be lost from different containers of water under specific conditions (e.g., cold water, high surface area to volume ratio). If you are able to use your model to make accurate predictions about the way water evaporates under different conditions, then you will be able to generate the evidence you need to convince others that the conceptual model you developed is valid or acceptable.

Connections to the Nature of Scientific Knowledge and Scientific Inquiry

As you work through your investigation, be sure to think about

- the difference between observations and inferences in science, and
- the nature and role of experiments in science.

Initial Argument

Once your group has finished collecting and analyzing your data, your group will need to develop an initial argument. Your initial argument needs to include a claim, evidence to support your claim, and a justification of the evidence. The *claim* is your group's answer to the guiding question. The *evidence* is an analysis and interpretation of your data. Finally, the *justification* of the evidence is why your group thinks the evidence matters. The justification of the evidence is important because scientists can use different kinds of evidence to support their claims. Your group will create your initial argument on a whiteboard. Your whiteboard should include all the information shown in Figure L12.4.

FIGURE L12.4

Argument presentation on a whiteboard

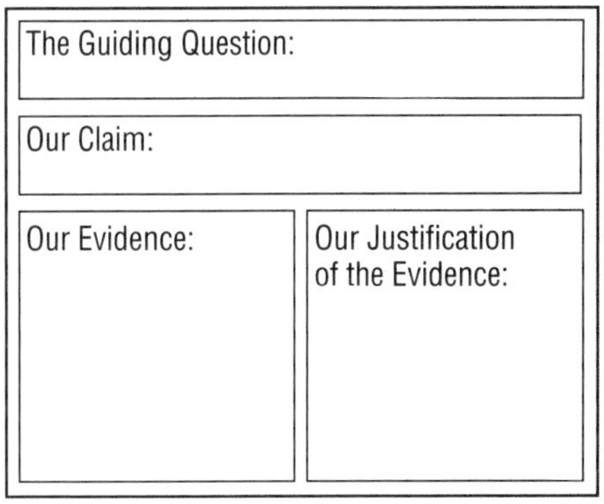

The Guiding Question:	
Our Claim:	
Our Evidence:	Our Justification of the Evidence:

Argumentation Session

The argumentation session allows all of the groups to share their arguments. One or two members of each group will stay at the lab station to share that group's argument, while the other members of the group go to the other lab stations to listen to and critique the other arguments. This is similar to what scientists do when they propose, support, evaluate, and refine new ideas during a poster session at a conference. If you are presenting your group's argument, your goal is to share your ideas and answer questions. You should also keep a record of the critiques and suggestions made by your classmates so you can use this feedback to make your initial argument stronger. You can keep track of specific critiques and suggestions for improvement that your classmates mention in the space below.

Critiques of our initial argument and suggestions for improvement:

If you are critiquing your classmates' arguments, your goal is to look for mistakes in their arguments and offer suggestions for improvement so these mistakes can be fixed. You should look for ways to make your initial argument stronger by looking for things that the other groups did well. You can keep track of interesting ideas that you see and hear during the argumentation in the space below. You can also use this space to keep track of any questions that you will need to discuss with your team.

Interesting ideas from other groups or questions to take back to my group:

Once the argumentation session is complete, you will have a chance to meet with your group and revise your initial argument. Your group might need to gather more data or design a way to test one or more alternative claims as part of this process. Remember, your goal at this stage of the investigation is to develop the best argument possible.

Report

Once you have completed your research, you will need to prepare an investigation report that consists of three sections. Each section should provide an answer for the following questions:

1. What question were you trying to answer and why?

2. What did you do to answer your question and why?

3. What is your argument?

Your report should answer these questions in two pages or less. You should write your report using a word processing application (such as Word, Pages, or Google Docs), if possible, to make it easier for you to edit and revise it later. You should embed any diagrams, figures, or tables into the document. Be sure to write in a persuasive style; you are trying to convince others that your claim is acceptable or valid.

Checkout Questions

Lab 12. Cycling of Water on Earth: Why Do the Temperature and the Surface Area to Volume Ratio of a Sample of Water Affect Its Rate of Evaporation?

1. How does the temperature of a water sample affect the rate of its evaporation?

2. How does the surface area to volume ratio of a water sample affect the rate of its evaporation?

3. A scientist has a collection of water samples. Each sample has the same volume of water, but the samples vary in their surface areas and temperatures. She is trying to compare the amount of evaporation in each sample.

Sample	Sample surface area (cm²)	Volume (cm³)	Sample temperature (°C)
A	6.25	20.0	25
B	10.00	20.0	65
C	12.56	20.0	25
D	3.14	20.0	65

a. After 45 minutes, which sample will have evaporated the most?

b. How do you know?

c. After 45 minutes, which sample will have evaporated the least?

d. How do you know?

4. This investigation was an experiment.

    a. I agree with this statement.

    b. I disagree with this statement.

Explain your answer, using an example from your investigation about the cycling of water on Earth.

5. "Sample A lost 5 grams of water due to evaporation" is an example of an observation.

    a. I agree with this statement.

    b. I disagree with this statement.

Explain your answer, using an example from your investigation about the cycling of water on Earth.

6. Scientists often track how matter moves into, out of, and within systems during an investigation. Explain why it is useful to do this, using an example from your investigation about the cycling of water on Earth.

7. Scientists often try to understand what controls the rate of change of a system. Explain what a rate of change in a system is and why it is useful to understand the factors that control a rate of change in a system, using an example from your investigation about the cycling of water on Earth.

Application Labs

LAB 13

Teacher Notes

Lab 13. Characteristics of Minerals: What Are the Identities of the Unknown Minerals?

Purpose

The purpose of this lab is for students to *apply* what they know about the disciplinary core idea (DCI) of Earth Materials and Systems to identify a set of unknown minerals. In addition, students have an opportunity to learn about the crosscutting concepts (CCs) of (a) Patterns and (b) Structure and Function. During the explicit and reflective discussion, students will also learn about (a) the difference between and data and evidence in science and (b) how scientists use different methods to answer different types of questions.

Important Earth and Space Science Content

Earth scientists classify rocks based on the type and proportion of the minerals that make up the rocks. Minerals are the inorganic solids that are the building blocks of rocks. Though there are thousands of known minerals, there are fewer than 50 types that make up rocks. Minerals themselves are, like every other type of matter, made of atoms. The *atomic composition* of a mineral refers to the different types of atoms found in it and the relative proportion of each type of atom. Dolomite (see Figure 13.1), for example, has the chemical composition of $CaMg(CO_3)_2$, whereas quartz (see Figure 13.2) has a chemical composition of SiO_2.

FIGURE 13.1 _____
Dolomite

FIGURE 13.2 _____
Quartz

The physical and chemical properties of a mineral refer to measurable or observable qualities or attributes that are used to distinguish between different minerals. Physical properties are descriptive characteristics of minerals; examples of these properties include

310

color, density, luster, and hardness. Chemical properties, in contrast, describe how a mineral interacts with other matter. For example, carbonate minerals such as calcite will react with an acid such as hydrochloric acid or vinegar, but silicates will not. Earth scientists can identify minerals based on their physical and chemical properties because every type of mineral has a unique set of physical and chemical properties that reflect the unique atomic composition of that mineral.

Students will be given a set of known minerals at the beginning of this investigation. They will then document, measure, or calculate at least three different chemical or physical properties for each mineral. Next, the students will return the known minerals and will be given a set of unknown minerals. The set of unknown minerals should consist of new samples of the minerals that the students already tested. The students will then be directed to design and carry out an investigation that will enable them to collect the data needed to determine the identity of the unknown minerals.

There are at least seven different properties that students can use to identify minerals during this investigation:

- *Color.* Minerals reflect different types of light, which we recognize as colors. The properties of a mineral give it its color, but minerals of the same type can be different colors. Quartz, for example, comes in many colors. Color is therefore not a reliable method to use for mineral identification.

- *Luster.* Minerals have an overall sheen, which can be metallic (looks like metal) or nonmetallic. Nonmetallic luster can be glossy, pearly, greasy, or dull. Minerals with metallic luster, however, may not always appear that way if they become tarnished.

- *Color of the powdered mineral.* Earth scientists use a streak test to determine the color of a powdered mineral. The streak test is valuable because many minerals occur in a variety of apparent colors, but all specimens of that mineral share a similar streak color. A sample of hematite, for example, can be black, red, brown, or silver in color; however, all samples of hematite produce a streak with a reddish color.

- *Hardness.* This property is the resistance of a material to being scratched. One of the most important tests for identifying mineral specimens is the Mohs Hardness Test, which compares the resistance of a mineral being scratched with the resistance of 10 reference minerals using the Mohs Hardness Scale (see Table 13.1). The test is useful because most

TABLE 13.1 _____

The Mohs Hardness Scale

Mineral	Hardness
Talc	1
Gypsum	2
Calcite	3
Fluorite	4
Apatite	5
Orthoclase	6
Quartz	7
Topaz	8
Corundum	9
Diamond	10

specimens of a given mineral are very close to the same hardness. This makes hardness a reliable diagnostic property for most minerals. Most of us, however, do not have samples of the 10 minerals on the Mohs Hardness Scale. Students can therefore use several different common materials to determine the relative hardness of an unknown mineral specimen. Glass, for example, has a known Mohs hardness of 5.5. If a mineral can scratch glass, it must be harder than 5.5. Streak plates are around 6.5. Nails are 4, pennies are 3, and a person's fingernail is 2.5.

- *How the mineral breaks.* Minerals that display cleavage break in flat, predictable shapes. Minerals that display fractures, in contrast, break unevenly or sometime crumble into smaller pieces that are not uniformly shaped.

- *Magnetism.* Some minerals, such as magnetite, are attracted to a magnet.

- *Reaction with an acid.* Some minerals, such as carbonates, fizz when an acid is added to them, whereas others, such as silicates, do not.

Timeline

The instructional time needed to complete this lab investigation is 200–280 minutes. Appendix 3 (p. 573) provides options for implementing this lab investigation over several class periods. Option E (280 minutes) should be used if students are unfamiliar with scientific writing, because this option provides extra instructional time for scaffolding the writing process. You can scaffold the writing process by modeling, providing examples, and providing hints as students write each section of the report. Option F (200 minutes) should be used if students are familiar with scientific writing and have developed the skills needed to write an investigation report on their own. In option F, students complete stage 6 (writing the investigation report) and stage 8 (revising the investigation report) as homework.

Materials and Preparation

The materials needed to implement this investigation are listed in Table 13.2. The consumables and equipment can be purchased from a science supply company such as Carolina, Flinn Scientific, or Ward's Science.

We recommend that the set of known minerals include 6–10 different samples (such as barite, calcite, feldspar, graphite, gypsum, halite, hematite, magnetite, mica, microcline, muscovite, olivine, pyrite, quartz, selenite, and talc) and the set of unknown minerals include at least 4 samples (such as halite, mica, quartz, and talc). The four samples in the set of unknown minerals must include only minerals that were included in the set of known minerals.

There are a number of options for creating the set of known minerals and the set of unknown minerals. You can purchase one or more different mineral collections from a science supply company and then use the minerals from the collections to create your own sets. Here are some good collections to use for this approach:

TABLE 13.2

Materials list for Lab 13

Item	Quantity
Consumables	
Water in squirt bottles	1 per group
Vinegar (acetic acid)	50 ml per group
Set of known minerals	1 per group
Set of unknown minerals	1 per group
Equipment and other materials	
Safety glasses or goggles	1 per student
Chemical-resistant apron	1 per student
Gloves	1 pair per student
Electronic or triple beam balance	1 per group
Magnifying glass	1 per group
Magnet	1 per group
Beaker, 250 ml	1 per group
Beaker, 400 ml	1 per group
Graduated cylinder, 100 ml	1 per group
Overflow can	1 per group
Pipette	1 per group
Ruler	1 per group
Streak plate (unglazed porcelain tile)	1 per group
Small piece of glass	1 per group
Penny	1 per group
Nail	1 per group
Computer or tablet with internet access	1 per group
Whiteboard, 2' × 3'*	1 per group
Lab Handout	1 per student
Peer-review guide and teacher scoring rubric	1 per student
Checkout Questions	1 per student

* As an alternative, students can use computer and presentation software such as Microsoft PowerPoint or Apple Keynote to create their arguments.

- Introduction to Minerals Study Kit from Carolina (item GEO2878)
- The Classification of Minerals Collection from Carolina (item GEO2194)
- Classroom Mineral Collection from Flinn Scientific (item AP4883)
- Introductory Mineral Collection from Flinn Scientific (item AP4880)
- Know Your Minerals Collection from Ward's Science (item 450210)
- Introductory Mineral Collection from Ward's Science (item 453300)

You can also purchase individual minerals in bulk and then use these minerals to create a set of known and unknown minerals. Whichever approach you decide to use, it is important for the samples in the set of unknown minerals to look different from the samples in the set of known minerals. To make the samples in each set look different from each other, simply change the amount of each mineral.

Be sure to use a set routine for distributing and collecting the materials during the lab investigation. One option is to set up the materials for each group at each group's lab station before class begins. This option works well when there is a dedicated section of the classroom for lab work and the materials are large and difficult to move (such as a stream table). A second option is to have all the materials on a table or cart at a central location. You can then assign a member of each group to be the "materials manager." This individual is responsible for collecting all the materials his or her group needs from the table or cart during class and for returning all the materials at the end of the class. This option works well when the materials are small and easy to move (such as stopwatches, metersticks, or thermometers). It also makes it easy to inventory the materials at the end of the class before students leave for the day.

Safety Precautions and Laboratory Waste Disposal

Remind students to follow all normal lab safety rules. In addition, tell students to take the following safety precautions:

- Follow safety precautions noted on safety data sheets for hazardous chemicals.
- Wear sanitized indirectly vented chemical-splash goggles and chemical-resistant, nonlatex aprons and gloves throughout the entire investigation (which includes setup and cleanup).
- Report and clean up any spills on the floor immediately to avoid a slip or fall hazard.
- Handle all glassware with care.
- When using the glass or streak plate, do not pick the plate up with your hands. Place it flat on the table. Otherwise, it may cut your hand.
- Handle nails with care. Nail ends are sharp and can cut or puncture skin.
- Do not place any materials in or around your mouth.

- Handle all glassware with care.

- Wash hands with soap and water when done collecting the data and after completing the lab.

The vinegar (acetic acid) may be rinsed down the drain with excess water according to the Flinn laboratory waste disposal method 26b. Information about laboratory waste disposal methods is included in the Flinn catalog and reference manual; you can request a free copy at *www.flinnsci.com/flinn-freebies/request-a-catalog*.

Topics for the Explicit and Reflective Discussion
Reflecting on the Use of Core Ideas and Crosscutting Concepts During the Investigation

Teachers should begin the explicit and reflective discussion by asking students to discuss what they know about the DCI they used during the investigation. The following are some important concepts related to the DCI of Earth Materials and Systems that students need to determine the identities of a set of unknown minerals:

- Minerals are the inorganic matter that makes up rocks.

- Minerals have specific physical and chemical properties.

- Physical properties of matter do not change based on the size or shape of the matter and can therefore be used to identify matter.

To help students reflect on what they know about these concepts, we recommend showing them two or three images using presentation software that help illustrate these important ideas. You can then ask the students the following questions to encourage students to share how they are thinking about these important concepts:

1. What do we see going on in this image?

2. Does anyone have anything else to add?

3. What might be going on that we can't see?

4. What are some things that we are not sure about here?

You can then encourage students to think about how CCs played a role in their investigation. There are at least two CCs that students need to determine the identities of a set of unknown minerals: (a) Patterns and (b) Structure and Function (see Appendix 2 [p. 569] for a brief description of these CCs). To help students reflect on what they know about these CCs, we recommend asking them the following questions:

1. Why do scientists look for and attempt to explain patterns in nature?

2. What patterns did you identify and use during your investigation? Why was that useful?

3. The way an object is shaped or structured determines many of its properties and how it functions. Why is it useful to think about the relationship between structure and function during an investigation?

4. How is the structure of a mineral related to its physical and chemical properties?

You can then encourage the students to think about how they used all these different concepts to help answer the guiding question and why it is important to use these ideas to help justify their evidence for their final arguments. Be sure to remind your students to explain why they included the evidence in their arguments and make the assumptions underlying their analysis and interpretation of the data explicit in order to provide an adequate justification of their evidence.

Reflecting on Ways to Design Better Investigations

It is important for students to reflect on the strengths and weaknesses of the investigation they designed during the explicit and reflective discussion. Students should therefore be encouraged to discuss ways to eliminate potential flaws, measurement errors, or sources of uncertainty in their investigations. To help students be more reflective about the design of their investigation and what they can do to make their investigations more rigorous in the future, you can ask the following questions:

1. What were some of the strengths of the way you planned and carried out your investigation? In other words, what made it scientific?

2. What were some of the weaknesses of the way you planned and carried out your investigation? In other words, what made it less scientific?

3. What rules can we make, as a class, to ensure that our next investigation is more scientific?

Reflecting on the Nature of Scientific Knowledge and Scientific Inquiry

This investigation can be used to illustrate two important concepts related to the nature of scientific knowledge and the nature of scientific inquiry: (a) the difference between data and evidence in science and (b) how scientists use different methods to answer different types of questions (see Appendix 2 [p. 569] for a brief description of these two concepts). Be sure to review these concepts during and at the end of the explicit and reflective discussion. To help students think about these concepts in relation to what they did during the lab, you can ask the following questions:

1. You had to talk about data and evidence during your investigation. Can you give me some examples of data and evidence from your investigation?

2. Can you work with your group to come up with a rule that you can use to decide if a piece of information is data or evidence? Be ready to share in a few minutes.

3. There is no universal step-by-step scientific method that all scientists follow. Why do you think there is no universal scientific method?

4. Think about what you did during this investigation. How would you describe the method you used to identify the set of unknown minerals? Why would you call it that?

You can also use presentation software or other techniques to encourage your students to think about these concepts. You can show examples of information from the investigation that are either data or evidence and ask students to classify each example and explain their thinking. You can also show one or more images of a "universal scientific method" that misrepresent the nature of scientific inquiry (see, e.g., *https://commons.wikimedia.org/wiki/File:The_Scientific_Method_as_an_Ongoing_Process.svg*) and ask students why each image is *not* a good representation of what scientists do to develop scientific knowledge. You can also ask students to suggest revisions to the image that would make it more consistent with the way scientists develop scientific knowledge.

Remind your students that, to be proficient in science, it is important that they understand what counts as scientific knowledge and how that knowledge develops over time.

Hints for Implementing the Lab

- We recommend including 6–10 samples in the set of known minerals and 4 samples in the set of unknown minerals in order to foster higher-quality argumentation during the lab. The more identities that student groups have to determine, the more opportunities there are for variation among groups that can lead to critical questioning and discussion during the argumentation session. However, if necessary for time or scheduling issues, the number of samples included in the set of known minerals and the set of unknown minerals can be decreased.

- Be sure to encourage groups to collect multiple types of data to develop a more complete set of evidence. Such work offers the opportunity to discuss the difference between data and evidence explicitly with students. Furthermore, using this type of analysis and comparison also provides opportunities to distinguish between observations (in this activity, the various physical or chemical property data collected) and inferences (the identity of the unknown is inferred through comparison of characteristics).

- We recommend that students use at least three different physical properties to identify the unknown samples, but you should not tell them which ones to use. The variation in approaches will foster better discussions during the argumentation session. We do recommend, however, that you insist that students be able to justify why they chose to uses a specific physical property over another.

- Do not tell students how to determine the volume of the samples; this is a methodological challenge associated with the lab and will provide an ideal opportunity to discuss measurement error and how choice of method can influence the accuracy of a measurement during the explicit and reflective discussion.

- Students often make mistakes during the data collection stage, but they should quickly realize these mistakes during the argumentation session. It will only take them a short period of time to re-collect data, and they should be allowed to do so. During the explicit and reflective discussion, students will also have the opportunity to reflect on and identify ways to improve the way they design investigations. This also offers an opportunity to discuss what scientists do when they realize that a mistake is made during a study.

Connections to Standards

Table 13.3 highlights how the investigation can be used to address specific (a) performance expectations from the *NGSS* and (b) *Common Core State Standards* in English language arts (*CCSS ELA*).

TABLE 13.3 _____

Lab 13 alignment with standards

NGSS performance expectation	Earth's systems • MS-ESS3-1: Construct a scientific explanation based on evidence for how the uneven distributions of Earth's mineral, energy, and groundwater resources are the result of past and current geoscience processes.
CCSS ELA—Reading in Science and Technical Subjects	Key ideas and details • CCSS.ELA-LITERACY.RST.6-8.1: Cite specific textual evidence to support analysis of science and technical texts. • CCSS.ELA-LITERACY.RST.6-8.2: Determine the central ideas or conclusions of a text; provide an accurate summary of the text distinct from prior knowledge or opinions. Craft and structure • CCSS.ELA-LITERACY.RST.6-8.4: Determine the meaning of symbols, key terms, and other domain-specific words and phrases as they are used in a specific scientific or technical context relevant to *grade 6–8 texts and topics*.

Continued

TABLE 13.3 (*continued*)

CCSS ELA—**Reading in Science and Technical Subjects** (*continued*)	Craft and structure (*continued*) • CCSS.ELA-LITERACY.RST.6-8.5: Analyze the structure an author uses to organize a text, including how the major sections contribute to the whole and to an understanding of the topic. • CCSS.ELA-LITERACY.RST.6-8.6: Analyze the author's purpose in providing an explanation, describing a procedure, or discussing an experiment in a text. Integration of knowledge and ideas • CCSS.ELA-LITERACY.RST.6-8.7: Integrate quantitative or technical information expressed in words in a text with a version of that information expressed visually (e.g., in a flowchart, diagram, model, graph, or table). • CCSS.ELA-LITERACY.RST.6-8.8: Distinguish among facts, reasoned judgment based on research findings, and speculation in a text. • CCSS.ELA-LITERACY.RST.6-8.9: Compare and contrast the information gained from experiments, simulations, video, or multimedia sources with that gained from reading a text on the same topic.
CCSS ELA—**Writing in Science and Technical Subjects**	Text types and purposes • CCSS.ELA-LITERACY.WHST.6-8.1: Write arguments focused on *discipline-specific content*. • CCSS.ELA-LITERACY.WHST.6-8.2: Write informative or explanatory texts, including the narration of historical events, scientific procedures/experiments, or technical processes. Production and distribution of writing • CCSS.ELA-LITERACY.WHST.6-8.4: Produce clear and coherent writing in which the development, organization, and style are appropriate to task, purpose, and audience. • CCSS.ELA-LITERACY.WHST.6-8.5: With some guidance and support from peers and adults, develop and strengthen writing as needed by planning, revising, editing, rewriting, or trying a new approach, focusing on how well purpose and audience have been addressed. • CCSS.ELA-LITERACY.WHST.6-8.6: Use technology, including the internet, to produce and publish writing and present the relationships between information and ideas clearly and efficiently. Range of writing • CCSS.ELA-LITERACY.WHST.6-8.10: Write routinely over extended time frames (time for reflection and revision) and shorter time frames (a single sitting or a day or two) for a range of discipline-specific tasks, purposes, and audiences.

Continued

TABLE 13.3 (*continued*)

***CCSS ELA*—Speaking and Listening**	Comprehension and collaboration • CCSS.ELA-LITERACY.SL.6-8.1: Engage effectively in a range of collaborative discussions (one-on-one, in groups, and teacher-led) with diverse partners on grade 6–8 topics, texts, and issues, building on others' ideas and expressing their own clearly. • CCSS.ELA-LITERACY.SL.6-8.2:* Interpret information presented in diverse media and formats (e.g., visually, quantitatively, orally) and explain how it contributes to a topic, text, or issue under study. • CCSS.ELA-LITERACY.SL.6-8.3:* Delineate a speaker's argument and specific claims, distinguishing claims that are supported by reasons and evidence from claims that are not. Presentation of knowledge and ideas • CCSS.ELA-LITERACY.SL.6-8.4:* Present claims and findings, sequencing ideas logically and using pertinent descriptions, facts, and details to accentuate main ideas or themes; use appropriate eye contact, adequate volume, and clear pronunciation. • CCSS.ELA-LITERACY.SL.6-8.5:* Include multimedia components (e.g., graphics, images, music, sound) and visual displays in presentations to clarify information. • CCSS.ELA-LITERACY.SL.6-8.6: Adapt speech to a variety of contexts and tasks, demonstrating command of formal English when indicated or appropriate.

* Only the standard for grade 6 is provided because the standards for grades 7 and 8 are similar. Please see *www.corestandards.org/ELA-Literacy/SL* for the exact wording of the standards for grades 7 and 8.

Lab Handout

Lab 13. Characteristics of Minerals: What Are the Identities of the Unknown Minerals?

Introduction

Rocks are made up of different types of minerals or other pieces of rock, which are made of minerals. Granite (Figure L13.1) and marble (Figure L13.2) are examples of different kinds of rock. Earth scientists group rocks into one of three categories:

- *Sedimentary* rocks are formed at the Earth's surface by the accumulation and cementation of fragments of sediments. Sandstone is an example of a sedimentary rock.

- *Igneous* rocks form through the cooling and solidification of magma or lava. Granite is an example of an igneous rock.

- *Metamorphic* rocks are produced when existing rocks are subjected to extreme temperature and pressure. Marble is an example of a metamorphic rock.

FIGURE L13.1 _____
A granite outcrop at Logan Rock, Cornwall, England

FIGURE L13.2 _____
Marble at a quarry in Carrara, Italy

Earth scientists use the mineral composition of a rock to classify it. Granite, for example, is made up minerals such as quartz, feldspar, and biotite; marble is made up of minerals called dolomite and calcite. Earth scientists must be able to determine the various types of minerals that are in a rock in order to identify it. Every mineral has a unique chemical composition. Dolomite (see Figure L13.3, p. 322), for example, has the chemical composition of $CaMg(CO_3)_2$, whereas quartz (see Figure L13.4, p. 322) has a chemical composition of SiO_2. The unique chemical composition of a mineral gives it a specific combination of chemical and physical properties. Earth scientists use these chemical and physical properties to identify a mineral.

Dolomite

Quartz

Chemical properties (see Figure L13.5) describe how a mineral interacts with other types of matter. Dolomite, for example, reacts with hydrochloric acid but quartz does not. *Physical properties* are descriptive characteristics of a mineral. Examples of physical properties include color, density, streak (whether the mineral streaks on a streak plate and the color of the powder), hardness (whether the mineral can scratch something with a known hardness, like glass or a nail), smell, how the mineral breaks (*cleavage* is when a mineral breaks evenly along a flat surface; *fracture* is when a mineral breaks apart roughly), and luster (whether the material appears metallic or nonmetallic). Some minerals will even attract magnets.

FIGURE L13.5 _____

How Earth scientists distinguish between different minerals

Minerals
Inorganic solids

Composition
Chemical makeup

Si O
O

Properties
Attributes or qualities
of minerals

Chemical
How a mineral will interact
with other types of matter

Physical
Characteristcs or
traits of a mineral

It is often challenging to determine the identity of an unknown mineral based on its chemical and physical properties. For example, if an Earth scientist has only a small amount of a mineral, he or she may not be able to conduct all the different types of tests that are needed because some tests may change the characteristics of the mineral during the process (such as when dolomite is mixed with an acid). It is also difficult to determine many of the physical properties of the sample, such as its density or its luster, when there is only a small amount of the substance, because taking measurements is harder. To complicate matters further, an unknown mineral may have an irregular shape, which can make it difficult to accurately measure its volume. Without knowing the mass and the volume of a substance, it is impossible to calculate its density. In this investigation, you will have an opportunity to learn about some of the challenges Earth scientists face when they need to identify an unknown mineral based on its chemical and physical properties and why it is important to make accurate measurements inside the laboratory.

Your Task

You will be given a set of known minerals. You will then document, measure, or calculate at least three different chemical or physical properties for each mineral. When you are done, you will return the known minerals to your teacher, who will then give you a set of unknown minerals. The set of unknowns will include samples of minerals that you tested. Your goal is to use what you know about the physical and chemical properties of matter, proportional relationships, and patterns to design and carry out an investigation that will enable you to collect the data you need to determine the identity of the unknown minerals.

The guiding question of this investigation is, *What are the identities of the unknown minerals?*

Materials

You may use any of the following materials during your investigation:

Consumables
- Water (in squirt bottles)
- Vinegar
- Set of known minerals
- Set of unknown minerals

Equipment
- Safety glasses or goggles (required)
- Chemical-resistant apron (required)
- Gloves (required)
- Electronic or triple beam balance
- Magnifying glass
- Magnet
- Beaker (250 ml)
- Beaker (400 ml)

- Graduated cylinder (100 ml)
- Overflow can
- Pipette
- Ruler
- Streak plate
- Small piece of glass
- Penny
- Nail

Safety Precautions

Follow all normal lab safety rules. In addition, take the following safety precautions:

- Follow safety precautions noted on safety data sheets for hazardous chemicals.

- Wear sanitized indirectly vented chemical-splash goggles and chemical-resistant, nonlatex aprons and gloves throughout the entire investigation (which includes setup and cleanup).

- Report and clean up any spills on the floor immediately to avoid a slip or fall hazard.

- Handle all glassware with care.

- When using the glass or streak plate, do not pick the plate up with your hands. Place it flat on the table. Otherwise, it may cut your hand.

- Handle nails with care. Nail ends are sharp and can cut or puncture skin.

- Do not place any materials in or around your mouth.

- Handle all glassware with care.

- Wash hands with soap and water when done collecting the data and after completing the lab.

Investigation Proposal Required?   ☐ Yes   ☐ No

Getting Started

To answer the guiding question, you will need to make several systematic observations of the known and unknown minerals. To accomplish this task, you must determine what type of data you need to collect, how you will collect it, and how you will analyze it.

To determine *what type of data you need to collect*, think about the following questions:

- Which three properties will you focus on as you make your systematic observations? The properties you choose to focus on can be chemical ones (reactions with other substances) or physical ones (e.g., color, density, hardness, streak).

- What information do you need to determine or calculate each of the chemical or physical properties?

- How will you determine if the physical properties of the various objects are the same or different?

To determine *how you will collect the data*, think about the following questions:

- What equipment will you need to collect the data you need?

- How will you make sure that your data are of high quality (i.e., how will you reduce error)?

- How will you keep track of and organize the data you collect?

To determine *how you will analyze the data*, think about the following questions:

- How might the unique chemical composition of a mineral (structure) be related to its unique chemical and physical properties (function)?

- What types of patterns might you look for as you analyze your data?

- What type of calculations will you need to make?

- What type of table or graph could you create to help make sense of your data?

Connections to the Nature of Scientific Knowledge and Scientific Inquiry

As you work through your investigation, be sure to think about

- the difference between data and evidence in science, and
- how scientists use different types of methods to answer different types of questions.

Initial Argument

Once your group has finished collecting and analyzing your data, your group will need to develop an initial argument. Your initial argument needs to include a claim, evidence to support your claim, and a justification of the evidence. The *claim* is your group's answer to the guiding question. The *evidence* is an analysis and interpretation of your data. Finally, the *justification* of the evidence is why your group thinks the evidence matters. The justification of the evidence is important because scientists can use different kinds of evidence to support their claims. Your group will create your initial argument on a whiteboard. Your whiteboard should include all the information shown in Figure L13.6.

FIGURE L13.6 _____

Argument presentation on a whiteboard

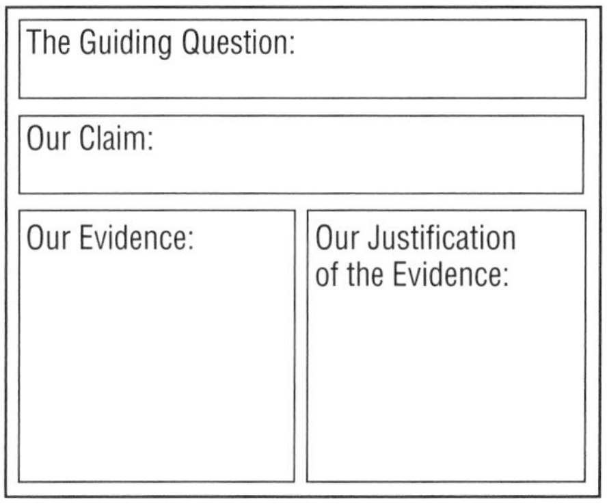

The Guiding Question:	
Our Claim:	
Our Evidence:	Our Justification of the Evidence:

Argumentation Session

The argumentation session allows all of the groups to share their arguments. One or two members of each group will stay at the lab station to share that group's argument, while the other members of the group go to the other lab stations to listen to and critique the other arguments. This is similar to what scientists do when they propose, support, evaluate, and refine new ideas during a poster session at a conference. If you are presenting your group's argument, your goal is to share your ideas and answer questions. You should also keep a record of the critiques and suggestions made by your classmates so you can use this feedback to make your initial argument stronger. You can keep track of specific critiques and suggestions for improvement that your classmates mention in the space below.

Critiques of our initial argument and suggestions for improvement:

If you are critiquing your classmates' arguments, your goal is to look for mistakes in their arguments and offer suggestions for improvement so these mistakes can be fixed. You should look for ways to make your initial argument stronger by looking for things that the other groups did well. You can keep track of interesting ideas that you see and hear during the argumentation in the space below. You can also use this space to keep track of any questions that you will need to discuss with your team.

Interesting ideas from other groups or questions to take back to my group:

Once the argumentation session is complete, you will have a chance to meet with your group and revise your initial argument. Your group might need to gather more data or design a way to test one or more alternative claims as part of this process. Remember, your goal at this stage of the investigation is to develop the best argument possible.

Report

Once you have completed your research, you will need to prepare an *investigation report* that consists of three sections. Each section should provide an answer for the following questions:

1. What question were you trying to answer and why?

2. What did you do to answer your question and why?

3. What is your argument?

Your report should answer these questions in two pages or less. You should write your report using a word processing application (such as Word, Pages, or Google Docs), if possible, to make it easier for you to edit and revise it later. You should embed any diagrams, figures, or tables into the document. Be sure to write in a persuasive style; you are trying to convince others that your claim is acceptable or valid.

Checkout Questions

Lab 13. Characteristics of Minerals: What Are the Identities of the Unknown Minerals?

1. Why is it possible to use physical properties to identify minerals found in rocks?

2. An Earth scientist has a rock she wants to identify and a list of known mineral characteristics. The rock is black with gold metallic specks. Depending on where she runs the streak test, streaks appear black. Its Mohs hardness is 6.2, and its density is 5.2. A table of known mineral characteristics is below.

Mineral	Color	Streak	Luster	Mohs Hardness	Density
Halite	Colorless or white when pure; impurities produce any color but usually yellow, gray, black, brown, red	White	Vitreous	2.5	2
Magnetite	Black to silver gray	Black	Metallic to submetallic	5–6.5	5.2
Muscovite	Thick specimens often appear to be black, brown, or silver in color; however, when split into thin sheets muscovite is colorless, sometimes with a tint of brown, yellow, green, or rose	White, often sheds tiny flakes	Pearly to vitreous	2.5–3	2.8–2.9
Pyrite	Brass-yellow	Greenish black to brownish black	Metallic	6–6.5	4.9–5.2

a. Which minerals does the rock likely contain?

b. How do you know?

3. There is no universal step-by step scientific method that all scientists follow.

a. I agree with this statement.

b. I disagree with this statement.

Explain your answer, using an example from your investigation about characteristics of minerals.

4. "The rock's color is red" is an example of data.

    a. I agree with this statement.

    b. I disagree with this statement.

Explain your answer, using an example from your investigation about characteristics of minerals.

5. Scientists often need to look for patterns that occur in the data they collect and analyze. Explain why identifying patterns are important, using an example from your investigation about characteristics of minerals.

6. In nature, the way something is structured often determines its function or places limits on what it can or cannot do. Explain why it is important to keep in mind the relationship between structure and function when attempting to collect or analyze data, using an example from your investigation about the characteristics of minerals.

Teacher Notes

Lab 14. Distribution of Natural Resources: Which Proposal for a New Copper Mine Maximizes the Potential Benefits While Minimizing the Potential Costs?

Purpose

The purpose of this lab is for students to *apply* what they know about the disciplinary core ideas (DCIs) of (a) Earth Materials and Systems, (b) Natural Resources, and (c) Human Impacts on Earth Systems to evaluate competing plans for a new copper mine based on cost-benefit ratio. In addition, students have an opportunity to learn about the crosscutting concepts (CCs) of (a) Patterns and (b) Cause and Effect: Mechanism and Explanation. During the explicit and reflective discussion, students will also learn about (a) how scientific knowledge changes over time and (b) the types of questions that scientists can investigate.

Important Earth and Space Science Content

A *natural resource* is a material or a substance that occurs in nature that people can use for economic gain. People depend on Earth's land, oceans, atmosphere, and biosphere for many different types of natural resources such as minerals, water, oil, forests, and fertile land. The amount of each type of natural resource on Earth, however, is limited, and many of these resources are not renewable or replaceable over the span of a human lifetime. In addition, natural resources are not evenly distributed on the planet because these resources were produced by different types of geologic or biological processes in the past. Many people, as a result, spend a great deal of time and money attempting to locate sites that contain a specific natural resource.

Once a source of a specific natural resource is located, people must find ways to extract and refine the resource before they can benefit from it. Every method used to extract or refine a resource, however, has economic, social, environmental, and geopolitical costs or risks associated with it that may or may not offset the potential benefits. New technologies and social regulations can change the balance of these factors, but the sustainability of human societies and the biodiversity that supports them require responsible management of natural resources.

Copper is an example of a natural resource. It is a valuable metal because of its unique physical properties. Copper is easily stretched, molded, and shaped; is resistant to corrosion; and is a good conductor of heat and electricity. People use it to make wires, pipes, appliances, consumer electronics, and automobiles. The United States used about 1.8 million metric tons of copper in 2016 (USGS 2017). Most of this copper was used for building construction, followed by transportation equipment, electric and electronic products, consumer and general products, and industrial machinery and equipment. The United States produced about 1.41 million metric tons of copper in 2016 (USGS 2017). In other words, the United States used more copper than it was able to produce by mining. Ninety-nine percent of all copper produced in the United States came from seven states: Arizona, New Mexico, Utah, Nevada, Montana, Michigan, and Missouri.

FIGURE 14.1

A map of copper in sediment-hosted and porphyry tracts around the world

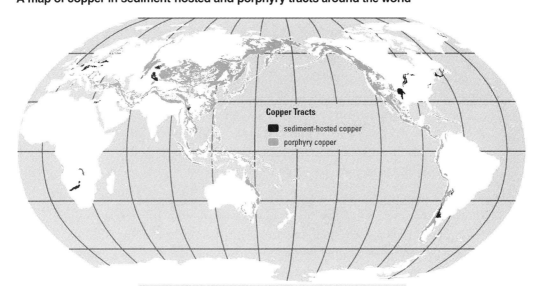

Copper, like most natural resources, is only located in a few sites around the world. These sites, known as *deposits,* are located within tracts (see Figure 14.1). A *tract* is a geographic area where the geology is consistent with the formation of a specific type of copper deposit (USGS 2014). Earth scientists classify copper deposits based on the geologic process that created them. *Porphyry copper deposits,* which contain about 60% of the world's copper, form when a large mass of molten rock cools and solidifies deep in the Earth's crust. These deposits are large and often contain between 100 million and 5 billion metric tons of copper-bearing rock called ore. The ore in these deposits, however, only contains 0.2%–1% copper by weight. Porphyry copper deposits are often located in East Asia and in the mountainous regions of western North and South America. *Sediment-hosted copper deposits,* in contrast, form due to the deposition and subsequent cementation of sediments. Sediment-hosted copper deposits contain about 20% of the world's copper. These deposits are smaller than porphyry deposits and usually only contain 1 million to 100 million metric tons of ore. The ore in these deposits, however, often contains 2%–6% copper by weight. People have found sediment-hosted copper deposits in Zambia, Zaire, Europe, central Asia, and the north central United States.

LAB 14

Copper deposits are often located 1–4 km below Earth's surface (Berger et al. 2008). People therefore rely on mining companies to access and refine the copper in a deposit. A mining project often lasts for decades and includes three major tasks ("Copper" 2017):

1. *Removing the ore from the deposit.* Before miners can access the ore in a deposit, they must first remove the rock and soil that covers it. Miners call this rock and soil overburden. A deposit that is deeper costs more in terms of time and resources to access than a deposit that is near the surface. Mining companies tend to use underground or open-pit mining methods to remove overburden and the ore.

2. *Separating the copper-based minerals from the other components of the ore.* This task is called milling. It is important because ore contains very little copper by weight. To separate the copper-based minerals from the other components of the ore, mining companies must first crush and grind the ore into a powder. Mining companies then use a method called froth flotation to separate copper-containing minerals from minerals that do not contain copper. The product of the froth flotation process is a solution called copper concentrate, which is about 25%–35% copper by weight.

3. *Making a pure copper metal that is usable by manufacturers.* Mining companies accomplish this task by first mixing the copper concentrate with a silica material called a flux and then heating the mixture to 1200°C. This process, which is called smelting, removes many impurities and produces a liquid of copper sulfide that is about 60% copper by weight. This liquid is called copper matte. The mining company then heats the copper matte to remove the sulfur and produce a material called blister copper, which is about 99% copper by weight. Next, the blister copper is refined and molded into sheets so it can go through a chemical process called electrolysis to remove the few remaining impurities in the metal. After the electrolysis process is complete, the resulting copper metal is about 99.95% pure. The mining company can then sell the pure copper by the pound to various manufacturing companies that make pipes, wire, and other goods.

Copper mining operations produce waste materials. There are five different types of mine waste material, which vary in their physical and chemical composition, their potential for environmental contamination, and how mine companies manage them over time:

- *Overburden,* which is the excess soil and rock that is removed to gain access to the mineral-rich ore. Overburden consists of acid-generating and non-acid-generating rock. Water that flows over or through acid-generating rock becomes more acidic. This acidic water can enter a nearby stream, river, or lake and cause environmental contamination.

- *Tailings,* which is a water-based slurry that is produced during the milling process. It consists of finely ground rock, mineral waste products, and toxic processing

334

chemicals. Mine companies typically store tailings in large artificial ponds. However, some companies are considering dumping tailings into oceans as an alternative method for disposal.

- *Slag,* which is a by-product of smelting. Slag consists of iron oxide and silicon dioxide. It is nontoxic, so mining companies can collect it and then use it to make concrete, roads, the grit used in sandblasters, and blocks for buildings.

- *Atmospheric emissions,* which include the dust that is produced when extracting and crushing the ore and the sulfur oxides that are generated as part of the smelting process. Atmospheric emissions vary in their potential for environmental contamination.

- *Mine water,* which is used to extract, crush, and grind ore and to control dust. It often contains dissolved minerals and metals or trace amounts of processing chemicals. This water can enter nearby streams, rivers, and lakes and cause environmental contamination.

In this lab investigation, students will evaluate several different proposals for a new copper mine. When evaluating design solutions or proposals such as these, it is important to take into account a range of constraints, including cost, safety, and aesthetics, and to consider the potential social, cultural, and environmental impacts as well as the potential benefits of the proposal. Students will place different values on the criteria that they identify and use to evaluate each proposal. For example, some students may place a higher value on the preservation of habitats than they do on the amount of copper that will be removed from a deposit, while others may place a higher value on extensive waste management plans for the mining project. Furthermore, there may be criteria that students identify and use that are not included in the Lab Handout, leading these students to look up information that other students will not. Therefore, there is no one correct answer to the guiding question of this investigation. The claims made by the students will differ based on the criteria they decide to use to conduct their benefit-and-cost analysis.

Timeline

The instructional time needed to complete this lab investigation is 220–280 minutes. Appendix 3 (p. 573) provides options for implementing this lab investigation over several class periods. Option A (280 minutes) should be used if students are unfamiliar with scientific writing, because this option provides extra instructional time for scaffolding the writing process. You can scaffold the writing process by modeling, providing examples, and providing hints as students write each section of the report. Option B (220 minutes) should be used if students are familiar with scientific writing and have developed the skills needed to write an investigation report on their own. In option B, students complete stage 6 (writing the investigation report) and stage 8 (revising the investigation report) as homework.

Materials and Preparation

The materials needed to implement this investigation are listed in Table 14.1. Students will need to use several different online resources during the investigation:

- Google Earth is available at *www.google.com/earth,* and Google Maps is available at *https://maps.google.com.* Students can use either one of these interactive maps to determine the location of a proposed mining project. They can also use these maps to learn more about the environment (e.g., rivers, lakes, forests) near the proposed mine site.

- The U.S. Geological Survey (USGS) Global Assessment of Undiscovered Copper Resources is available at *https://mrdata.usgs.gov/sir20105090z.* This interactive map and database allows students to learn more about the copper deposit at the site of a proposed mining operation.

- The USGS Mineral Commodity Summaries is available at *https://minerals.usgs. gov/minerals/pubs/mcs.* This online publication provides information about copper production, copper use, and the current market price of copper in the United States.

- Surf Your Watershed is available at *https://cfpub.epa.gov/surf/locate/index.cfm.* This Environmental Protection Agency (EPA) database provides information about the streams and lakes near a proposed mining operation. The database also includes recent findings from water quality monitoring tests.

- The U.S. Fish and Wildlife Service Environmental Conservation Online System (ECOS) is available at *https://ecos.fws.gov/ecp/report/table/critical-habitat. html.* This system includes an interactive map (click on "online mapper") that provides information about the location of protected habitats for threatened and endangered species throughout the United States.

You should access these websites and learn how the resources work before beginning the lab investigation. In addition, it is important to check if students can access and use the websites from a school computer or tablet, because some schools have set up firewalls and other restrictions on web browsing.

The copper mine proposals are in two files that can be downloaded from the book's Extras page at *www.nsta.org/adi-ess.* The Pebble Copper Mine Proposals file includes four different options (A–D) for a new mining project at the Pebble copper deposit in Alaska. The White Pine Copper Mine Proposals file includes four different options (A–D) for a new mining project at the White Pine copper deposit in Michigan. All of the mining proposals include the following information:

- the location of the mine,
- the mining method,

- the estimated life span of the mine,

- the estimated number of new jobs that will be created when the mine opens,

- the size of the site,

- the total amount of ore (rock with copper in it) to be removed from the site,

- the ore extraction rate,

- the amount of waste that will be produced,

- the waste management plan, and

- the expenses associated with operating and closing the mine.

You can have students evaluate just one set of proposals (e.g., options A–D for the Pebble Copper Mine) or more than one set (e.g., options A–D for Pebble Copper Mine and options A–D for the White Pine Copper Mine). To change the difficulty of the investigation based on the needs of your students, you can assign more or fewer sets of proposals for a new mining project.

TABLE 14.1

Materials list for Lab 14

Item	Quantity
Computer or tablet with internet access	1 per group
Pebble Copper Mine Proposals	1 per group
White Pine Copper Mine Proposals (optional)	1 per group
Whiteboard, 2' × 3'*	1 per group
Lab Handout	1 per student
Peer-review guide and instructor scoring rubric	1 per student
Checkout Questions	1 per student

* As an alternative, students can use computer and presentation software such as Microsoft PowerPoint or Apple Keynote to create their arguments.

Safety Precautions
Remind students to follow all normal lab safety rules.

Topics for the Explicit and Reflective Discussion
Reflecting on the Use of Core Ideas and Crosscutting Concepts During the Investigation
Teachers should begin the explicit and reflective discussion by asking students to discuss what they know about the DCIs they used during the investigation. The following are some important concepts related to the core idea of Earth's systems that students need

to be able to identify a location that that would be suitable to start a new copper mining project:

- People depend on Earth's land, oceans, atmosphere, and biosphere for many different natural resources.
- Natural resources are limited, and many are not renewable or replaceable over human lifetimes.
- Many natural resources are distributed unevenly around the planet as a result of past geologic processes.
- All forms of resource extraction have associated economic, social, environmental, and geopolitical costs and risks as well as benefits. New technologies and social regulations can change the balance of these factors.
- The sustainability of human societies and the biodiversity that supports them require responsible management of natural resources.
- When evaluating design solutions, it is important to take into account a range of constraints, including cost, safety, and aesthetics, and to consider the potential social, cultural, and environmental impacts.

To help students reflect on what they know about these concepts, we recommend showing them two or three images using presentation software that help illustrate these important ideas. You can then ask the students the following questions to encourage students to share how they are thinking about these important concepts:

1. What do we see going on in this image?

2. Does anyone have anything else to add?

3. What might be going on that we can't see?

4. What are some things that we are not sure about here?

You can then encourage students to think about how CCs played a role in their investigation. There are at least two CCs that students need to be able to identify a location that would be suitable to start a new copper mining project: (a) Patterns and (b) Cause and Effect: Mechanism and Explanation (see Appendix 2 [p. 569] for a brief description of these CCs). To help students reflect on what they know about these CCs, we recommend asking them the following questions:

1. Why do scientists look for and attempt to explain patterns in nature?

2. What patterns did you identify and use during your investigation? Why was that useful?

3. Why can cause-and-effect relationships be used to predict phenomena in natural systems?

4. What cause-and-effect relationships did you use during your investigation to make predictions? Why was that useful to do?

You can then encourage students to think about how they used all these different concepts to help answer the guiding question and why it is important to use these ideas to help justify their evidence for their final arguments. Be sure to remind your students to explain why they included the evidence in their arguments and make the assumptions underlying their analysis and interpretation of the data explicit in order to provide an adequate justification of their evidence.

Reflecting on Ways to Design Better Investigations

It is important for students to reflect on the strengths and weaknesses of the investigation they designed during the explicit and reflective discussion. Students should therefore be encouraged to discuss ways to eliminate potential flaws, measurement errors, or sources of uncertainty in their investigations. To help students be more reflective about the design of their investigation and what they can do to make them more rigorous in the future, you can ask the following questions:

1. What were some of the strengths of the way you planned and carried out your investigation? In other words, what made it scientific?

2. What were some of the weaknesses of the way you planned and carried out your investigation? In other words, what made it less scientific?

3. What rules can we make, as a class, to ensure that our next investigation is more scientific?

Reflecting on the Nature of Scientific Knowledge and Scientific Inquiry

This investigation can be used to illustrate two important concepts related to the nature of scientific knowledge and the nature of scientific inquiry: (a) how scientific knowledge changes over time and (b) the types of questions that scientists can investigate (see Appendix 2 [p. 569] for a brief description of these two concepts). Be sure to review these concepts during and at the end of the explicit and reflective discussion. To help students think about these concepts in relation to what they did during the lab, you can ask the following questions:

1. Scientific knowledge can and does change over time. Can you tell me why it changes?

2. Can you work with your group to come up with some examples of how scientific knowledge related to our understanding of natural resources has changed over time? Be ready to share in a few minutes.

3. Not all questions can be answered by science. Can you give me some examples of questions related to this investigation that can and cannot be answered by science?

4. Can you work with your group to come up with a rule that you can use to decide if a question can be answered by science or not? Be ready to share in a few minutes.

You can also use presentation software or other techniques to encourage your students to think about these concepts. You can show examples of how our thinking about natural resources has changed over time and ask students to discuss what they think led to those changes. You can also show one or more examples of questions that can be answered by science (e.g., What are the sources of greenhouse gases emissions? How do increased greenhouse gases in the atmosphere affect average surface temperature?) and cannot be answered by science (e.g., Should we increase taxes to help reduce carbon dioxide emissions? Who should be required to cut their greenhouse gas emissions?) and then ask students why each example is or is not a question that can be answered by science.

Remind your students that, to be proficient in science, it is important that they understand what counts as scientific knowledge and how that knowledge develops over time.

Hints for Implementing the Lab

- Learn how to use the online databases and interactive maps before the lab begins. It is important for you to know what information is available in these resources and how to use them so you can help students when they get stuck or confused. In addition, be sure to familiarize yourself with the data included in each resource.

- Allow the students to play with the online databases and online interactive maps as part of the tool talk before they begin to design their investigation. This gives students a chance to see what they can and cannot do with these resources.

- A group of three students per computer or tablet tends to work well. If more computers or tablets are available, you can also have each student work on a different computer or tablet so that individual members of a group can look up different information online and then share that information with the rest of the group. One student in a group, for example, could use the USGS Global Assessment of Undiscovered Copper Resources interactive map and database to look up the amount of copper in the deposit, while another student in the group uses the EPA Surf Your Watershed database to learn more about the watershed and a third student uses the U.S. Fish and Wildlife Service ECOS website to determine if there are endangered species in the area. This will speed up the data collection stage of the investigation. If you choose to use this option, make sure that the students are communicating with each other as they use these online resources during the investigation so that it continues to be a group activity rather than becoming an individual activity.

- The USGS Global Assessment of Undiscovered Copper Resources uses a unit called Megatonnes (Mt) when describing the size of a copper deposit (1 Mt = 1 million metric tons = 1,000,000,000 kg).

- Be sure that students record actual values rather than just attempting to describe what they see on the computer screen (e.g., there is lots of copper, there are endangered species that live in that area).

- Encourage students to use a decision matrix to determine which mine proposal is the best option given the potential benefits and costs associated with opening, operating, and closing a mine. A decision matrix includes rows for each proposal and columns that provide different evaluation criteria; see Table 14.2 (p. 342) for an example of a decision matrix. Evaluation criteria might include such things as potential value of extracted copper, amount of waste produced, waste management plan, and potential habitat loss. Students will need to determine which evaluation criteria to use based on what they know about the uneven distribution of natural resources and the economic, social, and environmental costs that are often associated with accessing natural resources. Once they have determined the evaluation criteria that they will use, they should rank order each proposal on each criterion. Direct students to give the best option the highest number (which is a 4 if there are four different mining proposals) and the worst option a 1. Students can then total the rankings across the different evaluation criteria, and the proposal with highest overall score is the best one.

- The best way to help students learn how to use a decision matrix is to provide "just-in-time" instruction. In other words, wait for students to need to determine how to evaluate the various proposals and then give a brief mini-lesson on how to set up and use a decision matrix. They will be much more interested in learning how to use a decision matrix if they know it will help solve a problem they are having or it will allow them to accomplish one of their goals.

- This is a good lab for students to make mistakes as they attempt to analyze the data. Students will quickly figure out what they did wrong during the argumentation session, and it will only take them a short period of time to reanalyze it data. It will also create an opportunity for students to reflect on and identify different ways to analyze data.

- This lab also provides an excellent opportunity to discuss how scientists must make choices about which data to use and how to analyze the data they have. Be sure to use this activity as a concrete example during the explicit and reflective discussion.

LAB 14

TABLE 14.2

An example of a decision matrix

Proposal	Evaluation criteria					Overall score
	Value of extracted copper	Amount of waste produced	Impact on local habitats	Waste management plan	Cost to operate	
Pebble A	4	1	1	4	1	11
Pebble B	3	4	4	1	3	15
Pebble C	2	2	3	2	2	11
Pebble D	1	3	2	3	4	13

Connections to Standards

Table 14.3 highlights how the investigation can be used to address specific (a) performance expectations from the *NGSS* and (b) *Common Core State Standards* in English language arts (*CCSS ELA*).

TABLE 14.3

Lab 14 alignment with standards

NGSS performance expectations	Earth's systems; Human impact • MS-ESS3-1: Construct a scientific explanation based on evidence for how the uneven distributions of Earth's mineral, energy, and groundwater resources are the result of past and current geoscience processes. • HS-ESS3-1: Construct an explanation based on evidence for how the availability of natural resources, occurrence of natural hazards, and changes in climate have influenced human activity • HS-ESS3-2: Evaluate competing design solutions for developing, managing, and utilizing energy and mineral resources based on cost-benefit ratios
CCSS ELA—Reading in Science and Technical Subjects	Key ideas and details • CCSS.ELA-LITERACY.RST.6-8.1: Cite specific textual evidence to support analysis of science and technical texts. • CCSS.ELA-LITERACY.RST.6-8.2: Determine the central ideas or conclusions of a text; provide an accurate summary of the text distinct from prior knowledge or opinions. • CCSS.ELA-LITERACY.RST.9-10.1: Cite specific textual evidence to support analysis of science and technical texts, attending to the precise details of explanations or descriptions.

Continued

TABLE 14.3 (*continued*)

CCSS ELA—Reading in Science and Technical Subjects (*continued*)	Key ideas and details (*continued*) • CCSS.ELA-LITERACY.RST.9-10.2: Determine the central ideas or conclusions of a text; trace the text's explanation or depiction of a complex process, phenomenon, or concept; provide an accurate summary of the text. • CCSS.ELA-LITERACY.RST.9-10.3: Follow precisely a complex multistep procedure when carrying out experiments, taking measurements, or performing technical tasks, attending to special cases or exceptions defined in the text. Craft and structure • CCSS.ELA-LITERACY.RST.6-8.4: Determine the meaning of symbols, key terms, and other domain-specific words and phrases as they are used in a specific scientific or technical context relevant to *grade 6–8 texts and topics*. • CCSS.ELA-LITERACY.RST.6-8.5: Analyze the structure an author uses to organize a text, including how the major sections contribute to the whole and to an understanding of the topic. • CCSS.ELA-LITERACY.RST.6-8.6: Analyze the author's purpose in providing an explanation, describing a procedure, or discussing an experiment in a text. • CCSS.ELA-LITERACY.RST.9-10.4: Determine the meaning of symbols, key terms, and other domain-specific words and phrases as they are used in a specific scientific or technical context relevant to *grade 9–10 texts and topics*. • CCSS.ELA-LITERACY.RST.9-10.5: Analyze the structure of the relationships among concepts in a text, including relationships among key terms (e.g., *force, friction, reaction force, energy*). • CCSS.ELA-LITERACY.RST.9-10.6: Analyze the author's purpose in providing an explanation, describing a procedure, or discussing an experiment in a text, defining the question the author seeks to address. Integration of knowledge and ideas • CCSS.ELA-LITERACY.RST.6-8.7: Integrate quantitative or technical information expressed in words in a text with a version of that information expressed visually (e.g., in a flowchart, diagram, model, graph, or table). • CCSS.ELA-LITERACY.RST.6-8.8: Distinguish among facts, reasoned judgment based on research findings, and speculation in a text. • CCSS.ELA-LITERACY.RST.6-8.9: Compare and contrast the information gained from experiments, simulations, video, or multimedia sources with that gained from reading a text on the same topic.

Continued

TABLE 14.3 (*continued*)

CCSS ELA—Reading in Science and Technical Subjects (*continued*)	Integration of knowledge and ideas (*continued*) • CCSS.ELA-LITERACY.RST.9-10.7: Translate quantitative or technical information expressed in words in a text into visual form (e.g., a table or chart) and translate information expressed visually or mathematically (e.g., in an equation) into words. • CCSS.ELA-LITERACY.RST.9-10.8: Assess the extent to which the reasoning and evidence in a text support the author's claim or a recommendation for solving a scientific or technical problem. • CCSS.ELA-LITERACY.RST.9-10.9: Compare and contrast findings presented in a text to those from other sources (including their own experiments), noting when the findings support or contradict previous explanations or accounts.
CCSS ELA—Writing in Science and Technical Subjects	Text types and purposes • CCSS.ELA-LITERACY.WHST.6-10.1: Write arguments focused on *discipline-specific content*. • CCSS.ELA-LITERACY.WHST.6-10.2: Write informative or explanatory texts, including the narration of historical events, scientific procedures/experiments, or technical processes. Production and distribution of writing • CCSS.ELA-LITERACY.WHST.6-10.4: Produce clear and coherent writing in which the development, organization, and style are appropriate to task, purpose, and audience. • CCSS.ELA-LITERACY.WHST.6-8.5: With some guidance and support from peers and adults, develop and strengthen writing as needed by planning, revising, editing, rewriting, or trying a new approach, focusing on how well purpose and audience have been addressed. • CCSS.ELA-LITERACY.WHST.6-8.6: Use technology, including the internet, to produce and publish writing and present the relationships between information and ideas clearly and efficiently. • CCSS.ELA-LITERACY.WHST.9-10.5: Develop and strengthen writing as needed by planning, revising, editing, rewriting, or trying a new approach, focusing on addressing what is most significant for a specific purpose and audience. • CCSS.ELA-LITERACY.WHST.9-10.6: Use technology, including the internet, to produce, publish, and update individual or shared writing products, taking advantage of technology's capacity to link to other information and to display information flexibly and dynamically.

Continued

TABLE 14.3 (*continued*)

CCSS ELA—**Writing in Science and Technical Subjects** (*continued*)	Range of writing • CCSS.ELA-LITERACY.WHST.6-10.10: Write routinely over extended time frames (time for reflection and revision) and shorter time frames (a single sitting or a day or two) for a range of discipline-specific tasks, purposes, and audiences.
CCSS ELA—**Speaking and Listening**	Comprehension and collaboration • CCSS.ELA-LITERACY.SL.6-8.1: Engage effectively in a range of collaborative discussions (one-on-one, in groups, and teacher-led) with diverse partners on grade 6–8 topics, texts, and issues, building on others' ideas and expressing their own clearly. • CCSS.ELA-LITERACY.SL.6-8.2:* Interpret information presented in diverse media and formats (e.g., visually, quantitatively, orally) and explain how it contributes to a topic, text, or issue under study. • CCSS.ELA-LITERACY.SL.6-8.3:* Delineate a speaker's argument and specific claims, distinguishing claims that are supported by reasons and evidence from claims that are not. • CCSS.ELA-LITERACY.SL.9-10.1: Initiate and participate effectively in a range of collaborative discussions (one-on-one, in groups, and teacher-led) with diverse partners on grade 9–10 topics, texts, and issues, building on others' ideas and expressing their own clearly and persuasively. • CCSS.ELA-LITERACY.SL.9-10.2: Integrate multiple sources of information presented in diverse media or formats (e.g., visually, quantitatively, orally) evaluating the credibility and accuracy of each source. • CCSS.ELA-LITERACY.SL.9-10.3: Evaluate a speaker's point of view, reasoning, and use of evidence and rhetoric, identifying any fallacious reasoning or exaggerated or distorted evidence. Presentation of knowledge and ideas • CCSS.ELA-LITERACY.SL.6-8.4:* Present claims and findings, sequencing ideas logically and using pertinent descriptions, facts, and details to accentuate main ideas or themes; use appropriate eye contact, adequate volume, and clear pronunciation. • CCSS.ELA-LITERACY.SL.6-8.5:* Include multimedia components (e.g., graphics, images, music, sound) and visual displays in presentations to clarify information. • CCSS.ELA-LITERACY.SL.6-8.6: Adapt speech to a variety of contexts and tasks, demonstrating command of formal English when indicated or appropriate.

Continued

TABLE 14.3 (*continued*)

***CCSS ELA*—Speaking and Listening** (*continued*)	Presentation of knowledge and ideas (*continued*) • CCSS.ELA-LITERACY.SL.9-10.4: Present information, findings, and supporting evidence clearly, concisely, and logically such that listeners can follow the line of reasoning and the organization, development, substance, and style are appropriate to purpose, audience, and task. • CCSS.ELA-LITERACY.SL.9-10.5: Make strategic use of digital media (e.g., textual, graphical, audio, visual, and interactive elements) in presentations to enhance understanding of findings, reasoning, and evidence and to add interest. • CCSS.ELA-LITERACY.SL.9-10.6: Adapt speech to a variety of contexts and tasks, demonstrating command of formal English when indicated or appropriate.

* Only the standard for grade 6 is provided because the standards for grades 7 and 8 are similar. Please see *www.corestandards.org/ELA-Literacy/SL* for the exact wording of the standards for grades 7 and 8.

References

Berger, B. R., R. A. Ayuso, J. C. Wynn, and R. R. Seal. 2008. Preliminary model of porphyry copper deposits. Open-File Report 2008–1321. U.S. Department of the Interior, U.S. Geological Survey. Available online at *https://pubs.usgs.gov/of/2008/1321/pdf/OF081321_508.pdf.*

Copper. 2017. *www.madehow.com/Volume-4/Copper.html.*

U.S. Geological Survey (USGS). 2014. Global mineral resource assessment: Estimate of undiscovered copper resources of the world, 2013. U.S. Department of the Interior, USGS. Available online at *https://pubs.usgs.gov/fs/2014/3004/pdf/fs2014-3004.pdf.*

U.S. Geological Survey (USGS). 2017. *Mineral commodity summaries.* Available online at *https://minerals.usgs.gov/minerals/pubs/commodity/copper/mcs-2017-coppe.pdf.*

Lab Handout

Lab 14. Distribution of Natural Resources: Which Proposal for a New Copper Mine Maximizes the Potential Benefits While Minimizing the Potential Costs?

Introduction

Copper is a useful metal because of its unique physical properties (see Figure L14.1). It is easily stretched, molded, and shaped; is resistant to corrosion; and is a good conductor of heat and electricity. People use copper to make the wires and pipes found in homes or businesses. In addition, manufacturing companies use copper to make the appliances (such as refrigerators and ovens) and consumer electronics (such as phones and computers) that we use in our homes every day. In addition, copper is an essential component in the motors, wiring, radiators, brakes, and bearings found in cars and trucks. The average car contains 1.5 kilometers (0.9 mile) of copper wire, and the total amount of copper in a vehicle can range from 20 kilograms (44 pounds) in a small car to 45 kilograms (99 pounds) in a large luxury or hybrid car (USGS 2009).

FIGURE L14.1

A sample of native copper extracted from a mine

Copper is a natural resource. Unfortunately, it is only located in a few specific sites around the world; these sites are called deposits. Earth scientists classify copper deposits based on the geologic process that created them. *Porphyry copper deposits*, which contain about 60% of the world's copper, form when a large mass of molten rock cools and solidifies deep in the Earth's crust. These deposits are large and often contain between 100 million to 5 billion metric tons of copper-bearing rock called ore. The ore in these deposits, however, only contains 0.2%–1% copper by weight. Porphyry copper deposits are often located in East Asia and in the mountainous regions of western North and South America. *Sediment-hosted copper deposits*, in contrast, form due to the deposition and subsequent cementation of sediments. Sediment-hosted copper deposits contain about 20% of the world's copper. These deposits are smaller than porphyry deposits and usually only contain 1 million to 100 million metric tons of ore. The ore in these deposits, however, often contains 2%–6% copper by weight. People have found sediment-hosted copper deposits in Zambia, Zaire, Europe, central Asia, and in the north central United States.

LAB 14

Most copper deposits, regardless of how they formed, have a definable boundary. An important component of the U.S. Geological Survey's (USGS) Mineral Resources Program is to identify the location and boundaries of untouched copper deposits, estimate the amount of copper that is likely in each one, and then share this information with scientists, other government agencies, and mining companies.

Copper deposits are often located far beneath Earth's surface. People therefore rely on mining companies to access and refine the copper in a deposit. A mining project often lasts for decades and include three major tasks. The first task in a mining project is to remove the ore from the deposit. Before miners can access the ore in a deposit, they must first remove the rock and soil that covers it. Miners call this rock and soil overburden. A deposit that is deeper costs more in terms of time and resources to access than a deposit that is near the surface. Mining companies tend to use underground or open-pit mining methods to remove overburden and the ore (see Figure L14.2 for an example of an open-pit copper mine).

The second major task is to separate the copper-based minerals from the ore. This task is called milling. This task is important because ore contains very little copper by weight, so

FIGURE L14.2 _____

An open-pit copper mine in Bisbee, Arizona

most of the minerals within the ore is considered waste by mining companies. To separate the copper-based minerals from the ore, mining companies must first crush and grind the ore into a powder. Mining companies then use a method called froth flotation to separate the copper-based minerals from minerals that do not contain copper. The result of the froth flotation process is a solution called copper concentrate, which is about 25%–35% copper by weight.

The third major task is to make a pure copper metal that is usable by manufacturers. Mining companies accomplish this task by first mixing the copper concentrate with silica and the heating the mixture to 1200°C. This process, which is called smelting, removes many impurities and produces a liquid of copper sulfide that is about 60% copper by weight. This liquid is called copper matte. The mining company then heats the copper matte to remove the sulfur and produce a material called blister copper, which is about 99% copper by weight. Next, the blister copper is refined and molded into sheets so it can go through a chemical process called electrolysis that removes the few remaining impurities in the metal. After the electrolysis is complete, the resulting copper metal is about 99.95% pure. The mining company can then sell the pure copper by the pound to various manufacturing companies.

Like the majority of human activities, accessing and refining copper produces waste materials. Waste is a general term for any material that currently has little or no economic value. There are different types of mine waste materials that vary in their physical and chemical composition, their potential for environmental contamination, and how mine companies manage them over time. There are five major categories of mine waste:

- *Overburden* is the excess soil and rock that is removed to gain access to the mineral-rich ore. Overburden consists of acid-generating and non-acid-generating rock. Water that flows over or through acid-generating rock becomes more acidic. This acidic water can enter a nearby stream, river, or lake and cause environmental contamination.

- *Tailings* is a water-based slurry that is produced when a mineral is separated from an ore. It consists of finely ground rock, mineral waste products, and toxic processing chemicals. Mine companies typically store tailings in large artificial ponds (see Figure L14.3, p. 350). However, some companies are considering dumping tailings into oceans as an alternative method for disposal.

- *Slag* is a by-product of smelting. It consists of iron oxide and silicon dioxide. Slag is nontoxic and can be used to make concrete, roads, the grit used in sandblasters, and blocks for buildings.

- *Atmospheric emissions* include dust and the sulfur oxides from the smelting process. Atmospheric emissions vary in their composition and potential for environmental contamination.

LAB 14

FIGURE L14.3

Aerial photograph of the Kennecott mine tailings storage pond near Salt Lake City, Utah

- *Mine water* is the water that is used to extract, crush, and grind ore and to control dust. It often contains dissolved minerals and metals or trace amounts of processing chemicals. Mine water can enter nearby streams, rivers, and lakes and cause environmental contamination.

Although many historic mining operations were not required to conduct their mining activities in ways that would reduce the negative impact on the environment, current federal and state regulations now require mining operations to use environmentally sound practices to minimize the effects of mineral development on human and ecosystem health. In this investigation, you will have an opportunity to examine several different proposals for developing and managing a copper mining operation to determine which one maximizes the benefits of mining copper while minimizing the potential costs. This is important to consider because the United States currently uses more copper (1.8 million metric tons in 2016) for construction and manufacturing than it produces (1.41 million tons in 2016). Although the United States can reduce its need for new copper mining projects by implementing more conservation, reuse, and recycling programs, it is also important to consider ways to minimize the impact of mining projects when new ones are needed.

Your Task

Use what you know about the uneven distribution of natural resources; the economic, social, and environmental costs associated with accessing natural resources; the importance of looking for patterns; and the nature of cause-and-effect relationships to identify the best proposal for starting a new copper mining project. Your assessment of each proposal must include an analysis of the potential benefits and costs associated with each proposal to determine which one has the highest benefit-to-cost ratio.

The guiding question of this investigation is, **Which proposal for a new copper mine maximizes the potential benefits while minimizing the potential costs?**

Materials

You may use the following resources during your investigation:

- Google Earth is an interactive map available at *www.google.com/earth,* and Google Maps is an interactive map available at *https://maps.google.com.*

- The U.S. Geological Survey (USGS) Global Assessment of Undiscovered Copper Resources is available at *https://mrdata.usgs.gov/sir20105090z.* This interactive map and database includes information about copper deposits all over the world.

- The USGS Mineral Commodity Summaries is available at *https://minerals.usgs.gov/minerals/pubs/mcs.* This online publication includes information about the market price of copper.

- Surf Your Watershed is available at *https://cfpub.epa.gov/surf/locate/index.cfm.* This Environmental Protection Agency (EPA) database provides information about stream water quality in different watersheds.

- The U.S. Fish and Wildlife Service Environmental Conservation Online System (ECOS) is available at *https://ecos.fws.gov/ecp/report/table/critical-habitat.html.* This system includes an interactive map that shows the location of protected habitats for threatened and endangered species.

Safety Precautions

Follow all normal lab safety rules.

Investigation Proposal Required?   ☐ Yes   ☐ No

Getting Started

Your teacher will give you several different copper mine proposals to evaluate. These proposals include different plans for accessing and refining the copper from a specific deposit within the United States. Mining companies must submit a proposal, which outlines their overall plan for developing and managing the extraction of a natural resource such as copper, to state and federal agencies for approval when they want to open a new mine. The overall plan for the mining project must be approved at both the state and federal level before the mining company can start digging. The copper mine proposals that you use during this investigation include the following information:

- Location of the mine
- Mining method
- Estimated life span of the mine
- Estimated number of new jobs that will be created when the mine opens
- Size of the site
- Total amount of ore (rock with copper in it) to be removed from the site
- Ore extraction rate
- Amount of waste that will be produced
- Waste management plan
- Expenses associated with opening, operating, and closing the mine

You will need to conduct a benefit-and-cost analysis of each proposal to determine which mine plan is the best one. A benefit-and-cost analysis requires the identification of the potential benefits of starting a new mining project and all the potential costs associated with opening, operating, and the closing the mine. Potential benefits associated with starting a new copper mine include such things as how much copper is in the deposit, how easy the deposit will be to access, how much ore that can be removed from the deposit, and the overall value of the copper on the open market. A mine can also have a positive impact on the local economy because it can create new jobs for people who live in that area. The potential costs, in contrast, include the amount of money needed to open, operate, and then close the mine, the negative impacts of mine waste on ecosystem health, and how the mine will be viewed by people who live in the area. You will therefore need to consider all of these issues, and potentially several others, as you conduct a benefit-and-cost analysis of each proposal during this investigation.

The first step in your benefit-and-cost analysis is to determine the location of each mine. To accomplish this step, you can enter the coordinates included with each proposal into Google Earth and/or Google Maps.

The second step in your benefit-and-cost analysis is to learn more about the copper deposit at a proposed location. You will need to use the USGS Global Assessment of Undiscovered Copper Resources database to accomplish this step. To use this database, simply zoom in on the location of the proposed mine on the interactive map. You can then click on the different deposits marked on the map until you find the name of the deposit you are interested in learning more about. You can then click on the name of the deposit to bring up information about it. As you use this resource, be sure to think about the following questions:

- What information will help you determine the potential benefits and costs of mining at this location?
- What would make one deposit more valuable than another deposit?
- How can you use mathematics to determine how much copper the mine could potentially produce based on the proposed amount of ore (rock with copper in it) that will be removed from the site?

The third step in your benefit-and-cost analysis is to determine the overall value of a proposed mine or the amount of revenue a mining company could potentially generate given the overall plan for a proposed mine. This is important to consider because a mine must be profitable to stay open. A mining company, in other words, must generate more money from selling the copper that it extracts from the mine than it spends to open, operate, and then close the mine. To determine how much money a mining company could make by selling the copper that it extracts from a mine, you can use the USGS Mineral Commodity Summaries. As you use this resource, be sure to think about the following questions:

- What information will help you determine the potential value of the copper in a deposit?
- How can you use mathematics to determine how much money a mining company could generate per day based on each copper mine proposal?
- What information will help you determine the potential cost of operating a proposed mine?
- How can you use mathematics to determine if a proposed mine will be profitable or not?

The fourth step in your benefit-and-cost analysis is to investigate the potential environmental impacts of each mine proposal. Mines can produce waste that can pollute water, contaminate soil, and destroy habitats. To determine potential environmental impacts of opening and operating mines, you will need to identify any streams, lakes, and important wildlife habitats around proposed mine sites. You can determine which streams and lakes are located near the proposed mine site by using Google Earth and/or Google Maps. You can also check the current water quality of these streams and lakes by accessing the Surf Your Watershed database. You can determine if there are any protected habitats of threatened or endangered species near a proposed mine by accessing the U.S. Fish and Wildlife Service Environmental Conservation Online System. As you use these resources, be sure to think about the following questions:

LAB 14

- What information will help you determine the potential environmental impacts of opening, operating, and closing a mine?

- What measurement scale or scales should you use to collect data?

- How can you describe or quantify the potential environmental impact of each proposed mine?

- What cause-and-effect relationships will you need to keep in mind as you use these resources?

The final step in your benefit-and-cost analysis is to choose between the different proposals based on the strengths and weakness of each one. To evaluate the trade-offs in each proposal fairly, you may want to create a decision matrix. A decision matrix includes rows for each proposal and columns that provide different evaluation criteria. Table L14.1 is an example of a decision matrix. Evaluation criteria might include such things as potential value of extracted copper, amount of waste produced, waste management plan, and potential habitat loss. You will need to determine which evaluation criteria to use based on what you know about the uneven distribution of natural resources and the economic, social, and environmental costs that are often associated with accessing natural resources. Once you have determined the evaluation criteria that your group will use, rank order each proposal on each criterion. Give the best option the highest number (which is a 4 if there are four different mining proposals) and the worst option a 1. You can then total the rankings, and the proposal with the highest overall score is the best one.

TABLE L14.1

An example of a decision matrix

Proposal	Evaluation criteria					Overall score
	1	2	3	4	5	
A						
B						
C						
D						

Connections to the Nature of Scientific Knowledge and Scientific Inquiry

As you work through your investigation, be sure to think about

- how scientific knowledge changes over time, and

- the types of questions that scientists can investigate.

Initial Argument

Once your group has finished collecting and analyzing your data, your group will need to develop an initial argument. Your initial argument needs to include a claim, evidence supporting your claim, and a justification of the evidence. The *claim* is your group's answer to the guiding question. The *evidence* is an analysis and interpretation of your data. Finally, the *justification* of the evidence is why your group thinks the evidence matters. The justification of the evidence is important because scientists can use different kinds of evidence to support their claims. Your group will create your initial argument on a whiteboard. Your whiteboard should include all the information shown in Figure L14.4.

FIGURE L14.4 _____

Argument presentation on a whiteboard

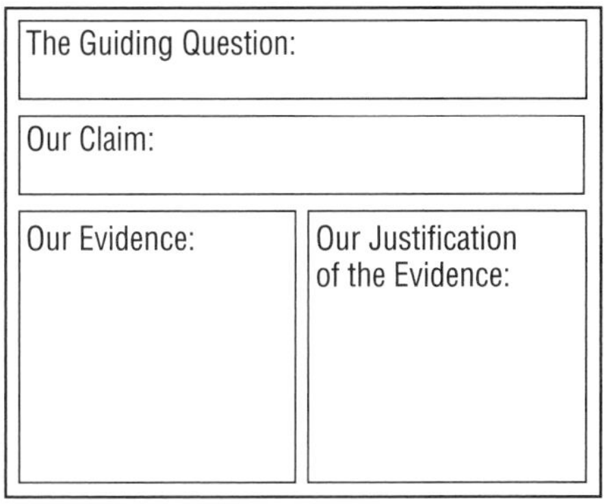

The Guiding Question:	
Our Claim:	
Our Evidence:	Our Justification of the Evidence:

Argumentation Session

The argumentation session allows all of the groups to share their arguments. One or two members of each group will stay at the lab station to share that group's argument, while the other members of the group go to the other lab stations to listen to and critique the other arguments. This is similar to what scientists do when they propose, support, evaluate, and refine new ideas during a poster session at a conference. If you are presenting your group's argument, your goal is to share your ideas and answer questions. You should also keep a record of the critiques and suggestions made by your classmates so you can use this feedback to make your initial argument stronger. You can keep track of specific critiques and suggestions for improvement that your classmates mention in the space below.

Critiques of our initial argument and suggestions for improvement:

If you are critiquing your classmates' arguments, your goal is to look for mistakes in their arguments and offer suggestions for improvement so these mistakes can be fixed. You should look for ways to make your initial argument stronger by looking for things that the other groups did well. You can keep track of interesting ideas that you see and hear during the argumentation in the space below. You can also use this space to keep track of any questions that you will need to discuss with your team.

Interesting ideas from other groups or questions to take back to my group:

Once the argumentation session is complete, you will have a chance to meet with your group and revise your initial argument. Your group might need to gather more data or design a way to test one or more alternative claims as part of this process. Remember, your goal at this stage of the investigation is to develop the best argument possible.

Report

Once you have completed your research, you will need to prepare an *investigation report* that consists of three sections. Each section should provide an answer for the following questions:

1. What question were you trying to answer and why?

2. What did you do to answer your question and why?

3. What is your argument?

Your report should answer these questions in two pages or less. You should write your report using a word processing application (such as Word, Pages, or Google Docs), if possible, to make it easier for you to edit and revise it later. You should embed any diagrams, figures, or tables into the document. Be sure to write in a persuasive style; you are trying to convince others that your claim is acceptable or valid.

Reference

U.S. Geological Survey (USGS). 2009. Copper—A metal for the ages. U.S. Department of the Interior, USGS. Available online at *https://pubs.usgs.gov/fs/2009/3031/FS2009-3031.pdf.*

LAB 14

Lab 14. Distribution of Natural Resources: Which Proposal for a New Copper Mine Maximizes the Potential Benefits While Minimizing the Potential Costs?

1. The map below shows the locations of the active mines producing copper in the United States as of 2003.

   a. Why are most copper mines located in the same region of the United States?

    b. What are some of the potential benefits of opening a new copper mine?

    c. What are some of the potential costs of opening a new copper mine?

2. Once something has been discovered in science, it is proven and will never change.

    a. I agree with this statement.

    b. I disagree with this statement.

Explain your answer, using an example from your investigation about the distribution of natural resources or other investigations you have done.

3. Scientists can only investigate certain types of questions.

   a. I agree with this statement.

   b. I disagree with this statement.

   Explain your answer, using an example from your investigation about the distribution of natural resources or other investigations you have done.

4. Identifying patterns is one of the important tasks that scientists carry out. Give an example of a pattern and why it was important for scientists to identify this pattern using information from your investigation about the distribution of natural resources.

5. Natural phenomena have causes, and uncovering causal relationships is a major activity of science. Explain why it is important to uncover causal relationships, using an example from your investigation about the distribution of natural resources.

SECTION 5
Weather and Climate

Introduction Labs

LAB 15

Lab 15. Air Masses and Weather Conditions: How Do the Motions and Interactions of Air Masses Result in Changes in Weather Conditions?

Purpose

The purpose of this lab is to *introduce* students to the disciplinary core idea (DCI) of Weather and Climate by having them determine how the motions and complex interactions of air masses result in changes in weather conditions. In addition, students have an opportunity to learn about the crosscutting concepts (CCs) of (a) Patterns and (b) Cause and Effect: Mechanism and Explanation. During the explicit and reflective discussion, students will also learn about (a) the difference between observations and inferences in science and (b) how scientists use different methods to answer different types of questions.

Important Earth and Space Science Content

An *air mass* is a large body of air that has relatively uniform temperature and moisture conditions at any given altitude. Air masses can be 1,000 km or more across, several kilometers thick, and extend through 20° or more of latitude. The atmospheric conditions within any air mass, as a result, will differ slightly from one side of the mass to the other. Atmospheric scientists expect to observe small differences in temperature and humidity levels within any air mass; however, these differences will be small in comparison to the atmospheric conditions in a different air mass. Air masses affect the weather of different regions as they move above the surface of the Earth over time.

An air mass forms when a portion of the air in the lower atmosphere stays over a relatively uniform region of Earth's surface, such as a large body of water, for several days, because air tends to assume the temperature and moisture conditions of the region beneath it. For example, air above a warm ocean will be warm and humid, air above a cold ocean will be cold and humid, and the air above land at higher latitudes is cold and dry. The area where an air mass forms is called its source region. A source region tends to be an area of high atmospheric pressure, which is the force exerted by the weight of a column of air above a given point, and light winds. The characteristics of an air mass, however, can change over time as it moves around the Earth. For example, an air mass that forms in the Arctic over land tends to consist of air that is very cold and dry. If the air mass moves over the ocean, it will pick up moisture. It will then be a cold and humid air mass rather than a cold and dry one.

Atmospheric scientists classify air masses based on their source region. Polar (P) and arctic (A) air masses originate in high latitudes near Earth's poles. Tropical (T) and Equatorial (E) air masses, in contrast, originate in low latitudes. The designation P, A, T, or E gives an indication of the temperature characteristics of an air mass. P and A indicate cold temperatures. and T and E indicate warm temperatures. Air masses are also classified based on the nature of the surface in the source region. Continental (c) air masses originate over land, and maritime (m) air masses form over water. The designation c or m therefore indicates the moisture characteristics of the air mass. Continental air masses tend to be dry, whereas maritime air tends to be humid. The basic types of air masses, according to this classification scheme, are continental polar (cP), continental arctic (cA), continental tropical (cT), continental equatorial (cE), maritime polar (mP), maritime arctic (mA), maritime tropical (mT), and maritime equatorial (mE). Figure 15.1 provides a map of the source region for these six common types of air masses.

FIGURE 15.1

Source regions of common air masses

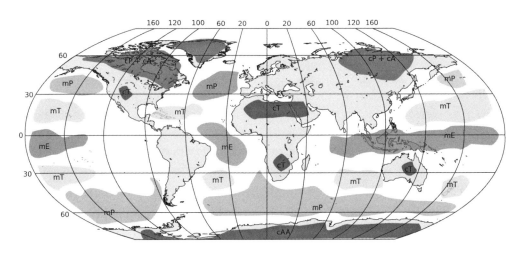

An air mass, as noted earlier, affects the weather of a given region as it moves into that area. The location where two air masses meet is called a front. Atmospheric scientists name fronts based on which type of air mass is moving into a specific location. There are three types of fronts: (a) a *cold front* occurs when a cold air mass moves into an area that was previously occupied by a warm air mass, (b) a *warm front* occurs when a warm air mass moves into an area that was occupied by a cold air mass, and (c) a *stationary front* is found in an area where a warm air mass and a cold air mass move past each other.

Cold fronts (see Figure 15.2, p. 366) are marked on a surface weather map with a solid blue line with triangles pointing in the direction of its movement. A cold front separates a cold and dry air mass from a warm and moist air mass. Cold fronts tend to move much faster than warm fronts. The cold air mass will push under the warm air mass as it moves because the cold air mass is denser than the warm air mass. In the Northern Hemisphere a cold front usually causes a shift of wind from the southwest to the northwest, whereas

LAB 15

FIGURE 15.2

Cold fronts are shown with triangles on weather maps; the triangles point to the direction that the air is moving

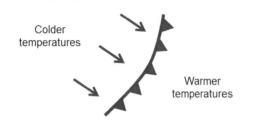

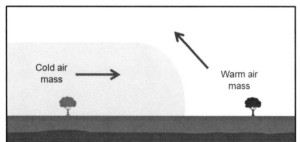

Note: A full-color version of this figure can be downloaded from the book's Extras page at *www.nsta.org/adi-ess*.

in the Southern Hemisphere a cold front usually causes a shift of wind from the northwest to the southwest.

There are a number of predictable changes in atmospheric conditions that are associated with the arrival of a cold front. Before a cold front arrives, the temperature tends to be warm, atmospheric pressure begins to decrease, there is often some light patchy rain, and the dew point is high and steady. The temperature drops suddenly when the front arrives. In addition, the atmospheric pressure will start to increase, there tend to be wind gusts that shift directions, and there is a sudden drop in the dew point. A narrow band of precipitation tends to move with the leading edge of a cold front. Depending on the amount of moisture in the air and the speed of the cold front, the amount of precipitation can range from light to heavy. Cold fronts often produce powerful thunderstorms in the summer months. The temperature in the region tends to stay cool after a cold front passes. The atmospheric pressure will continue to increase and the rain will clear. The dew point also begins to fall.

A warm front (see Figure 15.3) is marked on a surface weather map with a solid red line with semicircles pointing in the direction of its movement. A warm front separates a warm and humid air mass and a cold and dry air mass. The two air masses, however, do not mix when they interact with each other. Instead, the warm air mass will move over the colder air mass because it is less dense. The boundary between warm and cold air masses, as a result, tends to resemble a wedge with a gradual slope (see the illustration of a warm front in Figure 15.3). As the warm air mass moves over the mass of cold air, it cools and produces clouds and light to moderate precipitation over a large area. This precipitation tends to last for an extended period. Warm fronts tend to produce precipitation in this way because these air masses move slowly and the boundaries between warm and cold air masses have such a gradual slope. Warm fronts, however, will sometimes produce thunderstorms. This occurs when the overrunning warm air is humid and unstable and the temperature difference on the opposite sides of the air mass boundary is large. At the other extreme, a warm front associated with a dry and stable air mass will produce no precipitation at all.

There are several predictable changes in atmospheric conditions that are associated with the arrival of a warm front. The first sign of an approaching warm front is the appearance of cirrus clouds. These high clouds form as warm air climbs high above the cold air mass and cools. Cirrus clouds often appear 1,000 km or more ahead of the approaching warm front. Cirrostratus clouds appear as the warm front moves closer, which then blend into denser sheets of altostratus clouds. About 300 km ahead of the warm front, thicker stratus and nimbostratus clouds appear and it begins to rain or snow. The arrival of a warm front can also cause the wind direction to shift from the southeast to the southwest in the Northern Hemisphere and from the northeast to the northwest in the Southern Hemisphere. The temperature and the dew point tend to increase, and the atmospheric pressure decreases steadily.

A stationary front (see Figure 15.4) is a front that is not moving. Although the stationary boundary does not move, the warm and cold air masses along the boundary often move parallel to each other. Stationary fronts are marked on a weather map by alternating red and blue lines, with blue triangles and red semicircles facing opposite directions. Stationary fronts tend to be associated with heavy precipitation at a given location.

Timeline

The instructional time needed to complete this lab investigation is 270–330 minutes. Appendix 3 (p. 573) provides options for implementing this lab investigation over several class periods. Option G (330 minutes) should be used if students are unfamiliar with scientific writing, because this option provides

FIGURE 15.3

Warm fronts are shown with semicircles on weather maps; the semicircles indicate the direction that the air is moving

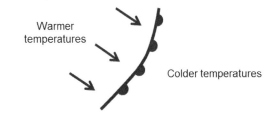

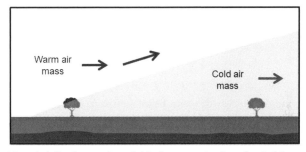

Note: A full-color version of this figure can be downloaded from the book's Extras page at *www.nsta.org/adi-ess*.

FIGURE 15.4

Stationary fronts are depicted with triangles and semicircles on weather maps

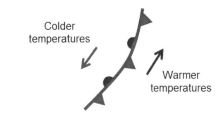

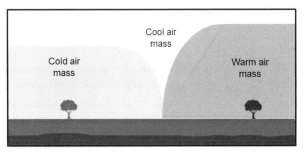

Note: A full-color version of this figure can be downloaded from the book's Extras page at *www.nsta.org/adi-ess*.

extra instructional time for scaffolding the writing process. You can scaffold the writing process by modeling, providing examples, and providing hints as students write each section of the report. Option H (270 minutes) should be used if students are familiar with scientific writing and have developed the skills needed to write an investigation report on their own. In option H, students complete stage 6 (writing the investigation report) and stage 8 (revising the investigation report) as homework. Both options give students time to test their models on day 3.

Materials and Preparation

The materials needed to implement this investigation are listed in Table 15.1. The National Oceanic and Atmospheric Administration (NOAA) / National Weather Service Weather Prediction Center maintains a record of daily weather maps for the United States dating back to 1871. You can access these weather maps at *www.wpc.ncep.noaa.gov/dwm/dwm.shtml*. These maps include information about the location of fronts, temperature, precipitation, and pressure. You should access the website and learn how to use it before beginning the lab investigation. In addition, it is important to check if students can access and use the website from a school computer or tablet, because some schools have set up firewalls and other restrictions on web browsing.

You can download Weather Map A, Weather Map B, Weather Conditions Table A, and Weather Conditions Table B from the book's online Extras page at *www.nsta.org/adi-ess*; these materials will allow students to test their conceptual models. You can also create your own versions of Weather Map A, Weather Map B, Weather Conditions Table A, and Weather Conditions B if you want students to examine weather conditions at specific cities or on specific dates.

To create your own version of Weather Maps A and B, you will first need to access the historical record of daily weather maps from the NOAA at *www.wpc.ncep.noaa.gov/dailywxmap/index.html*. Next, pick two different dates. The dates you pick do not matter, but we recommend picking dates that have some interesting atmospheric conditions. For each date, download the image of the map at the top of the page, which only shows the location of fronts and areas of high and low pressure, by right-clicking on the image and selecting "download image" or "save image as" (depending on your internet browser). Be sure to download the map that does *not* include weather station data. You can then add the map that you copied into a PowerPoint presentation. Next, mark two to four cities on each map by adding a circle to indicate the location of a city and a text box with the name of the city next to it. Label one map as Weather Map A and the other map as Weather Map B by adding a textbox. Then print out copies of each map so that each map prints out on a single piece of paper.

To create your own version of Weather Conditions Tables A and B, you can go to *www.wunderground.com/history* to find information about the maximum temperature, minimum temperature, barometric pressure, wind speed, wind direction, and amount of precipitation for the cities you marked on Weather Maps A and B. You can then create tables with this information in a Word or Excel document and print them out for students to use. We recommend that you include information about the weather conditions for each city for at least two different days, including the weather conditions for the day of the weather map and the day right after it.

TABLE 15.1

Materials list for Lab 15

Item	Quantity
Computer or tablet with internet access	1 per group
Weather Map A	1 per group
Weather Map B	1 per group
Weather Conditions Table A	1 per group
Weather Conditions Table B	1 per group
Investigation Proposal A (optional)	1 per group
Whiteboard, 2' × 3'*	1 per group
Lab Handout	1 per student
Peer-review guide and instructor scoring rubric	1 per student
Checkout Questions	1 per student

* As an alternative, students can use computer and presentation software such as Microsoft PowerPoint or Apple Keynote to create their arguments.

Safety Precautions
Remind students to follow all normal lab safety rules.

Topics for the Explicit and Reflective Discussion
Reflecting on the Use of Core Ideas and Crosscutting Concepts During the Investigation
Teachers should begin the explicit and reflective discussion by asking students to discuss what they know about the DCI they used during the investigation. The following are some important concepts related to the DCI of Weather and Climate that students need to determine how the motions and complex interactions of air masses result in changes in weather conditions:

- An air mass is a large body of air with generally uniform humidity and temperature.
- The source region of an air mass determines its characteristics.
- The area where two air masses meet is called a front.

- The motions and interaction of air masses results in changes in weather conditions.
- Weather at a specific location can be predicted based on current atmospheric conditions.

To help students reflect on what they know about these concepts, we recommend showing them two or three images using presentation software that help illustrate these important ideas. You can then ask the students the following questions to encourage students to share how they are thinking about these important concepts:

1. What do we see going on in this image?

2. Does anyone have anything else to add?

3. What might be going on that we can't see?

4. What are some things that we are not sure about here?

You can then encourage students to think about how CCs played a role in their investigation. There are at least two CCs that students need to determine how the motions and complex interactions of air masses result in changes in weather conditions: (a) Patterns and (b) Cause and Effect: Mechanism and Explanation (see Appendix 2 [p. 569] for a brief description of these CCs). To help students reflect on what they know about these CCs, we recommend asking them the following questions:

1. Why do scientists look for and attempt to explain patterns in nature?

2. What patterns did you identify and use during your investigation? Why was that useful?

3. Why can cause-and-effect relationships be used to predict phenomena in natural systems?

4. What cause-and-effect relationships did you use during your investigation to make predictions? Why was that useful to do?

You can then encourage students to think about how they used all these different concepts to help answer the guiding question and why it is important to use these ideas to help justify their evidence for their final arguments. Be sure to remind your students to explain why they included the evidence in their arguments and make the assumptions underlying their analysis and interpretation of the data explicit in order to provide an adequate justification of their evidence.

Reflecting on Ways to Design Better Investigations

It is important for students to reflect on the strengths and weaknesses of the investigation they designed during the explicit and reflective discussion. Students should therefore be encouraged to discuss ways to eliminate potential flaws, measurement errors, or sources of uncertainty in their investigations. To help students be more reflective about the design of their investigation and what they can do to make their investigations more rigorous in the future, you can ask the following questions:

1. What were some of the strengths of the way you planned and carried out your investigation? In other words, what made it scientific?

2. What were some of the weaknesses of the way you planned and carried out your investigation? In other words, what made it less scientific?

3. What rules can we make, as a class, to ensure that our next investigation is more scientific?

Reflecting on the Nature of Scientific Knowledge and Scientific Inquiry

This investigation can be used to illustrate two important concepts related to the nature of scientific knowledge and the nature of scientific inquiry: (a) the difference between observations and inferences in science and (b) how scientists use different methods to answer different types of questions (see Appendix 2 [p. 569] for a brief description of these two concepts). Be sure to review these concepts during and at the end of the explicit and reflective discussion. To help students think about these concepts in relation to what they did during the lab, you can ask the following questions:

- You made observations and inferences during your investigation. Can you give me some examples of these observations and inferences?

- Can you work with your group to come up with a rule that you can use to tell the difference between an observation and an inference? Be ready to share in a few minutes.

- There is no universal step-by-step scientific method that all scientists follow. Why do you think there is no universal scientific method?

- Think about what you did during this investigation. How would you describe the method you used to determine how the motions and complex interactions of air masses result in changes in weather conditions? Why would you call it that?

You can also use presentation software or other techniques to encourage your students to think about these concepts. You can show examples of information from the investigation that are either observations or inferences and ask students to classify each example and explain their thinking. You can also show one or more images of a "universal scientific method" that misrepresent the nature of scientific inquiry (see, e.g., *https://commons. wikimedia.org/wiki/File:The_Scientific_Method_as_an_Ongoing_Process.svg*) and ask students why each image is *not* a good representation of what scientists do to develop scientific knowledge. You can also ask students to suggest revisions to the image that would make it more consistent with the way scientists develop scientific knowledge.

Remind your students that, to be proficient in science, it is important that they understand what counts as scientific knowledge and how that knowledge develops over time.

Hints for Implementing the Lab

- Learn how to use NOAA database before the lab begins. It is important for you to know how to use the database so you can help students when they get stuck or confused.

- A group of three students per computer or tablet tends to work well.

- Allow the students to play with the database as part of the tool talk before they begin to design their investigation. This gives students a chance to see what information they can and cannot get on the website.

- Allowing students to design their own procedures for collecting data gives students an opportunity to try, to fail, and to learn from their mistakes. However, you can scaffold students as they develop their procedure by having them fill out an investigation proposal. These proposals provide a way for you to offer students hints and suggestions without telling them how to do it. You can also check the proposals quickly during a class period. We recommend using Investigation Proposal A for this lab.

- Encourage the students to look at multiple dates over a week or two to identify patterns in how weather conditions change over time. Remind them to focus on the fronts. They will need to keep track of the types of fronts they see and where the fronts are going over time and then relate that information to specific weather conditions at a specific location.

- You may want to do a quick mini-lesson during the tool talk (stage 1) to show students how to read weather maps if they are not familiar with them. If they have already learned how to read a weather map, you may need to do a "just-in-time" mini-lesson during stage 2 if students get stuck.

- Be sure that the students record actual values (e.g., temperature, wind speed and direction, amount of precipitation) as they collect data using the NOAA database, rather than just attempting to describe what they see on the computer screen (e.g., there was a lot of rain, there was a front over Texas).

- Students often make mistakes when developing their conceptual models and/ or initial arguments, but they should quickly realize these mistakes during the argumentation session. Be sure to allow students to revise their models and arguments at the end of the argumentation session. The explicit and reflective discussion will also give students an opportunity to reflect on and identify ways to improve how they develop and test models. This also offers an opportunity to discuss what scientists do when they realize a mistake is made.

Connections to Standards

Table 15.2 highlights how the investigation can be used to address specific (a) performance expectations from the *NGSS* and (b) *Common Core State Standards* in English language arts (*CCSS ELA*).

TABLE 15.2

Lab 15 alignment with standards

NGSS performance expectation	Weather and climate • MS-ESS2-5: Collect data to provide evidence for how the motions and complex interactions of air masses result in changes on weather conditions.
CCSS ELA—Reading in Science and Technical Subjects	Key ideas and details • CCSS.ELA-LITERACY.RST.6-8.1: Cite specific textual evidence to support analysis of science and technical texts. • CCSS.ELA-LITERACY.RST.6-8.2: Determine the central ideas or conclusions of a text; provide an accurate summary of the text distinct from prior knowledge or opinions. Craft and structure • CCSS.ELA-LITERACY.RST.6-8.4: Determine the meaning of symbols, key terms, and other domain-specific words and phrases as they are used in a specific scientific or technical context relevant to *grade 6–8 texts and topics*. • CCSS.ELA-LITERACY.RST.6-8.5: Analyze the structure an author uses to organize a text, including how the major sections contribute to the whole and to an understanding of the topic. • CCSS.ELA-LITERACY.RST.6-8.6: Analyze the author's purpose in providing an explanation, describing a procedure, or discussing an experiment in a text. Integration of knowledge and ideas • CCSS.ELA-LITERACY.RST.6-8.7: Integrate quantitative or technical information expressed in words in a text with a version of that information expressed visually (e.g., in a flowchart, diagram, model, graph, or table). • CCSS.ELA-LITERACY.RST.6-8.8: Distinguish among facts, reasoned judgment based on research findings, and speculation in a text.

Continued

TABLE 15.2 (*continued*)

CCSS ELA—**Reading in Science and Technical Subjects** (*continued*)	Integration of knowledge and ideas (*continued*) • CCSS.ELA-LITERACY.RST.6-8.9: Compare and contrast the information gained from experiments, simulations, video, or multimedia sources with that gained from reading a text on the same topic.
CCSS ELA—**Writing in Science and Technical Subjects**	Text types and purposes • CCSS.ELA-LITERACY.WHST.6-8.1: Write arguments focused on *discipline-specific content.* • CCSS.ELA-LITERACY.WHST.6-8.2: Write informative or explanatory texts, including the narration of historical events, scientific procedures/experiments, or technical processes. Production and distribution of writing • CCSS.ELA-LITERACY.WHST.6-8.4: Produce clear and coherent writing in which the development, organization, and style are appropriate to task, purpose, and audience. • CCSS.ELA-LITERACY.WHST.6-8.5: With some guidance and support from peers and adults, develop and strengthen writing as needed by planning, revising, editing, rewriting, or trying a new approach, focusing on how well purpose and audience have been addressed. • CCSS.ELA-LITERACY.WHST.6-8.6: Use technology, including the internet, to produce and publish writing and present the relationships between information and ideas clearly and efficiently. Range of writing • CCSS.ELA-LITERACY.WHST.6-8.10: Write routinely over extended time frames (time for reflection and revision) and shorter time frames (a single sitting or a day or two) for a range of discipline-specific tasks, purposes, and audiences.
CCSS ELA—**Speaking and Listening**	Comprehension and collaboration • CCSS.ELA-LITERACY.SL.6-8.1: Engage effectively in a range of collaborative discussions (one-on-one, in groups, and teacher-led) with diverse partners on grade 6–8 topics, texts, and issues, building on others' ideas and expressing their own clearly. • CCSS.ELA-LITERACY.SL.6-8.2:* Interpret information presented in diverse media and formats (e.g., visually, quantitatively, orally) and explain how it contributes to a topic, text, or issue under study. • CCSS.ELA-LITERACY.SL.6-8.3:* Delineate a speaker's argument and specific claims, distinguishing claims that are supported by reasons and evidence from claims that are not.

Continued

TABLE 15.2 (*continued*)

***CCSS ELA*—Speaking and Listening** (*continued*)	Presentation of knowledge and ideas
	• CCSS.ELA-LITERACY.SL.6-8.4:* Present claims and findings, sequencing ideas logically and using pertinent descriptions, facts, and details to accentuate main ideas or themes; use appropriate eye contact, adequate volume, and clear pronunciation.
	• CCSS.ELA-LITERACY.SL.6-8.5:* Include multimedia components (e.g., graphics, images, music, sound) and visual displays in presentations to clarify information.
	• CCSS.ELA-LITERACY.SL.6-8.6: Adapt speech to a variety of contexts and tasks, demonstrating command of formal English when indicated or appropriate.

* Only the standard for grade 6 is provided because the standards for grades 7 and 8 are similar. Please see *www.corestandards.org/ELA-Literacy/SL* for the exact wording of the standards for grades 7 and 8.

Lab Handout

Lab 15. Air Masses and Weather Conditions: How Do the Motions and Interactions of Air Masses Result in Changes in Weather Conditions?

Introduction

Meteorology is the study of the atmosphere. Meteorologists study the atmosphere so they can make accurate predictions about future weather conditions. In fact, meteorologists have generated detailed weather maps that include information about the atmosphere and current weather conditions in different regions of the United States for over a century. In the late 1800s, for example, newspapers printed a weather map every morning. An example of a U.S. daily weather map from 1899 can be seen in Figure L15.1. People relied on these weather maps to make predictions about the weather so they could make better decisions about what to do during the day, where to go, or what to wear. As technologies advanced, we developed faster ways to deliver up-to-date information about current weather conditions, including radios, live television broadcasts, and internet posts. Meteorologists are now able to consult many sources, such as computer models, real-time weather station data, and Doppler radar, to generate forecasts. These forecasts are often very accurate and can predict general daily weather up to 10 days in advance.

FIGURE L15.1

The U.S. Daily Weather Map for February 8, 1899

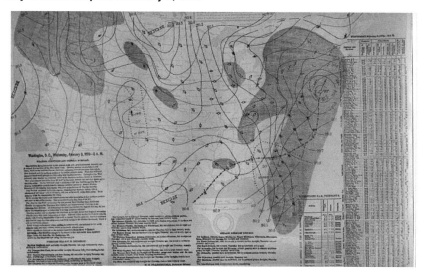

Although weather maps may look complicated, they simply display information about air masses and other atmospheric conditions. An *air mass* is a large body of air that has a relatively uniform temperature and humidity level. The curved lines on a weather map mark the boundaries of different air masses. The center of an air mass is marked with the letter *H* or *L,* which denotes whether the air mass has a high or low atmospheric pressure. The temperature of air affects the atmospheric pressure within an air mass. Warm air consists of molecules that are moving faster and more spread out compared with cold air. This makes hot air less dense than cold air.

When thinking about air masses, it is important to remember that air masses are three-dimensional; they spread out across a region (North, East, South, and West), while also extending upward from the surface of Earth. When an air mass warms at Earth's surface, it becomes less dense and begins to rise. As it rises farther away from Earth's surface, it cools, becomes denser, and sinks. When two air masses meet, the warmer, less dense air mass will rise above the colder, denser air mass.

The area where two air masses meet is called a front. Meteorologists categorize fronts based on the nature of the air mass that is moving into an area or how two or more air masses are interacting with each other. A cold front, for example, refers to instances when a cold air mass moves into an area that was previously occupied by a warm air mass. On a weather map, lines with shapes on them represent different types of fronts. A line with triangles is used to indicate the boundary and movement of a cold front (see Figure L15.2). A line with semicircles is used to indicate the boundary and movement of a warm front (see Figure L15.3). The shapes always point in the direction an air mass is moving. A third type of front is called a stationary front. Stationary fronts form in areas where warm air masses and cold air masses move past each other in opposite directions (see Figure L15.4, p. 378). The warm air mass is always on the side of the line without the semicircles, and the cold air mass is always on the side of the line without the triangles.

The interaction between two air masses can cause a change in weather conditions. Meteorologists therefore

FIGURE L15.2

Cold fronts are shown with triangles on weather maps; the triangles point to the direction that the air is moving

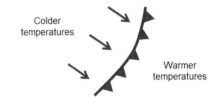

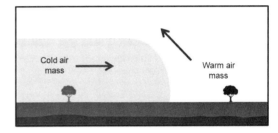

FIGURE L15.3

Warm fronts are shown with semicircles on weather maps; the semicircles indicate the direction that the air is moving

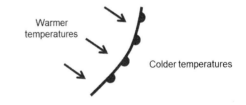

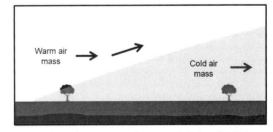

Note: A full-color version of Figures L15.2 and L15.3 can be downloaded from the book's Extras page at *www.nsta.org/adi-ess.*

LAB 15

Stationary fronts are depicted with triangles and semicircles on weather maps

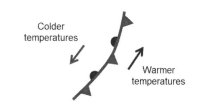

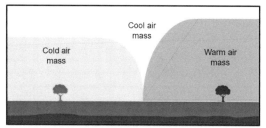

Note: A full-color version of this figure can be downloaded from the book's Extras page at *www.nsta.org/adi-ess.*

track the movement of air masses to make predictions about future weather conditions. In this investigation, you will have an opportunity to use historical weather maps and weather data from several different regions to learn how the motions and complex interactions of air masses are related to changes in weather conditions. Your goal is to develop a conceptual model that you can use to not only explain how the motions and interactions of air masses result in specific weather conditions but also predict how weather conditions will change over time in a given area.

Your Task

Develop a conceptual model that can be used to explain weather conditions based on the movement and interactions of air masses. Your conceptual model must be based on what we know about the weather, the kinds of air masses found in the atmosphere, the importance of looking for patterns in nature, and cause-and-effect relationships. Once you have developed your model, you will need to test it to see if you can use it to make accurate predictions about the weather conditions in different cities on a given date.

The guiding question of this investigation is, *How do the motions and interactions of air masses result in changes in weather conditions?*

Materials

You may use any of the following materials during your investigation:

Equipment
- Computer or tablet with internet access

Other Resources
- Weather Map A (use to test your model)
- Weather Conditions Table A (use to test your model)
- Weather Map B (use to test your model)
- Weather Conditions Table B (use to test your model)

Safety Precautions

Follow all normal lab safety rules.

Investigation Proposal Required? ☐ Yes ☐ No

Getting Started

The first step in this investigation is to analyze an existing data set to determine how the movement or interaction of different kinds of air masses is related to specific weather conditions. To accomplish this goal, you will need to examine several different historical weather maps and look for patterns that you can use to explain and predict changes in weather conditions. You can access U.S. Daily Weather Maps from the National Oceanic and Atmospheric Administration (NOAA) / National Weather Service Weather Prediction Center at *www.wpc.ncep.noaa.gov/dwm/dwm.shtml*.

Once you have identified patterns in the historical weather maps, you can develop your conceptual model. A conceptual model is an idea or set of ideas that explains what causes a particular phenomenon in nature. People often use words, images, and arrows to describe a conceptual model. Your conceptual model needs to be able to explain changes in weather conditions based on the movement and interactions of air masses. The model also needs to be consistent with what we know about what causes changes in atmospheric pressure, the nature of wind, and the cycling of water on Earth.

The last step in this investigation is to test your model. To accomplish this goal, you can make predictions about the weather conditions at several different cities using the information found on Weather Maps A and B. Your teacher will identify the cities that you will need to include in your predictions. You can then use Weather Conditions Tables A and B to determine if your predictions were accurate. If you are able to use your model to make accurate predictions about the weather conditions in different cities, then you will be able to generate the evidence you need to convince others that the conceptual model you developed is valid or acceptable.

Connections to the Nature of Scientific Knowledge and Scientific Inquiry

As you work through your investigation, be sure to think about

- the difference between observations and inferences in science, and
- how scientists use different methods to answer different types of questions.

Initial Argument

Once your group has finished collecting and analyzing your data, your group will need to develop an initial argument. Your initial argument needs to include a claim, evidence to support your claim, and a justification of the evidence. The *claim* is your group's answer to the guiding question. The *evidence* is an analysis and interpretation of your data. Finally, the *justification* of the evidence is why your group thinks the evidence matters. The justification of the evidence is important because scientists can use different kinds of evidence to support their claims. Your group will create your initial argument on a whiteboard. Your

LAB 15

FIGURE L15.5

Argument presentation on a whiteboard

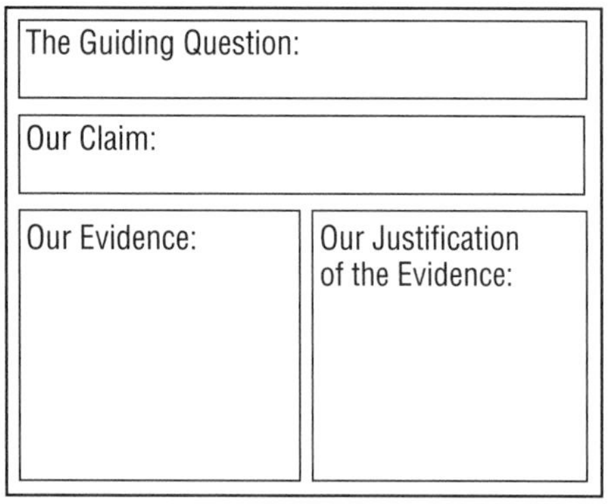

The Guiding Question:	
Our Claim:	
Our Evidence:	Our Justification of the Evidence:

whiteboard should include all the information shown in Figure L15.5.

Argumentation Session

The argumentation session allows all of the groups to share their arguments. One or two members of each group will stay at the lab station to share that group's argument, while the other members of the group go to the other lab stations to listen to and critique the other arguments. This is similar to what scientists do when they propose, support, evaluate, and refine new ideas during a poster session at a conference. If you are presenting your group's argument, your goal is to share your ideas and answer questions. You should also keep a record of the critiques and suggestions made by your classmates so you can use this feedback to make your initial argument stronger. You can keep track of specific critiques and suggestions for improvement that your classmates mention in the space below.

Critiques of our initial argument and suggestions for improvement:

If you are critiquing your classmates' arguments, your goal is to look for mistakes in their arguments and offer suggestions for improvement so these mistakes can be fixed. You should look for ways to make your initial argument stronger by looking for things that the other groups did well. You can keep track of interesting ideas that you see and hear during the argumentation in the space below. You can also use this space to keep track of any questions that you will need to discuss with your team.

Interesting ideas from other groups or questions to take back to my group:

Once the argumentation session is complete, you will have a chance to meet with your group and revise your initial argument. Your group might need to gather more data or design a way to test one or more alternative claims as part of this process. Remember, your goal at this stage of the investigation is to develop the best argument possible.

Report

Once you have completed your research, you will need to prepare an *investigation report* that consists of three sections. Each section should provide an answer for the following questions:

1. What question were you trying to answer and why?

2. What did you do to answer your question and why?

3. What is your argument?

Your report should answer these questions in two pages or less. You should write your report using a word processing application (such as Word, Pages, or Google Docs), if possible, to make it easier for you to edit and revise it later. You should embed any diagrams, figures, or tables into the document. Be sure to write in a persuasive style; you are trying to convince others that your claim is acceptable or valid.

Checkout Questions

Lab 15. Air Masses and Weather Conditions: How Do the Motions and Interactions of Air Masses Result in Changes in Weather Conditions?

1. What are air masses and how do meteorologists classify them?

2. A meteorologist is given the weather map below and is asked to predict the upcoming weather for the four cities marked with the letters A, B, C, and D in black circles.

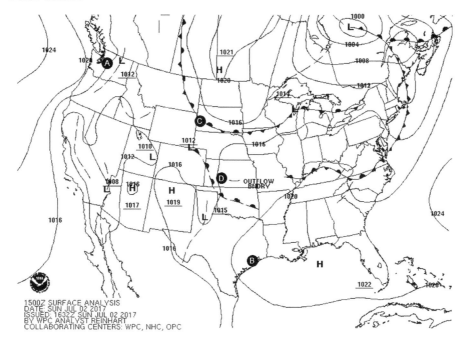

a. Which city or cities will probably experience rain in the near future?

b. How do you know?

3. The statement "an air mass with low pressure is replacing one with high pressure" is an example of an observation.

   a. I agree with this statement.
   b. I disagree with this statement.

   Explain your answer, using an example from your investigation about air masses and weather conditions.

4. Scientists, regardless of their discipline, follow the same step-by step method to answer questions about natural phenomena.

   a. I agree with this statement.
   b. I disagree with this statement.

   Explain your answer, using an example from your investigation about air masses and weather conditions.

5. Scientists often need to look for patterns during an investigation. Explain why identifying patterns is important in science, using an example from your investigation about air masses and weather conditions.

6. Scientists often attempt to uncover a cause-and-effect relationship as part of an investigation. Explain what a cause-and-effect relationship is and why these relationships are so important in science, using an example from your investigation about air masses and weather conditions.

LAB 16

Teacher Notes

Lab 16. Surface Materials and Temperature Change: How Does the Nature of the Surface Material Covering a Specific Location Affect Heating and Cooling Rates at That Location?

Purpose

The purpose of this lab is to *introduce* students to the disciplinary core ideas (DCIs) of (a) Earth Materials and Systems and (b) Weather and Climate by having them determine the relationship between the materials covering an area and the rate at which the temperature of that area changes over time. In addition, students have an opportunity to learn about the crosscutting concepts (CCs) of (a) Cause and Effect: Mechanism and Explanation and (b) Stability and Change. During the explicit and reflective discussion, students will also learn about (a) the difference between laws and theories in science and (b) the types of questions that scientists can investigate.

Important Earth and Space Science Content

Temperature is used to describe the average *kinetic energy* of the particles (atoms or molecules) that make up an object. *Heat,* on the other hand, is the transfer of thermal energy from one object to another or within an object due to a temperature difference. When thermal energy transfers into an object, the temperature of that object will increase; when thermal energy transfers out of an object, the temperature of that object will decrease. The direction of thermal energy transfer is always from a warmer object to a cooler one (or from the higher-temperature part of object to the cooler-temperature part of that object). The rate of thermal energy transfer depends on the temperature difference between two objects (or parts of an object). There are three ways thermal energy can move into, out of, or within an object: conduction, convection, and radiation.

Conduction is the transfer of thermal energy into, out of, or through an object due to the collision of the particles (atoms or molecules) within that object. The particles that make up an object are constantly in motion. These particles, as a result, have kinetic energy. Kinetic energy is transferred from one particle to another particle when they bump into or collide with each other. An object that consists of particles that have greater kinetic energy will transfer thermal energy to an object that consists of particles with lower kinetic energy until there is no longer a temperature difference between the two objects. Similarly, a part of an object that consists of particles with greater kinetic energy will transfer thermal energy to another part of that object that consists of particles with lower kinetic energy until there is no longer a temperature difference within that object. An example of conduction is when you hold one end of an iron nail in a flame. Thermal energy transfers from the high-temperature flame to the end of the nail. The particles at the end of the nail start moving faster. The particles at the end of the nail will collide with nearby particles in the nail that have lower kinetic energy. The particles with lower kinetic energy will start moving faster as a result. Over time, thermal energy moves through the entire length of the nail, and it will become too hot to hold. In this case, the nail is at a higher temperature than your hand, and so thermal energy will transfer from the particles in the nail to the particles that make up your hand. Similarly, when you hold a piece of ice, thermal energy transfers from your hand (the higher-temperature object) to the molecules in the piece of ice (the lower-temperature object).

Convection is the transfer of thermal energy due to the mass movement or circulation of particles within a fluid. Fluids are liquids (like lakes or oceans) and gases (such as air). When a fluid is heated, the part of the fluid that is closest to a heat source will increase in temperature, expand, becomes less dense, and rise. Cooler parts of the fluid then move in and take the place of the particles in the fluid that were near the heat source. This movement of warmer and cooler parts of the fluid creates convection currents that transfer thermal energy within the fluid over time. An example of convection is when food is cooked in an oven. When the oven is turned on, the coils inside the oven increase in temperature. The hot coils warm the air surrounding them. The air around the coils, as a result, expands, becomes less dense, and rises. The cooler air in the oven then migrates toward the hot coils and is warmed. The convection currents stir the air and transfers thermal energy from the coils throughout the oven.

Radiation is the transfer of thermal energy through electromagnetic waves. Unlike conduction and convection, radiation can transfer thermal energy in a vacuum. All objects at any temperature above absolute zero emit electromagnetic waves. The average frequency of the electromagnetic waves emitted by an object is directly proportional to the absolute temperature of that object. The Sun, for example, has a very high temperature so it emits high-frequency visible light and ultraviolet waves. The Earth, in contrast, is relatively cool. It emits lower-frequency infrared waves. Higher-frequency electromagnetic waves have shorter wavelengths and carry more energy than lower-frequency electromagnetic waves.

All objects, including people, furniture, and appliances, emit electromagnetic waves in a mixture of frequencies (because temperature corresponds to a mixture of molecular kinetic energies). Objects at room temperature mostly emit low-frequency infrared electromagnetic waves. When your skin absorbs higher-frequency infrared or visible light waves, it feels hot to you. Common sources of higher-frequency visible light that will feel hot to you include burning embers in a fireplace, a burner on a stovetop, and a lightbulb filament.

When an electromagnetic wave strikes an object, thermal energy can transfer into the object. The transfer of thermal energy will increase the kinetic energy of some of the atoms or molecules within that object and result in an increase in object's temperature. The greater the amount of thermal energy that is transferred to the object, the greater the increase in the kinetic energy of the particles within that object and the hotter the object will become. The amount of thermal energy that will transfer into an object depends on three factors:

- *The nature of the electromagnetic waves striking the object.* Waves that carry more energy can transfer more thermal energy to an object in a given unit of time.

- *The surface area of the object.* The greater the surface area for electromagnetic waves to strike, the greater the amount of thermal energy that will be transferred to an object in a given unit of time.

- *The albedo of the object (or the fraction of the total electromagnetic radiation that is reflected off the surface of the object).* The greater the albedo of an object, the less thermal energy that the object will absorb in a given unit of time.

It is important to note that radiation can heat only the surface of an opaque object, because electromagnetic waves can only penetrate a small distance into these objects. Thermal energy transfers within an opaque object through conduction or convection. For example, if you were to leave a metal bar outside on a clear summer day, thermal energy would transfer into the part of the metal facing the Sun because of radiation. The thermal energy would then move through the bar as molecules collide with other molecules and transfer kinetic energy.

For a given amount of energy transferring into an object, its temperature change depends on two factors. First, the *mass* of the object being heated will influence the temperature change. The more massive the object is, the more energy it requires to increase the temperature of the whole object. When the energy flowing into the object is constant, the more massive the object, the smaller the energy increase. The second factor is the *specific heat* of an object, which is defined as the amount of energy needed to increase the temperature of 1 g of a substance by 1°C. All substances have different specific heats. The higher the specific heat, the more energy is required to flow into the substance to increase the temperature by 1oC. When the amount of the energy flowing into an object is held constant, the higher the specific heat of the object, the smaller the temperature change will be. For example, the specific heat of lead is 0.128 J/g•°C, which means that it takes 0.128 J of energy to increase the temperature of 1 g of lead by 1°C. The specific heat of water, in contrast, is 4.186 J/g•°C. It therefore takes 4.186 J of energy to increase the temperature of 1 g of water by the same amount.

In this lab, students will place different materials in Styrofoam cups and place the cups under a heat lamp. They will then collect data on the temperature change of each material. Depending on the mass of each material and the surface area of the material facing the light source, students will get slightly different results. However, they should find that the sand and the concrete heat the fastest and the water heats the slowest.

Surface Materials and Temperature Change

How Does the Nature of the Surface Material Covering a Specific Location Affect Heating and Cooling Rates at That Location?

Timeline

The instructional time needed to complete this lab investigation is 220–280 minutes. Appendix 3 (p. 573) provides options for implementing this lab investigation over several class periods. Option A (280 minutes) should be used if students are unfamiliar with scientific writing, because this option provides extra instructional time for scaffolding the writing process. You can scaffold the writing process by modeling, providing examples, and providing hints as students write each section of the report. Option B (220 minutes) should be used if students are familiar with scientific writing and have developed the skills needed to write an investigation report on their own. In option B, students complete stage 6 (writing the investigation report) and stage 8 (revising the investigation report) as homework.

Materials and Preparation

The materials needed to implement this investigation are listed in Table 16.1 (p. 390). The consumables and equipment can be purchased from a science supply company such as Carolina, Flinn Scientific, or Ward's Science. Concrete and grass turf can be purchased from a local hardware store.

You should prepare the concrete samples before the lab begins. For each concrete sample, mix a small amount of concrete following the instructions on the packaging. We suggest mixing the concrete in a Styrofoam cup similar to the ones you will provide to students during this lab. Before you let the concrete dry, place a drinking straw in the concrete that is lubricated in vegetable or canola oil. When the concrete dries, remove the straw from the concrete. This will provide a hole for the students to place a thermometer to measure the temperature change of the concrete.

Be sure to use a set routine for distributing and collecting the materials during the lab investigation. One option is to set up the materials for each group at each group's lab station before class begins. This option works well when there is a dedicated section of the classroom for lab work and the materials are large and difficult to move (such as a stream table). A second option is to have all the materials on a table or cart at a central location. You can then assign a member of each group to be the "materials manager." This individual is responsible for collecting all the materials his or her group needs from the table or cart during class and for returning all the materials at the end of the class. This option works well when the materials are small and easy to move (such as stopwatches, metersticks, or thermometers). It also makes it easy to inventory the materials at the end of the class before students leave for the day.

LAB 16

TABLE 16.1

Materials list for Lab 16

Item	Quantity
Consumables	
Water (at room temperature)	250 ml per group
Soil	250 g per group
Dark sand	250 g per group
Light sand	250 g per group
Concrete	1 sample per group
Sod (cut into 10 cm × 10 cm squares)	1 per group
Equipment and other materials	
Safety glasses or goggles	1 per student
Chemical-resistant apron	1 per student
Gloves	1 pair per student
Styrofoam cups	5 per group
Electronic or triple beam balance	1 per group
Infrared lamp and reflector	1–2 per group
Partial immersion (nonmercury) thermometers	5 per group
Digital or laser thermometer (optional)	1 per group
Graduated cylinder, 250 ml	1 per group
Support stand	1 per group
Ruler	1 per group
Investigation Proposal C (optional)	1 per group
Whiteboard, 2' × 3'*	1 per group
Lab Handout	1 per student
Peer-review guide and teacher scoring rubric	1 per student
Checkout Questions	1 per student

* As an alternative, students can use computer and presentation software such as Microsoft PowerPoint or Apple Keynote to create their arguments.

Safety Precautions and Laboratory Waste Disposal

Remind students to follow all normal lab safety rules. In addition, tell the students to take the following safety precautions:

- Wear sanitized indirectly vented chemical-splash goggles and chemical-resistant, nonlatex aprons, and gloves throughout the entire investigation (which includes setup and cleanup).

- Use only a GFCI-protected electrical receptacle for the lamp to prevent or reduce risk of shock.

- Handle the infrared lamp with care; it can get hot enough to burn skin.

- Do not spill or splash water on the hot lamp bulb—this can crack glass and form a projectile.

- Report and clean up spills immediately, and avoid walking in areas where water has been spilled.

- Wash hands with soap and water when done collecting the data and after completing the lab.

Waste for this lab will be minimal. You may choose to collect the soil, sand, and concrete samples after use for later reuse, or all the samples can be disposed of directly into the trash.

Topics for the Explicit and Reflective Discussion

Reflecting on the Use of Core Ideas and Crosscutting Concepts During the Investigation

Teachers should begin the explicit and reflective discussion by asking students to discuss what they know about the DCIs they used during the investigation. The following are some important concepts related to the DCIs of (a) Earth Materials and Systems and (b) Weather and Climate that students need to determine the relationship between the materials covering an area and the rate at which the temperature of that area changes over time:

- Temperature is the average kinetic energy of the atoms or molecules that make up an object.

- Heat is a form of energy that transfers among particles in an object or a system.

- Heat can transfer by conduction, convection, or radiation.

- Materials have unique physical properties.

To help students reflect on what they know about these concepts, we recommend showing them two or three images using presentation software that help illustrate these important ideas. You can then ask the students the following questions to encourage students to share how they are thinking about these important concepts:

1. What do we see going on in this image?

2. Does anyone have anything else to add?

3. What might be going on that we can't see?

4. What are some things that we are not sure about here?

You can then encourage students to think about how CCs played a role in their investigation. There are at least two CCs that students need to determine the relationship between the materials covering an area and the rate at which the temperature of that area changes over time: (a) Cause and Effect: Mechanism and Explanation and (b) Stability and Change (see Appendix 2 [p. 569] for a brief description of these CCs). To help students reflect on what they know about these CCs, we recommend asking them the following questions:

1. Why can cause-and-effect relationships be used to predict phenomena in natural systems?

2. What cause-and-effect relationships did you use during your investigation to make predictions? Why was that useful to do?

3. Why is it important to think about factors that control or affect the rate of change in system? How can we measure a rate of change?

4. Which factor(s) might have controlled the rate at which materials heated up and cooled down in your investigation? What did exploring these factors allow you to do?

You can then encourage the students to think about how they used all these different concepts to help answer the guiding question and why it is important to use these ideas to help justify their evidence for their final arguments. Be sure to remind your students to explain why they included the evidence in their arguments and make the assumptions underlying their analysis and interpretation of the data explicit in order to provide an adequate justification of their evidence.

Reflecting on Ways to Design Better Investigations

It is important for students to reflect on the strengths and weaknesses of the investigation they designed during the explicit and reflective discussion. Students should therefore be encouraged to discuss ways to eliminate potential flaws, measurement errors, or sources of uncertainty in their investigations. To help students be more reflective about the design of their investigation and what they can do to make their investigations more rigorous in the future, you can ask the following questions:

1. What were some of the strengths of the way you planned and carried out your investigation? In other words, what made it scientific?

2. What were some of the weaknesses of the way you planned and carried out your investigation? In other words, what made it less scientific?

3. What rules can we make, as a class, to ensure that our next investigation is more scientific?

Reflecting on the Nature of Scientific Knowledge and Scientific Inquiry

This investigation can be used to illustrate two important concepts related to the nature of scientific knowledge and the nature of scientific inquiry: (a) the difference between laws and theories in science and (b) the types of questions that scientists can investigate (see Appendix 2 [p. 569] for a brief description of these two concepts). Be sure to review these concepts during and at the end of the explicit and reflective discussion. To help students think about these concepts in relation to what they did during the lab, you can ask the following questions:

- Laws and theories are different in science. The amount of energy needed to raise the temperature of an object can be determined using the equation $Q = c \bullet m \bullet \Delta T$. Is this statement an example of a law or a theory? Why?

- Can you work with your group to come up with a rule that you can use to decide if something is a law or a theory? Be ready to share in a few minutes.

- Not all questions can be answered by science. Can you give me some examples of questions related to this investigation that can and cannot be answered by science?

- Can you work with your group to come up with a rule that you can use to decide if a question can or cannot be answered by science? Be ready to share in a few minutes.

You can also use presentation software or other techniques to encourage your students to think about these concepts. You can show examples of laws (such as $Q = c \bullet m \bullet \Delta T$) and theories (such as *objects are made of atoms or particles that are constantly in motion*) and ask students to indicate if they think each example is a law or a theory and explain their thinking. You can also show one or more examples of questions that can be answered by science (e.g., What are the sources of greenhouse gases emissions? How does increased greenhouse gases in the atmosphere affect average surface temperature?) and cannot be answered by science (e.g., Should we increase taxes to help reduce carbon dioxide emissions? Who should be required to cut their greenhouse gas emissions?) and then ask students why each example is or is not a question that can be answered by science.

Remind your students that, to be proficient in science, it is important that they understand what counts as scientific knowledge and how that knowledge develops over time.

LAB 16

Hints for Implementing the Lab

- Allow the students to become familiar with the equipment as part of the tool talk before they begin to design their experiments. This gives students a chance to see what they can and cannot do with the equipment.

- Allowing students to design their own procedures for collecting data gives students an opportunity to try, to fail, and to learn from their mistakes. However, you can scaffold students as they develop their procedure by having them fill out an investigation proposal. These proposals provide a way for you to offer students hints and suggestions without telling them how to do it. You can also check the proposals quickly during a class period. We recommend using Investigation Proposal C for this lab.

- Investigation Proposal C works best for the experiment that the students will conduct during this investigation because students can describe a different relationship between the independent and dependent variable for each hypothesis. For example, materials that cover the surface of cities might (a) heat up and cool down more quickly than materials that cover rural areas (hypothesis 1), (b) heat up and cool down more slowly than materials that cover rural areas (hypothesis 2), or (c) heat up and cool down at the same rate as materials that cover rural areas (hypothesis 3).

- Students do not need to use a large mass of each material during their experiments. The more mass students use for each material, the slower the rate of change will be when investigating each material. Because you will be making the concrete samples for students, we recommend making each sample of concrete to have a mass between 150 g and 250 g. This way, students who choose to control for the mass of the samples will not require more than 250 g of each sample.

- Make sure that all of the materials start at room temperature. *Note:* There might not be sufficient time for each material (sand, soil, etc.) to return to room temperature between classes. This means you might need to have a separate set of sand, soil, water, and concrete for each class.

- If you have access to laser thermometers, students can easily collect the temperature of the surface of the material with the laser thermometer while using more conventional thermometers to measure the temperature of the material beneath the surface.

- Students often make mistakes during the data collection stage, but they should quickly realize these mistakes during the argumentation session. It will only take them a short period of time to re-collect data, and they should be allowed to do so. During the explicit and reflective discussion, students will also have the opportunity to reflect on and identify ways to improve the way they design investigations (especially how they attempt to control variables as part of an experiment). This also offers an opportunity to discuss what scientists do when they realize that a mistake is made during a study.

Connections to Standards

Table 16.2 highlights how the investigation can be used to address specific (a) performance expectations from the *NGSS* and (b) *Common Core State Standards* in English language arts (*CCSS ELA*).

TABLE 16.2

Lab 16 alignment with standards

***NGSS* performance expectation**	Weather and climate • MS-ESS2-6: Develop and use a model to describe how unequal heating and rotation of the Earth cause patterns of atmospheric and oceanic circulation that determine regional climates.
***CCSS ELA*—Reading in Science and Technical Subjects**	Key ideas and details • CCSS.ELA-LITERACY.RST.6-8.1: Cite specific textual evidence to support analysis of science and technical texts. • CCSS.ELA-LITERACY.RST.6-8.2: Determine the central ideas or conclusions of a text; provide an accurate summary of the text distinct from prior knowledge or opinions. Craft and structure • CCSS.ELA-LITERACY.RST.6-8.4: Determine the meaning of symbols, key terms, and other domain-specific words and phrases as they are used in a specific scientific or technical context relevant to *grade 6–8 texts and topics*. • CCSS.ELA-LITERACY.RST.6-8.5: Analyze the structure an author uses to organize a text, including how the major sections contribute to the whole and to an understanding of the topic. • CCSS.ELA-LITERACY.RST.6-8.6: Analyze the author's purpose in providing an explanation, describing a procedure, or discussing an experiment in a text. Integration of knowledge and ideas • CCSS.ELA-LITERACY.RST.6-8.7: Integrate quantitative or technical information expressed in words in a text with a version of that information expressed visually (e.g., in a flowchart, diagram, model, graph, or table). • CCSS.ELA-LITERACY.RST.6-8.8: Distinguish among facts, reasoned judgment based on research findings, and speculation in a text.

Continued

TABLE 16.2 (*continued*)

***CCSS ELA*—Reading in Science and Technical Subjects** (*continued*)	Integration of knowledge and ideas (*continued*) • CCSS.ELA-LITERACY.RST.6-8.9: Compare and contrast the information gained from experiments, simulations, video, or multimedia sources with that gained from reading a text on the same topic.
***CCSS ELA*—Writing in Science and Technical Subjects**	Text types and purposes • CCSS.ELA-LITERACY.WHST.6-8.1: Write arguments focused on *discipline-specific content*. • CCSS.ELA-LITERACY.WHST.6-8.2: Write informative or explanatory texts, including the narration of historical events, scientific procedures/experiments, or technical processes. Production and distribution of writing • CCSS.ELA-LITERACY.WHST.6-8.4: Produce clear and coherent writing in which the development, organization, and style are appropriate to task, purpose, and audience. • CCSS.ELA-LITERACY.WHST.6-8.5: With some guidance and support from peers and adults, develop and strengthen writing as needed by planning, revising, editing, rewriting, or trying a new approach, focusing on how well purpose and audience have been addressed. • CCSS.ELA-LITERACY.WHST.6-8.6: Use technology, including the internet, to produce and publish writing and present the relationships between information and ideas clearly and efficiently. Range of writing • CCSS.ELA-LITERACY.WHST.6-8.10: Write routinely over extended time frames (time for reflection and revision) and shorter time frames (a single sitting or a day or two) for a range of discipline-specific tasks, purposes, and audiences.
***CCSS ELA*—Speaking and Listening**	Comprehension and collaboration • CCSS.ELA-LITERACY.SL.6-8.1: Engage effectively in a range of collaborative discussions (one-on-one, in groups, and teacher-led) with diverse partners on grade 6–8 topics, texts, and issues, building on others' ideas and expressing their own clearly. • CCSS.ELA-LITERACY.SL.6-8.2:* Interpret information presented in diverse media and formats (e.g., visually, quantitatively, orally) and explain how it contributes to a topic, text, or issue under study. • CCSS.ELA-LITERACY.SL.6-8.3:* Delineate a speaker's argument and specific claims, distinguishing claims that are supported by reasons and evidence from claims that are not.

Continued

TABLE 16.2 (*continued*)

CCSS ELA—**Speaking and Listening** (*continued*)	Presentation of knowledge and ideas • CCSS.ELA-LITERACY.SL.6-8.4:* Present claims and findings, sequencing ideas logically and using pertinent descriptions, facts, and details to accentuate main ideas or themes; use appropriate eye contact, adequate volume, and clear pronunciation. • CCSS.ELA-LITERACY.SL.6-8.5:* Include multimedia components (e.g., graphics, images, music, sound) and visual displays in presentations to clarify information. • CCSS.ELA-LITERACY.SL.6-8.6: Adapt speech to a variety of contexts and tasks, demonstrating command of formal English when indicated or appropriate.

* Only the standard for grade 6 is provided because the standards for grades 7 and 8 are similar. Please see *www. corestandards.org/ELA-Literacy/SL* for the exact wording of the standards for grades 7 and 8.

Lab Handout

Lab 16. Surface Materials and Temperature Change: How Does the Nature of the Surface Material Covering a Specific Location Affect Heating and Cooling Rates at That Location?

Introduction

Inner cities and suburbs tend to be much warmer than rural areas as a result of land use and human activities. Scientists call this phenomenon the urban heat island effect. Take Atlanta, Georgia, as an example. Figure L16.1 shows two Landsat satellite images of Atlanta taken on September 28, 2000. Image A is a true-color picture of Atlanta, where trees and other vegetation are dark green and roads or buildings are different shades of gray. Image B, in contrast, is a map of land surface temperature. In image B, cooler temperatures are yellow and hotter temperatures are red. Downtown Atlanta is in the center of both images. On this day in 2000, the temperature in the areas in and around downtown Atlanta reached 30°C (86°F), while some of the less densely developed areas outside of the city only reached 20°C (68°F). Las Vegas, Nevada, also experiences a significant urban heat island effect. On hot summer days, it can be 13°C (24°F) warmer in downtown Las Vegas than it is in the surrounding desert. Downtown Las Vegas also has, on average, 22 more days that are above 32°C (90°F) each year when compared with the surrounding rural areas. Most major cities in the United States, including Dallas, Phoenix, New York, Los Angeles, Denver, and Washington, D.C., experience a significant urban heat island effect.

FIGURE L16.1 _____

Two Landsat satellite images of Atlanta: (a) a true-color picture of the city and (b) an image showing the differences in temperature for the city and areas around the city in the afternoon

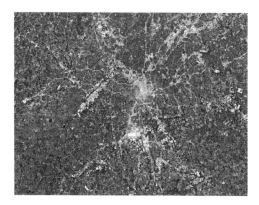

a

b

Note: A full-color version of this figure can be downloaded from the book's Extras page at *www.nsta.org/adi-ess*.

In our everyday conversations, we often use the terms *temperature* and *heat* interchangeably. In science, however, these two terms have different meanings. *Temperature* is used to describe the average kinetic energy of the atoms or molecules that make up an object. *Heat*, on the other hand, is the transfer of thermal energy into, within, or out of an object. There are three ways thermal energy can transfer into, within, or out of an object: conduction, convection, and radiation.

- *Conduction* is the transfer of thermal energy due to the collision of the atoms or molecules within an object or between two objects in contact. Thermal energy always transfers from an object or area of higher temperature to an object or area of lower temperature.

- *Convection* is the transfer of thermal energy due to the mass movement or circulation of particles within a fluid. Fluids are liquids (like lakes or oceans) and gases (such as air).

- *Radiation* is the transfer of thermal energy through electromagnetic waves. An example of radiation is what happens to a car when it sits in the sunlight on a hot summer day. The car absorbs sunlight, and the temperature of the car increases. As more sunlight is absorbed over time, the temperature of the car increases as well.

When thermal energy transfers into an object, the temperature of the object will increase; when heat transfers out of an object, the temperature will decrease. However, not all objects will undergo the same change in temperature when the same amount of thermal energy is added to them. For example, adding 1 joule (J) of thermal energy to a 1-kilogram (kg) sample of lead will cause the piece of lead to increase in temperature by about 8°C. Adding 1 J of thermal energy to a 1 kg sample of water, however, will only increase the temperature of the water by approximately 0.2°C.

There are many potential explanations for the urban heat island effect. First, cities have more people living in them. Some scientists have therefore speculated that a higher concentration of people using air conditioners may be causing the urban heat island effect because air conditioners remove heat from the air in buildings and transfer it outside. Other scientists have suggested that car exhaust is causing the urban heat island effect because the gases in exhaust can trap thermal energy, and there are many more cars in cities than there are in rural areas. Finally, other scientists suggest that the materials we use to build roads, homes, and other buildings are the source of the urban heat island effect. Scientists studying this possibility note that in cities that experience an urban heat island effect, there tends to be a much higher concentration of concrete inside the city when compared with the area surrounding the city, which tends to be covered by naturally occurring materials such as plants, sand, or water.

LAB 16

Your Task

Use what you know about heat and temperature, cause-and-effect relationships, and stability and change in systems to plan and carry out an investigation that will allow you to determine the relationship between the materials covering an area and the rate at which the temperature of that area changes over time. This investigation will aid you in understanding the underlying cause of urban heat islands.

The guiding question of this investigation is, *How does the nature of the surface material covering a specific location affect heating and cooling rates at that location?*

Materials

You may use any of the following materials during your investigation:

Consumables
- Water
- Soil
- Dark sand
- Light sand
- Concrete
- Sod

Equipment
- Safety glasses or goggles (required)
- Chemical-resistant apron (required)
- Gloves (required)
- Styrofoam cups
- Electronic or triple beam balance
- Infrared lamp and reflector

- Partial immersion (nonmercury) thermometers
- Digital or laser thermometer (optional)
- Graduated cylinder (250 ml)
- Support stand
- Ruler

Safety Precautions

Follow all normal lab safety rules. In addition, take the following safety precautions:

- Wear sanitized indirectly vented chemical-splash goggles and chemical-resistant, nonlatex aprons, and gloves throughout the entire investigation (which includes setup and cleanup).

- Use only a GFCI-protected electrical receptacle for the lamp to prevent or reduce risk of shock.

- Handle the infrared lamp with care; it can get hot enough to burn skin.

- Do not spill or splash water on the hot lamp bulb—this can crack glass and form a projectile.

- Report and clean up spills immediately, and avoid walking in areas where water has been spilled.

- Wash hands with soap and water when done collecting the data and after completing the lab.

Investigation Proposal Required?   ☐ Yes   ☐ No

Surface Materials and Temperature Change

*How Does the Nature of the Surface Material Covering a Specific Location Affect Heating and
Cooling Rates at That Location?*

Getting Started

To answer the guiding question, you will need to design and carry out an experiment.
Figure L16.2 shows how you can use a heat lamp to warm different types of materials, such
as soil, water, sand, concrete, or sod (grass).
Before you begin to design your experiment
using this equipment, think about what type of
data you need to collect, how you will collect
the data, and how you will analyze the data.

To determine *what type of data you need to
collect,* think about the following questions:

- What are the components of the
  system you are studying?
- Which factor(s) might control the rate
  of change in this system?
- How will you measure how quickly
  the materials heat up (or rate of
  change)?
- How will you measure how quickly
  the materials cool down (or rate of
  change)?

FIGURE L16.2

How to use a heat lamp to warm different types of materials

To determine *how you will collect the data,* think about the following questions:

- What conditions need to be satisfied to establish a cause-and-effect relationship?
- What will serve as your independent variable and dependent variables?
- How will you vary the independent variable while holding the other variables
  constant?
- How will you make sure the amount of each material is the same?
- How will you make sure that your data are of high quality (i.e., how will you
  reduce error)?
- How will you keep track of and organize the data you collect?

To determine *how you will analyze the data,* think about the following questions:

- What type of calculations will you need to make?
- How could you use mathematics to document a difference between conditions?
- What type of table or graph could you create to help make sense of your data?
- How will you determine if rates of change are the same or different?

LAB 16

Connections to the Nature of Scientific Knowledge and Scientific Inquiry

As you work through your investigation, be sure to think about

- the difference between laws and theories in science, and
- the types of questions that scientists can investigate.

Initial Argument

Once your group has finished collecting and analyzing your data, your group will need to develop an initial argument. Your initial argument needs to include a claim, evidence to support your claim, and a justification of the evidence. The *claim* is your group's answer to the guiding question. The *evidence* is an analysis and interpretation of your data. Finally, the *justification* of the evidence is why your group thinks the evidence matters. The justification of the evidence is important because scientists can use different kinds of evidence to support their claims. Your group will create your initial argument on a whiteboard. Your whiteboard should include all the information shown in Figure L16.3.

FIGURE L16.3

Argument presentation on a whiteboard

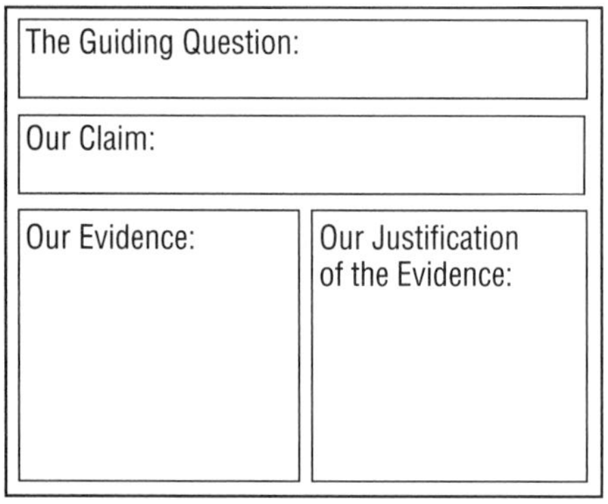

The Guiding Question:	
Our Claim:	
Our Evidence:	Our Justification of the Evidence:

Argumentation Session

The argumentation session allows all of the groups to share their arguments. One or two members of each group will stay at the lab station to share that group's argument, while the other members of the group go to the other lab stations to listen to and critique the other arguments. This is similar to what scientists do when they propose, support, evaluate, and refine new ideas during a poster session at a conference. If you are presenting your group's argument, your goal is to share your ideas and answer questions. You should also keep a record of the critiques and suggestions made by your classmates so you can use this feedback to make your initial argument stronger. You can keep track of specific critiques and suggestions for improvement that your classmates mention in the space below.

Critiques of our initial argument and suggestions for improvement:

If you are critiquing your classmates' arguments, your goal is to look for mistakes in their' arguments and offer suggestions for improvement so these mistakes can be fixed. You should look for ways to make your initial argument stronger by looking for things that the other groups did well. You can keep track of interesting ideas that you see and hear during the argumentation in the space below. You can also use this space to keep track of any questions that you will need to discuss with your team.

Interesting ideas from other groups or questions to take back to my group:

LAB 16

Once the argumentation session is complete, you will have a chance to meet with your group and revise your initial argument. Your group might need to gather more data or design a way to test one or more alternative claims as part of this process. Remember, your goal at this stage of the investigation is to develop the best argument possible.

Report

Once you have completed your research, you will need to prepare an *investigation report* that consists of three sections. Each section should provide an answer for the following questions:

1. What question were you trying to answer and why?

2. What did you do to answer your question and why?

3. What is your argument?

Your report should answer these questions in two pages or less. You should write your report using a word processing application (such as Word, Pages, or Google Docs), if possible, to make it easier for you to edit and revise it later. You should embed any diagrams, figures, or tables into the document. Be sure to write in a persuasive style; you are trying to convince others that your claim is acceptable or valid.

Checkout Questions

Lab 16. Surface Materials and Temperature Change: How Does the Nature of the Surface Material Covering a Specific Location Affect Heating and Cooling Rates at That Location?

1. If all objects are in the sunlight for the same time, why do some objects increase in temperature more than others?

2. Using data from your lab, explain how the design of cities contributes to heat islands. Make a recommendation for an urban planner on how to reduce the degree to which a city is a heat island.

3. An experiment is one possible method for answering a question in science. In an experiment, scientists develop a systematic plan for recording data. They do not manipulate or change any variables in order to answer their questions.

   a. I agree with this description of an experiment.
   b. I disagree with this description of an experiment.

   Explain your answer, using an example from your investigation about surface materials and temperature change.

4. Theories and laws are both important in science. Theories provide explanations for why phenomena occur, and laws provide descriptions of phenomena.

   a. I agree with this statement.
   b. I disagree with this statement.

   Explain your answer, using an example from your investigation about surface materials and temperature change.

5. Natural phenomena have causes, and uncovering causal relationships is a major activity of science. Explain what a causal relationship is and why it important to identify causal relationships in science, using an example from your investigation about surface materials and temperature change.

6. In science, it is important to understand what factors influence rates of change in a system. Explain why this is so important, using an example from your investigation about surface materials and temperature change.

LAB 17

Teacher Notes

Lab 17. Factors That Affect Global Temperature: How Do Cloud Cover and Greenhouse Gas Concentration in the Atmosphere Affect the Surface Temperature of Earth?

Purpose

The purpose of this lab is to *introduce* students to the disciplinary core idea (DCI) of Weather and Climate by having them determine how the temperature of Earth responds to changes in the amount of cloud cover and the concentration of various greenhouse gases in the atmosphere. In addition, students have an opportunity to learn about the crosscutting concepts (CCs) of (a) Energy and Matter: Flows, Cycles, and Conservation; and (b) Stability and Change. During the explicit and reflective discussion, students will also learn about (a) how scientific knowledge changes over time and (b) the types of questions that scientists can investigate.

Important Earth and Space Science Content

The Sun radiates energy. Some of that energy makes its way to Earth, where it is (a) absorbed by the atmosphere, (b) reflected by the atmosphere back into space, or (c) absorbed by Earth's surface. Earth's surface is warmed by the solar energy that it has absorbed. The Earth radiates some of that energy back into the atmosphere where it is (a) transferred into space, (b) absorbed by the atmosphere, or (b) radiated back to Earth to be absorbed (see Figure 17.1). As a result, Earth's surface is considerably warmer than it would be without an atmosphere. Earth's average surface temperature is about 33°C warmer than a black body (an ideal body that absorbs all radiation but does not have an atmosphere) having the approximate same size of Earth (for more information, see "Earth Fact Sheet" at *https://nssdc.gsfc.nasa.gov/planetary/factsheet/earthfact.html*). Scientists call this warming the *greenhouse effect*.

The gases in the atmosphere that are most responsible for the greenhouse effect are water vapor, carbon dioxide, methane, nitrous oxide, and ozone. Clouds also contribute to the greenhouse effect because they absorb and emit infrared radiation. The greenhouse effect is important to the success of life on Earth. Without the greenhouse effect, Earth would not be warm enough to support life as we know it. The Earth system, which includes the surface and the atmosphere, currently absorbs an average of about 340 watts of solar power per square meter over the course of the year (NASA n.d.). The Earth system also emits about the same amount of infrared energy into space. The average global surface temperature, as a result, tends to be stable over time.

Human influence has amplified the greenhouse effect because humans have significantly increased the amount of greenhouse gases present in the atmosphere. As this has

FIGURE 17.1

The greenhouse effect

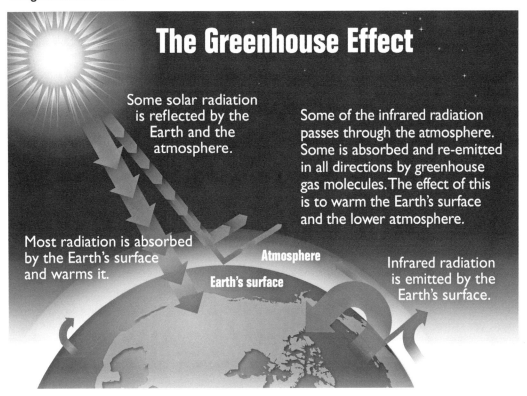

Note: A full-color version of this figure can be downloaded from the book's Extras page at *www.nsta.org/adi-ess.*

occurred, Earth has experienced an unprecedented warming trend. The majority of the scientific community agrees that the warming trend is directly related to humans burning fossil fuels along with other anthropogenic causes, such as deforestation, agriculture, production of organic fertilizers, decomposing organic matter in landfills, and use of chlorofluorocarbons (CFCs) in refrigerators and air conditioners. The average global temperature of Earth has increased approximately 0.8°C (1.4°F) within the past 100 years as less energy leaves Earth's atmosphere than enters it (NASA Goddard Institute for Space Studies 2016). Earth's oceans have absorbed over 90% of this additional energy. This additional energy absorbed by the oceans has caused their overall water temperature to increase slightly and, as a result, has influenced global weather patterns, sea levels, and terrestrial and aquatic ecosystems.

Timeline

The instructional time needed to complete this lab investigation is 170–230 minutes. Appendix 3 (p. 573) provides options for implementing this lab investigation over several class periods. Option C (230 minutes) should be used if students are unfamiliar with scientific writing, because this option provides extra instructional time for scaffolding the writing process. You can scaffold the writing process by modeling, providing examples, and providing hints as students write each section of the report. Option D (170 minutes) should be used if students are familiar with scientific writing and have developed the skills needed to write an investigation report on their own. In option D, students complete stage 6 (writing the investigation report) and stage 8 (revising the investigation report) as homework.

Materials and Preparation

The materials needed to implement this investigation are listed in Table 17.1. *The Greenhouse Effect* simulation, which was developed by PhET Interactive Simulations, University of Colorado (*http://phet.colorado.edu*), is available at *https://phet.colorado.edu/en/simulation/legacy/greenhouse*. It is free to use and can be run online using an internet browser. You should access the website and learn how the simulation works before beginning the lab investigation. In addition, it is important to check if students can access and use the simulation from a school computer or tablet, because some schools have set up firewalls and other restrictions on web browsing. You can also download an app-based version of PhET that includes this simulation. The app-based version currently only works on Apple products and is available at the Apple App Store.

TABLE 17.1

Materials list for Lab 17

Item	Quantity
Computer with internet access (or tablet with PhET app)	1 per group
Investigation Proposal C (optional)	1 per group
Whiteboard, 2' × 3'*	1 per group
Lab Handout	1 per student
Peer-review guide and teacher scoring rubric	1 per student
Checkout Questions	1 per student

* As an alternative, students can use computer and presentation software such as Microsoft PowerPoint or Apple Keynote to create their arguments.

Safety Precautions

Remind students to follow all normal lab safety rules.

Topics for the Explicit and Reflective Discussion

Reflecting on the Use of Core Ideas and Crosscutting Concepts During the Investigation

Teachers should begin the explicit and reflective discussion by asking students to discuss what they know about the DCI they used during the investigation. The following are some important concepts related to the DCI of Weather and Climate that students need to determine how the temperature of Earth responds to changes in the amount of cloud cover and the concentration of carbon dioxide in the atmosphere:

- All matter in the universe radiates energy across a range of wavelengths in the electromagnetic spectrum.
- The energy from the sunlight increases the temperature of the surface.
- The atmosphere traps some of this infrared radiation before it can escape into space.
- Many gases that are found naturally in the Earth's atmosphere, including carbon dioxide, methane, and nitrous oxide, are called greenhouse gases because these gases are able to trap infrared energy in the atmosphere.
- Any change to the Earth system that affects how much energy enters or leaves the system can cause a significant change in Earth's average global temperature.

To help students reflect on what they know about these concepts, we recommend showing them two or three images using presentation software that help illustrate these important ideas. You can then ask the students the following questions to encourage students to share how they are thinking about these important concepts:

1. What do we see going on in this image?
2. Does anyone have anything else to add?
3. What might be going on that we can't see?
4. What are some things that we are not sure about here?

You can then encourage students to think about how CCs played a role in their investigation. There are at least two CCs that students need to determine how the temperature of Earth responds to changes in the amount of cloud cover and the concentration of carbon dioxide in the atmosphere: (a) Energy and Matter: Flows, Cycles, and Conservation; and (b) Stability and Change (see Appendix 2 [p. 569] for a brief description of these CCs). To help students reflect on what they know about these CCs, we recommend asking them the following questions:

1. Why is it important to track how energy flows into, out of, or within a system during an investigation?

2. How did you track the flow of energy in the system you were studying? What did tracking the flow of energy allow you to do during your investigation?

3. Why is it important to think about what controls or affects the rate of change in system? How can we measure a rate of change?

4. Which factor(s) might have controlled the rate of temperature change in your investigation? What did exploring these factors allow you to do?

You can then encourage students to think about how they used all these different concepts to help answer the guiding question and why it is important to use these ideas to help justify their evidence for their final arguments. Be sure to remind your students to explain why they included the evidence in their arguments and make the assumptions underlying their analysis and interpretation of the data explicit as to provide an adequate justification of their evidence.

Reflecting on Ways to Design Better Investigations

It is important for students to reflect on the strengths and weaknesses of the investigation they designed during the explicit and reflective discussion. Students should therefore be encouraged to discuss ways to eliminate potential flaws, measurement errors, or sources of uncertainty in their investigations. To help students be more reflective about the design of their investigation and what they can do to make their investigations more rigorous in the future, you can ask the following questions:

1. What were some of the strengths of the way you planned and carried out your investigation? In other words, what made it scientific?

2. What were some of the weaknesses of the way you planned and carried out your investigation? In other words, what made it less scientific?

3. What rules can we make, as a class, to ensure that our next investigation is more scientific?

Reflecting on the Nature of Scientific Knowledge and Scientific Inquiry

This investigation can be used to illustrate two important concepts related to the nature of scientific knowledge and the nature of scientific inquiry: (a) how scientific knowledge changes over time and (b) the types of questions that scientists can investigate (see Appendix 2 [p. 569] for a brief description of these two concepts). Be sure to review these concepts during and at the end of the explicit and reflective discussion. To help students think about these concepts in relation to what they did during the lab, you can ask the following questions:

- Scientific knowledge can and does change over time. Can you tell me why it changes?

- Can you work with your group to come up with some examples of how scientific knowledge related to our understanding of average global temperature change has changed over time? Be ready to share in a few minutes.

- Not all questions can be answered by science. Can you give me some examples of questions related to this investigation that can and cannot be answered by science?

- Can you work with your group to come up with a rule that you can use to decide if a question can or cannot be answered by science? Be ready to share in a few minutes.

You can also use presentation software or other techniques to encourage your students to think about these concepts. You can show examples of how our thinking about average global temperature change has changed over time and ask students to discuss what they think led to those changes. You can also show one or more examples of questions that can be answered by science (e.g., What are the sources of greenhouse gases emissions? How does increased greenhouse gases in the atmosphere affect average global surface temperature) and cannot be answered by science (e.g., What is the best way to reduce carbon dioxide emissions? Who should be required to cut their greenhouse gas emissions?) and then ask students why each example is or is not a question that can be answered by science.

Remind your students that, to be proficient in science, it is important that they understand what counts as scientific knowledge and how that knowledge develops over time.

Hints for Implementing the Lab

- Learn how to use the online simulation before the lab begins. It is important for you to know how to use the simulation so you can help students when they get stuck or confused.

- A group of three students per computer or tablet tends to work well.

- Allow the students to play with the simulation as part of the tool talk before they begin to design their investigation. This gives students a chance to see what they can and cannot do with the simulation.

- Allowing students to design their own procedures for collecting data gives students an opportunity to try, to fail, and to learn from their mistakes. However, you can scaffold students as they develop their procedure by having them fill out an investigation proposal. These proposals provide a way for you to offer students hints and suggestions without telling them how to do it. You can also check the proposals quickly during a class period. We recommend using Investigation Proposal C for this lab.

- Be sure that students record actual values (e.g., greenhouse gas composition, number of photons, or temperature) and are not just attempting to hand draw what they see on the computer screen.

- Students often make mistakes during the data collection stage, but they should quickly realize these mistakes during the argumentation session. It will only take them a short period of time to re-collect data, and they should be allowed to do so. During the explicit and reflective discussion, students will also have the opportunity to reflect on and identify ways to improve the way they design investigations (especially how they attempt to control variables as part of an experiment). This also offers an opportunity to discuss what scientists do when they realize that a mistake is made during a study.

Connections to Standards

Table 17.2 highlights how the investigation can be used to address specific (a) performance expectations from the *NGSS* and (b) *Common Core State Standards* in English language arts (*CCSS ELA*).

TABLE 17.2

Lab 17 alignment with standards

NGSS performance expectations	Weather and climate • MS-ESS3-5: Ask questions to clarify evidence of the factors that have caused the rise in global temperatures over the past century. • HS-ESS2-4: Use a model to describe how variations in the flow of energy into and out of Earth's systems result in changes in climate.
CCSS ELA—Reading in Science and Technical Subjects	Key ideas and details • CCSS.ELA-LITERACY.RST.6-8.1: Cite specific textual evidence to support analysis of science and technical texts. • CCSS.ELA-LITERACY.RST.6-8.2: Determine the central ideas or conclusions of a text; provide an accurate summary of the text distinct from prior knowledge or opinions. • CCSS.ELA-LITERACY.RST.9-10.1: Cite specific textual evidence to support analysis of science and technical texts, attending to the precise details of explanations or descriptions. • CCSS.ELA-LITERACY.RST.9-10.2: Determine the central ideas or conclusions of a text; trace the text's explanation or depiction of a complex process, phenomenon, or concept; provide an accurate summary of the text.

Continued

TABLE 17.2 (*continued*)

***CCSS ELA*—Reading in Science and Technical Subjects** (*continued*)	Key ideas and details (*continued*) • CCSS.ELA-LITERACY.RST.9-10.3: Follow precisely a complex multistep procedure when carrying out experiments, taking measurements, or performing technical tasks, attending to special cases or exceptions defined in the text. Craft and structure • CCSS.ELA-LITERACY.RST.6-8.4: Determine the meaning of symbols, key terms, and other domain-specific words and phrases as they are used in a specific scientific or technical context relevant to *grade 6–8 texts and topics*. • CCSS.ELA-LITERACY.RST.6-8.5: Analyze the structure an author uses to organize a text, including how the major sections contribute to the whole and to an understanding of the topic. • CCSS.ELA-LITERACY.RST.6-8.6: Analyze the author's purpose in providing an explanation, describing a procedure, or discussing an experiment in a text. • CCSS.ELA-LITERACY.RST.9-10.4: Determine the meaning of symbols, key terms, and other domain-specific words and phrases as they are used in a specific scientific or technical context relevant to *grade 9–10 texts and topics*. • CCSS.ELA-LITERACY.RST.9-10.5: Analyze the structure of the relationships among concepts in a text, including relationships among key terms (e.g., *force, friction, reaction force, energy*). • CCSS.ELA-LITERACY.RST.9-10.6: Analyze the author's purpose in providing an explanation, describing a procedure, or discussing an experiment in a text, defining the question the author seeks to address. Integration of knowledge and ideas • CCSS.ELA-LITERACY.RST.6-8.7: Integrate quantitative or technical information expressed in words in a text with a version of that information expressed visually (e.g., in a flowchart, diagram, model, graph, or table). • CCSS.ELA-LITERACY.RST.6-8.8: Distinguish among facts, reasoned judgment based on research findings, and speculation in a text. • CCSS.ELA-LITERACY.RST.6-8.9: Compare and contrast the information gained from experiments, simulations, video, or multimedia sources with that gained from reading a text on the same topic. • CCSS.ELA-LITERACY.RST.9-10.7: Translate quantitative or technical information expressed in words in a text into visual form (e.g., a table or chart) and translate information expressed visually or mathematically (e.g., in an equation) into words.

Continued

TABLE 17.2 (*continued*)

CCSS ELA—**Reading in Science and Technical Subjects** (*continued*)	Integration of knowledge and ideas (*continued*) • CCSS.ELA-LITERACY.RST.9-10.8: Assess the extent to which the reasoning and evidence in a text support the author's claim or a recommendation for solving a scientific or technical problem. • CCSS.ELA-LITERACY.RST.9-10.9: Compare and contrast findings presented in a text to those from other sources (including their own experiments), noting when the findings support or contradict previous explanations or accounts.
CCSS ELA—**Writing in Science and Technical Subjects**	Text types and purposes • CCSS.ELA-LITERACY.WHST.6-10.1: Write arguments focused on *discipline-specific content.* • CCSS.ELA-LITERACY.WHST.6-10.2: Write informative or explanatory texts, including the narration of historical events, scientific procedures/experiments, or technical processes. Production and distribution of writing • CCSS.ELA-LITERACY.WHST.6-10.4: Produce clear and coherent writing in which the development, organization, and style are appropriate to task, purpose, and audience. • CCSS.ELA-LITERACY.WHST.6-8.5: With some guidance and support from peers and adults, develop and strengthen writing as needed by planning, revising, editing, rewriting, or trying a new approach, focusing on how well purpose and audience have been addressed. • CCSS.ELA-LITERACY.WHST.6-8.6: Use technology, including the internet, to produce and publish writing and present the relationships between information and ideas clearly and efficiently. • CCSS.ELA-LITERACY.WHST.9-10.5: Develop and strengthen writing as needed by planning, revising, editing, rewriting, or trying a new approach, focusing on addressing what is most significant for a specific purpose and audience. • CCSS.ELA-LITERACY.WHST.9-10.6: Use technology, including the internet, to produce, publish, and update individual or shared writing products, taking advantage of technology's capacity to link to other information and to display information flexibly and dynamically. Range of writing • CCSS.ELA-LITERACY.WHST.6-10.10: Write routinely over extended time frames (time for reflection and revision) and shorter time frames (a single sitting or a day or two) for a range of discipline-specific tasks, purposes, and audiences.

Continued

TABLE 17.2 (*continued*)

CCSS ELA—Speaking and Listening	Comprehension and collaboration • CCSS.ELA-LITERACY.SL.6-8.1: Engage effectively in a range of collaborative discussions (one-on-one, in groups, and teacher-led) with diverse partners on grade 6–8 topics, texts, and issues, building on others' ideas and expressing their own clearly. • CCSS.ELA-LITERACY.SL.6-8.2:* Interpret information presented in diverse media and formats (e.g., visually, quantitatively, orally) and explain how it contributes to a topic, text, or issue under study. • CCSS.ELA-LITERACY.SL.6-8.3:* Delineate a speaker's argument and specific claims, distinguishing claims that are supported by reasons and evidence from claims that are not. • CCSS.ELA-LITERACY.SL.9-10.1: Initiate and participate effectively in a range of collaborative discussions (one-on-one, in groups, and teacher-led) with diverse partners on grade 9–10 topics, texts, and issues, building on others' ideas and expressing their own clearly and persuasively. • CCSS.ELA-LITERACY.SL.9-10.2: Integrate multiple sources of information presented in diverse media or formats (e.g., visually, quantitatively, orally) evaluating the credibility and accuracy of each source. • CCSS.ELA-LITERACY.SL.9-10.3: Evaluate a speaker's point of view, reasoning, and use of evidence and rhetoric, identifying any fallacious reasoning or exaggerated or distorted evidence. Presentation of knowledge and ideas • CCSS.ELA-LITERACY.SL.6-8.4:* Present claims and findings, sequencing ideas logically and using pertinent descriptions, facts, and details to accentuate main ideas or themes; use appropriate eye contact, adequate volume, and clear pronunciation. • CCSS.ELA-LITERACY.SL.6-8.5:* Include multimedia components (e.g., graphics, images, music, sound) and visual displays in presentations to clarify information. • CCSS.ELA-LITERACY.SL.6-8.6: Adapt speech to a variety of contexts and tasks, demonstrating command of formal English when indicated or appropriate. • CCSS.ELA-LITERACY.SL.9-10.4: Present information, findings, and supporting evidence clearly, concisely, and logically such that listeners can follow the line of reasoning and the organization, development, substance, and style are appropriate to purpose, audience, and task.

Continued

TABLE 17.2 (*continued*)

CCSS ELA—Speaking and Listening (*continued*)	Presentation of knowledge and ideas (*continued*)
	• CCSS.ELA-LITERACY.SL.9-10.5: Make strategic use of digital media (e.g., textual, graphical, audio, visual, and interactive elements) in presentations to enhance understanding of findings, reasoning, and evidence and to add interest. • CCSS.ELA-LITERACY.SL.9-10.6: Adapt speech to a variety of contexts and tasks, demonstrating command of formal English when indicated or appropriate.

* Only the standard for grade 6 is provided because the standards for grades 7 and 8 are similar. Please see *www.corestandards.org/ELA-Literacy/SL* for the exact wording of the standards for grades 7 and 8.

References

National Aeronautics and Space Administration (NASA). n.d. Earth's energy budget. *https://earthobservatory.nasa.gov/Features/EnergyBalance/page4.php*.

National Aeronautics and Space Administration (NASA) Goddard Institute for Space Studies. 2016. GISS surface temperature analysis. *https://data.giss.nasa.gov/gistemp/graphs_v3*.

Lab Handout

Lab 17. Factors That Affect Global Temperature: How Do Cloud Cover and Greenhouse Gas Concentration in the Atmosphere Affect the Surface Temperature of Earth?

Introduction

All matter in the universe radiates energy across a range of wavelengths in the electromagnetic spectrum. Hotter objects tend to emit radiation with shorter wavelengths than cooler objects. The hottest objects in the universe, as a result, mostly emit gamma rays and x-rays. Cooler objects, in contrast, emit mostly longer-wavelength radiation, including visible light, infrared (IR), microwaves, and radio waves. The surface of the Sun has a temperature of about 5500°C or about 10000°F. At that temperature, most of the energy the Sun radiates is visible and near-IR light.

When sunlight first reaches Earth, some of it is reflected back out into space and some is absorbed by the atmosphere. The rest of the sunlight travels through the atmosphere and then hits the surface of Earth. The energy from the sunlight is absorbed by the surface and warms it. All objects, including Earth's surface, emit (or give off) IR radiation. The hotter an object is, the more IR radiation it emits. The amount of IR radiation emitted by Earth's surface therefore increases as it warms. The atmosphere traps some of this IR radiation before it can escape into space. The trapped IR radiation in the atmosphere helps keep the temperature of Earth warmer than it would be without the atmosphere. Scientists call the warming of the atmosphere that is caused by trapped IR radiation the *greenhouse effect* (see Figure L17.1, p. 420). Many gases that are found naturally in Earth's atmosphere, including water vapor, carbon dioxide, methane, nitrous oxide, and ozone, are called greenhouse gases because these gases are able to trap IR energy in the atmosphere.

The amount of energy that enters and leaves the Earth system is directly related to the average global temperature of the Earth. The Earth system, which includes the surface and the atmosphere, currently absorbs an average of about 340 watts of solar power per square meter over the course of the year (NASA n.d.). The Earth system also emits about the same amount of IR energy into space. The average global surface temperature, as a result, tends to be stable over time. However, if something were to change the amount of energy that enters or leaves this system, then the flow of energy would be unbalanced and the average global temperature would change in response. Therefore, any change to the Earth system that affects how much energy enters or leaves the system can cause a significant change in Earth's average global temperature.

The average global surface temperature of Earth has increased approximately 0.8°C (1.4°F) over the last 100 years (NASA Goddard Institute for Space Studies 2016). There are at least two potential explanations for this observation. One explanation is that the

FIGURE L17.1

The greenhouse effect

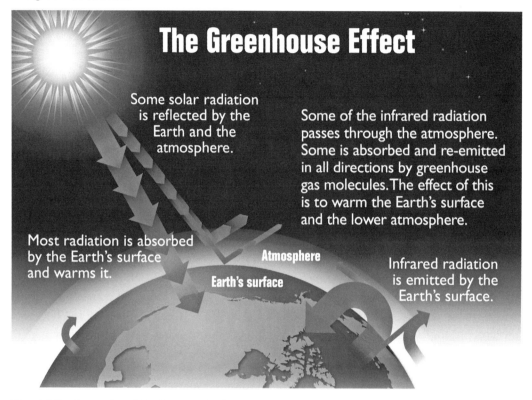

Note: A full-color version of this figure can be downloaded from the book's Extras page at *www.nsta.org/adi-ess.*

average global surface temperature of Earth normally increases and decreases over time and the current increase in temperature is just a normal part of this cycle. These changes could be to due to differences in the Sun's brightness, Milankovitch cycles (small variations in the shape of Earth's orbit and its axis of rotation that occur over thousands of years), or an increase or decrease in cloud cover (clouds form when water vapor in the air condenses into water droplets or ice). This explanation, however, does not account for the rapid increase in the average global surface temperature. An alternative explanation, which is the consensus view of the scientific community, is that humans have caused the rapid increase in the average global surface temperature of Earth by adding large amounts of greenhouse gases to the atmosphere. The addition of greenhouse gases magnifies the greenhouse effect. The atmosphere, as a result, traps more IR radiation and emits less IR energy out into space.

Before you can evaluate the merits of these two explanations for the observed change in average global surface temperature, it is important for you to understand how energy from the Sun interacts with the surface of the Earth and the various components of the atmosphere, such as clouds. You will therefore need to learn more about the relationships between surface temperature, cloud cover, and greenhouse gas levels in the atmosphere.

Your Task

Use a computer simulation and what you know about stability and change and the importance of tracking how energy flows into, within, and out of systems to determine how the temperature of Earth responds to changes in the amount of cloud cover and the concentration of carbon dioxide in the atmosphere.

The guiding question of this investigation is, *How do cloud cover and greenhouse gas concentration in the atmosphere affect the surface temperature of Earth?*

Materials

You will use an online simulation called *The Greenhouse Effect* to conduct your investigation; the simulation is available at *https://phet.colorado.edu/en/simulation/legacy/greenhouse.*

Safety Precautions

Follow all normal lab safety rules.

Investigation Proposal Required?   ☐ Yes   ☐ No

Getting Started

The *Greenhouse Effect* computer simulation models how energy flows into, within, and out of the Earth system and records changes in average global temperature over time (see Figure L17.2, p. 422). It shows the surface of Earth as a green strip. Above the green strip there is a blue atmosphere and black space at the top. Yellow dots stream downward representing photons of sunlight. Red dots represent photons of IR light that are emitted by the surface of Earth and travel toward space. The greenhouse gas concentration in the atmosphere, including amounts of water vapor (H_2O), carbon dioxide (CO_2), methane (CH_4), and nitrous oxide (N_2O), can be changed so it reflects the current level of these gases, the level in 1750, the level during the last ice age, or a level of your choice. Clouds can also be added or removed from the atmosphere. Greenhouse gases block IR light (energy) that is emitted by Earth's surface. Clouds can block sunlight and IR photons.

To answer the research question, you must determine what type of data you need to collect, how you will collect it, and how will you analyze it. To determine *what type of data you need to collect*, think about the following questions:

- What are the boundaries and components of the system you are studying?
- How do the components of the system interact with each other?
- When is this system stable, and under which conditions does it change?

FIGURE L17.2

A screenshot from *The Greenhouse Effect* simulation

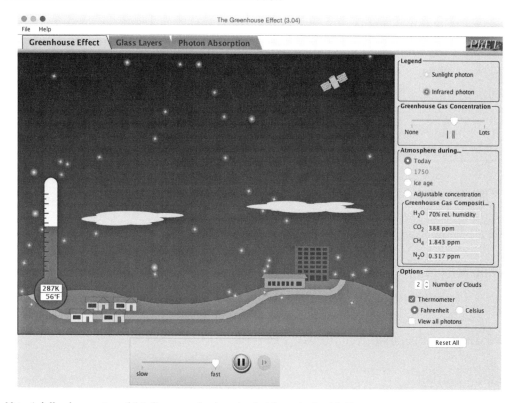

Note: A full-color version of this figure can be downloaded from the book's Extras page at *www.nsta.org/adi-ess*.

- Which factor(s) might control the rate of change in this system?
- How can you describe the components of the system quantitatively?
- How could you keep track of changes in this system quantitatively?
- How can you track how energy flows into, out of, or within this system?

To determine *how you will collect your data,* think about the following questions:

- What type of measurements or observations will you need to record during your investigation?
- How often will you need to make these measurements or observations?
- What will serve as your dependent variable?
- What will serve as a control condition?
- What types of treatment conditions will you need to set up?
- How many trials will you need to run in each condition?

- How long will you let the simulation run before you collect data?
- How will you keep track of and organize the data you collect?

To determine *how you will analyze the data*, think about the following questions:

- What types of patterns might you look for as you analyze your data?
- How could you use mathematics to describe a change over time?
- How could you use mathematics to document a difference between treatment and control conditions?
- What type of calculations will you need to make?
- What type of graph could you create to help make sense of your data?

Connections to the Nature of Scientific Knowledge and Scientific Inquiry

As you work through your investigation, be sure to think about

- how scientific knowledge can change over time, and
- the types of questions that scientists can investigate.

Initial Argument

Once your group has finished collecting and analyzing your data, your group will need to develop an initial argument. Your initial argument needs to include a claim, evidence to support your claim, and a justification of the evidence. The *claim* is your group's answer to the guiding question. The *evidence* is an analysis and interpretation of your data. Finally, the *justification* of the evidence is why your group thinks the evidence matters. The justification of the evidence is important because scientists can use different kinds of evidence to support their claims. Your group will create your initial argument on a whiteboard. Your whiteboard should include all the information shown in Figure L17.3.

FIGURE L17.3

Argument presentation on a whiteboard

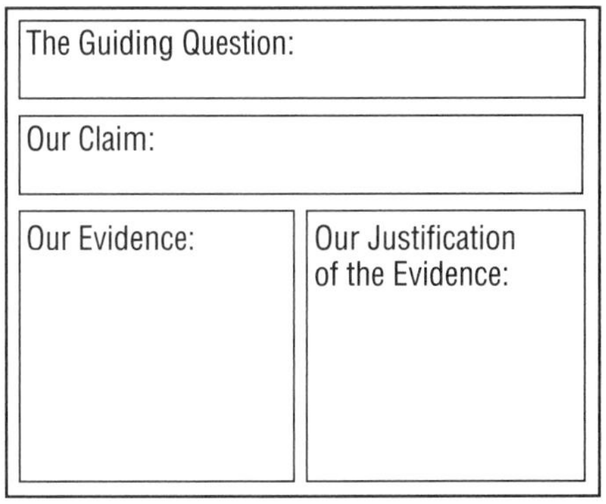

The Guiding Question:	
Our Claim:	
Our Evidence:	Our Justification of the Evidence:

Argumentation Session

The argumentation session allows all of the groups to share their arguments. One or two members of each group will stay at the lab station to share that group's argument, while the other members of the group go to the other lab stations to listen to and critique the other arguments. This is similar to what scientists do when they propose, support, evaluate, and refine new ideas during a poster session at a conference. If you are presenting your group's argument, your goal is to share your ideas and answer questions. You should also keep a record of the critiques and suggestions made by your classmates so you can use this feedback to make your initial argument stronger. You can keep track of specific critiques and suggestions for improvement that your classmates mention in the space below.

Critiques of our initial argument and suggestions for improvement:

If you are critiquing your classmates' arguments, your goal is to look for mistakes in their arguments and offer suggestions for improvement so these mistakes can be fixed. You should look for ways to make your initial argument stronger by looking for things that the other groups did well. You can keep track of interesting ideas that you see and hear during the argumentation in the space below. You can also use this space to keep track of any questions that you will need to discuss with your team.

Interesting ideas from other groups or questions to take back to my group:

Once the argumentation session is complete, you will have a chance to meet with your group and revise your initial argument. Your group might need to gather more data or design a way to test one or more alternative claims as part of this process. Remember, your goal at this stage of the investigation is to develop the best argument possible.

Report

Once you have completed your research, you will need to prepare an *investigation report* that consists of three sections. Each section should provide an answer for the following questions:

1. What question were you trying to answer and why?

2. What did you do to answer your question and why?

3. What is your argument?

Your report should answer these questions in two pages or less. You should write your report using a word processing application (such as Word, Pages, or Google Docs), if possible, to make it easier for you to edit and revise it later. You should embed any diagrams, figures, or tables into the document. Be sure to write in a persuasive style; you are trying to convince others that your claim is acceptable or valid.

References

National Aeronautics and Space Administration (NASA). n.d. Earth's energy budget. *https://earthobservatory.nasa.gov/Features/EnergyBalance/page4.php.*

National Aeronautics and Space Administration (NASA) Goddard Institute for Space Studies. 2016. GISS surface temperature analysis. *https://data.giss.nasa.gov/gistemp/graphs_v3.*

Lab 17. Factors That Affect Global Temperature: How Do Cloud Cover and Greenhouse Gas Concentration in the Atmosphere Affect the Surface Temperature of Earth?

1. Continued use of fossil fuels adds more greenhouse gases to the atmosphere each year. Does this have an impact on short-term temperature changes? In other words, will tomorrow be hotter than today because there will be more greenhouse gases in the atmosphere tomorrow?

2. Does human activity contribute to increased cloud cover? Explain your answer.

3. There are several different types of questions that scientists can ask as part of their investigations.

    a. I agree with this statement.

    b. I disagree with this statement.

    Explain your answer, using an example from your investigation about greenhouse gases and cloud cover.

4. Scientific knowledge can change over time.

    a. I agree with this statement.

    b. I disagree with this statement.

Explain your answer, using an example from your investigation about greenhouse gases and cloud cover.

5. In science, it is critical to understand what makes a system stable or unstable and what controls rates of change in a system. Explain whether the system determining Earth's temperature is stable or unstable. Also, indicate what controls rates of change in Earth's temperature.

6. In science, it is important to track how energy moves into, within, and out of a system. Explain how energy moves into, out of, and within Earth's surface and atmosphere. It may help to draw a diagram. Also, indicate if there is a net gain or loss in energy for Earth's surface and for Earth's atmosphere.

LAB 18

Teacher Notes

Lab 18. Carbon Dioxide Levels in the Atmosphere: How Has the Concentration of Atmospheric Carbon Dioxide Changed Over Time?

Purpose

The purpose of this lab is to *introduce* students to the disciplinary core ideas (DCIs) of (a) Human Impacts on Earth Systems and (b) Global Climate Change by having them determine whether carbon dioxide (CO_2) levels and average global temperature are changing at a different rate than they have in the past. In addition, students have an opportunity to learn about the crosscutting concepts (CCs) of (a) Scale, Proportion, and Quantity; and (b) Stability and Change. During the explicit and reflective discussion, students will also learn about (a) how scientific knowledge changes over time and (b) the assumptions made by scientists about order and consistency in nature.

Important Earth and Space Science Content

The Earth is warming at an unprecedented rate. According to an ongoing temperature analysis conducted by scientists at the National Aeronautics and Space Administration (NASA) Goddard Institute for Space Studies, the average global temperature on Earth has increased by about 0.8°C (1.4°F) since 1880 (see Figure 18.1 and *https://data.giss.nasa. gov/gistemp/graphs_v3*). Scientists agree that this unprecedented rate of warming is due to anthropogenic causes, primarily the increase of greenhouse gases from the burning of fossil fuels. Greenhouse gases affect the amount of energy that is retained by Earth. High-energy waves from the Sun penetrate Earth's atmosphere, warming the oceans and surface. Earth then radiates infrared energy at a lower wavelength. Some of that energy escapes Earth's atmosphere, but much of it is trapped. This effect, called the greenhouse effect, is essential to life on Earth. However, an increase in greenhouse gases exacerbates this effect, causing the widespread rapid warming trend.

Carbon dioxide is one such greenhouse gas released during many human activities; 82% of greenhouse gas released by the United States is CO_2 (EPA 2017). Most greenhouse gas emissions are the by-products of transportation or the production of electricity. The Environmental Protection Agency (2017) reports that from 1990 to 2015, U.S. CO_2 emissions increased 6%. Though data for CO_2 emissions in the recent past are easily accessible, historical data are more difficult to obtain. To determine the amount of CO_2 present in years in Earth's distant past, scientists drill ice cores (see Figure 18.2). Ice cores are obtained from ice sheets in Antarctica or Greenland, where the influence of human activity is small. Scientists examine the trapped gas bubbles in each layer of the ice cores. Because the law of superposition states that the oldest layers within an ice core sample are found at the

FIGURE 18.1

Changes in average global temperature over time

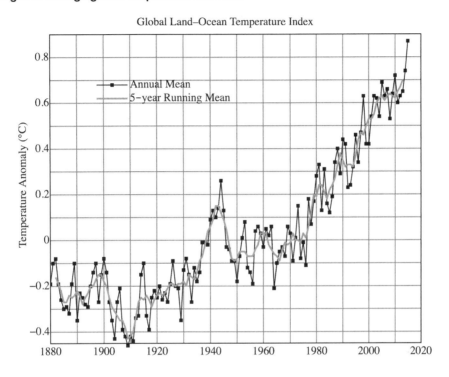

Global Land–Ocean Temperature Index

- Annual Mean
- 5-year Running Mean

(Y-axis: Temperature Anomaly (°C), from −0.4 to 0.8)
(X-axis: 1880, 1900, 1920, 1940, 1960, 1980, 2000, 2020)

bottom of the sample, scientists can count the layers in an ice core sample to determine the age of each layer.

In this investigation, students will analyze a data set of atmospheric CO_2 measurements and temperature anomalies between 1880 and 2016 and atmospheric CO_2 levels dating back 416,000 years which are based on measurements taken from ice cores. Students may examine the rate of increase of CO_2 parts per million for the period of time before and after the Industrial Revolution. The water molecules that make up the ice can be analyzed to get information about temperatures, and other trapped particles like pollen can give clues about other indicators of climate at that time. The rate of increase in atmospheric CO_2 levels after the Industrial Revolution is significantly greater than that before the industrial revolution.

FIGURE 18.2

An example of an ice core

Although Earth has experienced periods of time when the average global temperature rose and fell, the *rate* of this most recent change is the reason scientists are concerned. A few hundred years is an insignificant amount of time on a geologic time scale, and a temperature increase this great in such a relatively small amount of time is truly exceptional within Earth's known history. The rate at which CO_2 has increased within Earth's atmosphere is correlated with the rate at which Earth's average temperature has increased.

Timeline

The instructional time needed to complete this lab investigation is 170–230 minutes. Appendix 3 (p. 573) provides options for implementing this lab investigation over several class periods. Option C (230 minutes) should be used if students are unfamiliar with scientific writing, because this option provides extra instructional time for scaffolding the writing process. You can scaffold the writing process by modeling, providing examples, and providing hints as students write each section of the report. Option D (170 minutes) should be used if students are familiar with scientific writing and have developed the skills needed to write an investigation report on their own. In option D, students complete stage 6 (writing the investigation report) and stage 8 (revising the investigation report) as homework.

Materials and Preparation

The materials needed to implement this investigation are listed in Table 18.1. The Average Global Temperature and Ice Core CO2 Data Excel file can be downloaded from the book's Extras page at *www.nsta.org/adi-ess*. It can be loaded onto student computers before the investigation, e-mailed to students, or uploaded to a class website that students can access. It is important that Excel be available on the computers that students will use so they can analyze the data set using the tools built into the spreadsheet application. It is also important for you to look over the file before the investigation begins so you can learn how the data in the file are organized. This will enable you to give students suggestions on how to analyze the data. Students also need access to visualizations of trends in sea ice levels, CO_2 levels, and global temperatures provided by NASA at *http://climate.nasa.gov/interactives/climate-time-machine*.

TABLE 18.1

Materials list for Lab 18

Item	Quantity
Computer with Excel (or other spreadsheet application) and internet access	1 per group
Graphing calculator or computer with graphing software (optional)	1 per group
Average Global Temperature and Ice Core CO2 Data Excel file	1 per group
Investigation Proposal A (optional)	1 per group
Whiteboard, 2' × 3'*	1 per group
Lab Handout	1 per student
Peer-review guide and teacher scoring rubric	1 per student
Checkout Questions	1 per student

* As an alternative, students can use computer and presentation software such as Microsoft PowerPoint or Apple Keynote to create their arguments.

Safety Precautions

Remind students to follow all normal lab safety rules.

Topics for the Explicit and Reflective Discussion

Reflecting on the Use of Core Ideas and Crosscutting Concepts During the Investigation

Teachers should begin the explicit and reflective discussion by asking students to discuss what they know about the DCIs they used during the investigation. The following are some important concepts related to the DCIs of (a) Human Impacts on Earth Systems and (b) Global Climate Change that students need to determine how the temperature of Earth responds to changes in the amount of cloud cover and the concentration of CO_2 in the atmosphere:

- The layers of an ice core sample correspond to time periods.

- Based on the law of superposition, the oldest layers within an ice core sample are found at the bottom of the sample.

- Dissolved CO_2 levels in an ice core layer correspond to the amount of CO_2 present in the atmosphere during the time the ice core layer was created.

To help students reflect on what they know about these concepts, we recommend showing them two or three images using presentation software that help illustrate these important ideas. You can then ask the students the following questions to encourage students to share how they are thinking about these important concepts:

1. What do we see going on in this image?

2. Does anyone have anything else to add?

3. What might be going on that we can't see?

4. What are some things that we are not sure about here?

You can then encourage students to think about how CCs played a role in their investigation. There are at least two CCs that students need to determine how the temperature of Earth responds to changes in the amount of cloud cover and the concentration of CO_2 in the atmosphere: (a) Scale, Proportion, and Quantity; and (b) Stability and Change (see Appendix 2 [p. 569] for a brief description of these CCs). To help students reflect on what they know about these CCs, we recommend asking them the following questions:

1. Why is it important to consider what measurement scale or scales to use during an investigation? Why is useful to look for proportional relationships when analyzing data?

2. What measurement scale or scales did you use during your investigation? What did that allow you to do? Did you attempt to look for proportional relationships when you were analyzing your data? Why or why not?

3. Why is it important to think about what controls or affects the rate of change in system? How can we measure a rate of change?

4. Which factor(s) might have controlled the rate of change in atmospheric CO_2 in your investigation? What did exploring these factors allow you to do?

You can then encourage students to think about how they used all these different concepts to help answer the guiding question and why it is important to use these ideas to help justify their evidence for their final arguments. Be sure to remind your students to explain why they included the evidence in their arguments and make the assumptions underlying their analysis and interpretation of the data explicit in order to provide an adequate justification of their evidence.

Reflecting on Ways to Design Better Investigations

It is important for students to reflect on the strengths and weaknesses of the investigation they designed during the explicit and reflective discussion. Students should therefore be encouraged to discuss ways to eliminate potential flaws, measurement errors, or sources of uncertainty in their investigations. To help students be more reflective about the design of their investigation and what they can do to make their investigations more rigorous in the future, you can ask the following questions:

1. What were some of the strengths of the way you planned and carried out your investigation? In other words, what made it scientific?

2. What were some of the weaknesses of the way you planned and carried out your investigation? In other words, what made it less scientific?

3. What rules can we make, as a class, to ensure that our next investigation is more scientific?

Reflecting on the Nature of Scientific Knowledge and Scientific Inquiry

This investigation can be used to illustrate two important concepts related to the nature of scientific knowledge and the nature of scientific inquiry: (a) how scientific knowledge changes over time and (b) the assumptions made by scientists about order and consistency in nature (see Appendix 2 [p. 569] for a brief description of these two concepts). Be sure to review these concepts during and at the end of the explicit and reflective discussion. To help students think about these concepts in relation to what they did during the lab, you can ask the following questions:

- Scientific knowledge can and does change over time. Can you tell me why it changes?

- Can you work with your group to come up with some examples of how scientific knowledge related to our understanding of average global temperature change has changed over time? Be ready to share in a few minutes.

- Scientists assume that natural laws operate today as they did in the past and that they will continue to do so in the future. Why do you think this assumption is important?

- Think about what you were trying to do during this investigation. What would you have had to do differently if you could not assume natural laws operate today as they did in the past?

You can also use presentation software or other techniques to encourage your students to think about these concepts. You can show examples of how our thinking about average global temperature change has changed over time and ask students to discuss what they think led to those changes. You can also show images of different scientific laws (such as the law of universal gravitation, the law of conservation of mass, or the law of superposition) and ask students if they think these laws have been the same throughout Earth's history. Then ask them to think about what scientists would need to do to be able to study the past if laws are not consistent through time and space.

Remind your students that, to be proficient in science, it is important that they understand what counts as scientific knowledge and how that knowledge develops over time.

LAB 18

Hints for Implementing the Lab

- Examine the Average Global Temperature and Ice Core CO2 Data Excel file before the lab begins. It is important for you to know what is included in the file and how you can analyze these data so you can help students when they get stuck or confused.

- A group of three students per computer tends to work well.

- Allow the students to play with the Excel file as part of the tool talk before they begin to design their investigation. This gives students a chance to see what they can and cannot do with the data in Excel (or another spreadsheet application).

- Allowing students to decide how to analyze the data in the Excel file gives students an opportunity to try, to fail, and to learn from their mistakes. However, you can scaffold students as they attempt to decide what to do by having them fill out an investigation proposal. These proposals provide a way for you to offer students hints and suggestions without telling them how to do it. You can also check the proposals quickly during a class period. For this lab we suggest using Investigation Proposal A.

- The best way to help students to learn how to use Excel (or other spreadsheet application) is to provide "just-in-time" instruction. In other words, wait for students to get stuck and then give a brief mini-lesson on how to use a specific tool in the application based on what students are trying to do. They will be much more interested in learning about how to use the tools in the application if they know it will help solve a problem they are having or it will allow them to accomplish one of their goals.

- Students often make mistakes as they attempt to analyze the data, but they should quickly realize these mistakes during the argumentation session. It will only take them a short period of time to reanalyze the data, and they should be allowed to do so. During the explicit and reflective discussion, students will also have the opportunity to reflect on and identify different ways to analyze data. This also offers an opportunity to discuss what scientists do when they realize that a mistake is made during a study.

- This lab provides an excellent opportunity to discuss how scientists must make choices about how to analyze the data they have and how the choice of analysis reflects the nature of the question they are trying to answer. Be sure to use this activity as a concrete example during the explicit and reflective discussion.

- This lab also provides an excellent opportunity to discuss how scientists identify a signal (a pattern or trend) from the noise (measurement error) in their data. Be sure to use this activity as a concrete example during the explicit and reflective discussion.

Connections to Standards

Table 18.2 highlights how the investigation can be used to address specific (a) performance expectations from the *NGSS* and (b) *Common Core State Standards* in English language arts (*CCSS ELA*).

TABLE 18.2

Lab 18 alignment with standards

***NGSS* performance expectation**	Weather and climate • MS-ESS3-5: Ask questions to clarify evidence of the factors that have caused the rise in global temperatures over the past century.
***CCSS ELA*—Reading in Science and Technical Subjects**	Key ideas and details • CCSS.ELA-LITERACY.RST.6-8.1: Cite specific textual evidence to support analysis of science and technical texts. • CCSS.ELA-LITERACY.RST.6-8.2: Determine the central ideas or conclusions of a text; provide an accurate summary of the text distinct from prior knowledge or opinions. Craft and structure • CCSS.ELA-LITERACY.RST.6-8.4: Determine the meaning of symbols, key terms, and other domain-specific words and phrases as they are used in a specific scientific or technical context relevant to *grade 6–8 texts and topics*. • CCSS.ELA-LITERACY.RST.6-8.5: Analyze the structure an author uses to organize a text, including how the major sections contribute to the whole and to an understanding of the topic. • CCSS.ELA-LITERACY.RST.6-8.6: Analyze the author's purpose in providing an explanation, describing a procedure, or discussing an experiment in a text. Integration of knowledge and ideas • CCSS.ELA-LITERACY.RST.6-8.7: Integrate quantitative or technical information expressed in words in a text with a version of that information expressed visually (e.g., in a flowchart, diagram, model, graph, or table). • CCSS.ELA-LITERACY.RST.6-8.8: Distinguish among facts, reasoned judgment based on research findings, and speculation in a text. • CCSS.ELA-LITERACY.RST.6-8.9: Compare and contrast the information gained from experiments, simulations, video, or multimedia sources with that gained from reading a text on the same topic.

Continued

TABLE 18.2 (*continued*)

***CCSS ELA*—Writing in Science and Technical Subjects**	Text types and purposes • CCSS.ELA-LITERACY.WHST.6-8.1: Write arguments focused on discipline-specific content. • CCSS.ELA-LITERACY.WHST.6-8.2: Write informative or explanatory texts, including the narration of historical events, scientific procedures/experiments, or technical processes. Production and distribution of writing • CCSS.ELA-LITERACY.WHST.6-8.4: Produce clear and coherent writing in which the development, organization, and style are appropriate to task, purpose, and audience. • CCSS.ELA-LITERACY.WHST.6-8.5: With some guidance and support from peers and adults, develop and strengthen writing as needed by planning, revising, editing, rewriting, or trying a new approach, focusing on how well purpose and audience have been addressed. • CCSS.ELA-LITERACY.WHST.6-8.6: Use technology, including the internet, to produce and publish writing and present the relationships between information and ideas clearly and efficiently. Range of writing • CCSS.ELA-LITERACY.WHST.6-8.10: Write routinely over extended time frames (time for reflection and revision) and shorter time frames (a single sitting or a day or two) for a range of discipline-specific tasks, purposes, and audiences.
***CCSS ELA*—Speaking and Listening**	Comprehension and collaboration • CCSS.ELA-LITERACY.SL.6-8.1: Engage effectively in a range of collaborative discussions (one-on-one, in groups, and teacher-led) with diverse partners on grade 6–8 topics, texts, and issues, building on others' ideas and expressing their own clearly. • CCSS.ELA-LITERACY.SL.6-8.2:* Interpret information presented in diverse media and formats (e.g., visually, quantitatively, orally) and explain how it contributes to a topic, text, or issue under study. • CCSS.ELA-LITERACY.SL.6-8.3:* Delineate a speaker's argument and specific claims, distinguishing claims that are supported by reasons and evidence from claims that are not. Presentation of knowledge and ideas • CCSS.ELA-LITERACY.SL.6-8.4:* Present claims and findings, sequencing ideas logically and using pertinent descriptions, facts, and details to accentuate main ideas or themes; use appropriate eye contact, adequate volume, and clear pronunciation.

Continued

TABLE 18.2 (*continued*)

CCSS ELA—**Speaking and Listening** (*continued*)	Presentation of knowledge and ideas (*continued*) • CCSS.ELA-LITERACY.SL.6-8.5:* Include multimedia components (e.g., graphics, images, music, sound) and visual displays in presentations to clarify information. • CCSS.ELA-LITERACY.SL.6-8.6: Adapt speech to a variety of contexts and tasks, demonstrating command of formal English when indicated or appropriate.

* Only the standard for grade 6 is provided because the standards for grades 7 and 8 are similar. Please see *www.corestandards.org/ELA-Literacy/SL* for the exact wording of the standards for grades 7 and 8.

Reference

U.S. Environmental Protection Agency (EPA). 2017. Carbon dioxide emissions. *www.epa.gov/ghgemissions/overview-greenhouse-gases*.

Lab Handout

Lab 18. Carbon Dioxide Levels in the Atmosphere: How Has the Concentration of Atmospheric Carbon Dioxide Changed Over Time?

Introduction

There has been a lot of discussion about climate in recent years. This discussion usually focuses on average global temperature. In the United States some states have had above-average temperatures, some states have had below-average temperatures, and some states have had had near-average temperatures over the last 100 years. Figure L18.1 shows decadal temperature anomalies, or how the decadal average temperature for each state differs from the 20th-century average during three different decades. According to an ongoing temperature analysis conducted by scientists at the National Aeronautics and Space Administration (NASA) Goddard Institute for Space Studies, the average global temperature on Earth has increased by about 0.8°Celsius (1.4°Fahrenheit) since 1880 (see *https://data.giss.nasa.gov/gistemp/graphs_v3*).

A major contributing factor to global temperature is the concentration of carbon dioxide (CO_2) in the atmosphere. CO_2 is one type of greenhouse gas. Greenhouse gases trap heat from the Sun and warm the surface of Earth. Without greenhouse gases in the atmosphere, Earth would be too cold for humans to survive. As the concentration of greenhouse gases in the atmosphere increases, the temperature of Earth's surface will also increase.

Climate experts agree that human activity has significantly increased the amount of CO_2 in the atmosphere, leading to an overall rise in global temperatures. Some people, however,

FIGURE L18.1

Decadal average temperature maps

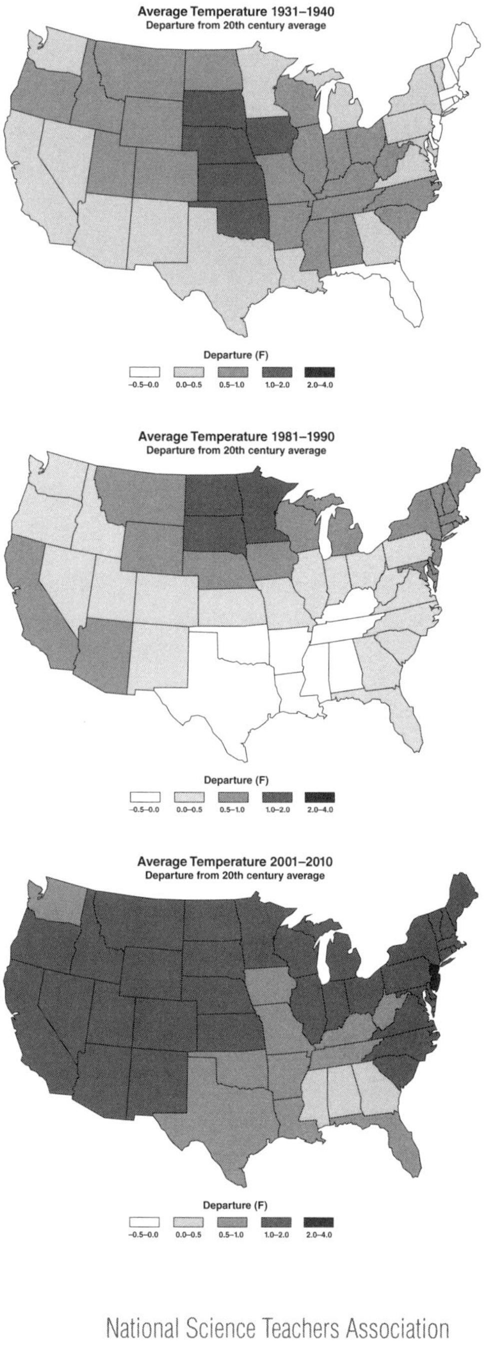

Average Temperature 1931–1940
Departure from 20th century average

Departure (F)
-0.5–0.0   0.0–0.5   0.5–1.0   1.0–2.0   2.0–4.0

Average Temperature 1981–1990
Departure from 20th century average

Departure (F)
-0.5–0.0   0.0–0.5   0.5–1.0   1.0–2.0   2.0–4.0

Average Temperature 2001–2010
Departure from 20th century average

Departure (F)
-0.5–0.0   0.0–0.5   0.5–1.0   1.0–2.0   2.0–4.0

still question whether this increase is primarily due to human activity or to a natural process that causes climate change. There is research that shows global temperatures and atmospheric CO_2 levels have increased and decreased in a cyclical pattern for at least 650,000 years (Etheridge et al. 1998).

Before you can evaluate the merits of alternative explanations for the observed increase in average global temperature, it is important to understand how CO_2 levels have changed over Earth's history. You will therefore need to learn more about historical patterns of CO_2 levels.

Your Task

Analyze long-term historical data to determine whether CO_2 levels and average global temperature are changing at a different rate than they have in the past. Your goal is to use what you know about climate, patterns, and stability and change in systems to determine if human activity has made a significant change in global CO_2 levels and thus global temperature.

The guiding question of this investigation is: *How has the concentration of atmospheric carbon dioxide changed over time?*

Materials

You may use the following resources during your investigation:

- Average Global Temperature and Ice Core CO2 Data Excel file: This file provides information about changes in average global temperature over time and atmospheric CO_2 levels based on ice core samples that date back to 416,000 years before the present time.

- Climate Time Machine: This NASA website provides visualizations of current trends in sea ice levels, CO_2 levels, and global temperatures at *http://climate.nasa. gov/interactives/climate-time-machine*.

Safety Precautions

Be sure to follow all normal lab safety rules.

Investigation Proposal Required? ☐ Yes ☐ No

Getting Started

Scientists use some clever data sources to gain insight into Earth's history. One such data source is an ice core sample. To obtain an ice core sample, scientists drill down into a glacier or ice sheet and bring out a long cylindrical piece of ice (see Figure L18.2, p. 440). Scientists can then count the layers in the ice core sample and determine how many years

FIGURE L18.2

An example of an ice core

ago each layer was on the surface of Earth. The ability to determine the age of layers by counting them is based on the law of superposition, which states that the oldest layers in a geologic sample are found at the bottom of the sample.

Scientists can also analyze the tiny air bubbles that are trapped in the ice at each layer of an ice core sample to determine the amounts of different gases that were in the atmosphere at the time that layer was created. When scientists make these measurements, they assume that natural laws operate today as they did in the past and that they will continue to do so in the future. Scientists therefore assume that the dissolved CO_2 levels in an ice core layer correspond to the amount of CO_2 present in the atmosphere at the time the ice was made, just like dissolved CO_2 levels in fresh ice match the CO_2 levels in the current atmosphere.

The Average Global Temperature and Ice Core CO2 Data Excel file includes information about the atmospheric CO_2 concentration and changes in average global temperature over time. The file includes two tabs:

1. The first tab, which is called "CO2 and Temp 1880-2016," includes atmospheric CO_2 levels and average global temperature anomalies from 1880 to 2016. The term *temperature anomaly* means the difference from the long-term average. A positive anomaly value indicates that the observed temperature was warmer than the long-term average, and a negative anomaly indicates that the observed temperature was cooler than the long-term average. Scientists calculate and report temperature anomalies because they more accurately describe climate variability than absolute temperatures do, and these anomalies make it easier to find patterns in temperature trends. The yearly temperature anomaly values come from the National Oceanic and Atmospheric Administration's National Centers for Environmental Information (see *www.ncdc.noaa.gov/monitoring-references/faq/anomalies.php*), the 1880–2004 atmospheric CO_2 levels come from Etheridge et al. (2010), and the 2005–2016 atmospheric CO_2 levels come from NASA's Global Climate Change website (see *https://climate.nasa.gov/vital-signs/carbon-dioxide*).

2. The second tab, which is called "CO2 Before 1880," includes atmospheric CO_2 levels dating back 416,000 years based on measurements taken from ice cores by Etheridge et al. (1998).

You can use the data from the Excel file to see how CO_2 concentrations and global temperature typically change over a very long time scale. As you analyze these data, think about the following questions:

- Will you need to analyze some data separately from others?
- What types of patterns might you look for as you analyze your data?
- What type of diagram could you create to help make sense of your data?
- How could you use mathematics to describe a change over time or if there is a relationship between variables?
- What type of graph could you create to help make sense of your data?

You can also use the visualizations on NASA's Climate Time Machine web page to examine how some of Earth's key climate indicators have changed in the recent past. This web page provides satellite pictures of the annual Arctic sea ice minimums dating back to 1979. At the end of each summer, the sea ice cover reaches its minimum extent, leaving what is called the perennial ice cover. The Climate Time Machine also shows global changes in the concentration and distribution of CO_2 in the atmosphere dating back to 2002 at an altitude range of 1.9–8 miles. The yellow-to-red regions indicate higher concentrations of CO_2, while the blue-to-green areas indicate lower concentrations, measured in parts per million. Finally, and perhaps most important, the Climate Time Machine provides a color-coded map that shows how global surface temperatures have changed dating back to 1884. Dark blue indicates areas cooler than average, and dark red indicates areas warmer than average.

Connections to the Nature of Scientific Knowledge and Scientific Inquiry

As you work through your investigation, be sure to think about

- how scientific knowledge can change over time, and
- the assumptions made by scientists about order and consistency in nature.

Initial Argument

Once your group has finished collecting and analyzing your data, your group will need to develop an initial argument. Your initial argument needs to include a claim, evidence to support your claim, and a justification of the evidence. The *claim* is your group's answer to the guiding question. The *evidence* is an analysis and interpretation of your data. Finally, the *justification* of the evidence is why your group thinks the evidence matters. The justification of the evidence is important because scientists can use different kinds of evidence to support their claims. Your group will create your initial argument on a whiteboard. Your whiteboard should include all the information shown in Figure L18.3 (p. 442).

LAB 18

FIGURE L18.3 _____

Argument presentation on a whiteboard

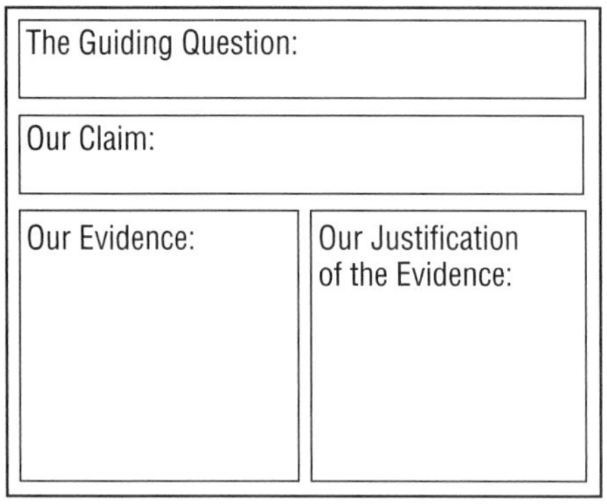

The Guiding Question:	
Our Claim:	
Our Evidence:	Our Justification of the Evidence:

Argumentation Session

The argumentation session allows all of the groups to share their arguments. One or two members of each group will stay at the lab station to share that group's argument, while the other members of the group go to the other lab stations to listen to and critique the other arguments. This is similar to what scientists do when they propose, support, evaluate, and refine new ideas during a poster session at a conference. If you are presenting your group's argument, your goal is to share your ideas and answer questions. You should also keep a record of the critiques and suggestions made by your classmates so you can use this feedback to make your initial argument stronger. You can keep track of specific critiques and suggestions for improvement that your classmates mention in the space below.

Critiques of our initial argument and suggestions for improvement:

If you are critiquing your classmates' arguments, your goal is to look for mistakes in their arguments and offer suggestions for improvement so these mistakes can be fixed. You should look for ways to make your initial argument stronger by looking for things that the other groups did well. You can keep track of interesting ideas that you see and hear during the argumentation in the space below. You can also use this space to keep track of any questions that you will need to discuss with your team.

Interesting ideas from other groups or questions to take back to my group:

Once the argumentation session is complete, you will have a chance to meet with your group and revise your initial argument. Your group might need to gather more data or design a way to test one or more alternative claims as part of this process. Remember, your goal at this stage of the investigation is to develop the best argument possible.

Report

Once you have completed your research, you will need to prepare an *investigation report* that consists of three sections. Each section should provide an answer for the following questions:

1. What question were you trying to answer and why?

2. What did you do to answer your question and why?

3. What is your argument?

Your report should answer these questions in two pages or less. You should write your report using a word processing application (such as Word, Pages, or Google Docs), if possible, to make it easier for you to edit and revise it later. You should embed any diagrams, figures, or tables into the document. Be sure to write in a persuasive style; you are trying to convince others that your claim is acceptable or valid.

References

Etheridge, D. M, L. P. Steele, R. L. Langenfelds, R. J. Francey, J. Barnola, V. I. and Morgan. 1998. Historical CO_2 records from the Law Dome DE08, DE08-2, and DSS ice cores. In *Trends: A compendium of data on global change*. Oak Ridge, TN: U.S. Department of Energy, Oak Ridge National Laboratory, Carbon Dioxide Information Analysis Center. Available at *www.co2.earth/co2-ice-core-data*.

Etheridge, et al. 2010. Law Dome Ice Core 2000-Year CO2, CH4, and N2O Data. IGBP PAGES/World Data Center for Paleoclimatology Data Contribution Series 2010-070. Boulder, CO: NOAA/NCDC Paleoclimatology Program. Available at *ftp://ftp.ncdc.noaa.gov/pub/data/paleo/icecore/antarctica/law/law2006.txt*.

LAB 18

Lab 18. Carbon Dioxide Levels in the Atmosphere: How Has the Concentration of Atmospheric Carbon Dioxide Changed Over Time?

1. Sketch a graph of the how the concentration of atmospheric carbon dioxide has changed over time.

2. A scientist collects yearly global average temperature and compiles it into the table below.

Year	Average temperature
1880	56.8°F
1900	56.9°F
1920	56.7°F
1930	56.9°F
1940	57.3°F
1960	57.1°F
1980	57.6°F
2000	57.9°F
2010	58.4°F

a. What is the rate of change for the time period between1880 and 1930 and the time period between 1960 and 2010?

b. Are the rates of change significantly different from one another?

c. How do you know?

d. What additional information would you need to determine whether the global climate is stable?

3. Once new scientific knowledge is developed, it will not be abandoned or modified in light of new evidence.

   a. I agree with this statement.

   b. I disagree with this statement.

Explain your answer, using an example from your investigation about carbon dioxide levels in the atmosphere.

4. Science assumes that objects and events in natural systems occur in consistent patterns that are understandable through measurement and observation.

   a. I agree with this statement.
   b. I disagree with this statement.

   Explain your answer, using an example from your investigation about carbon dioxide levels in the atmosphere.

5. It is critical for scientists to be able to recognize what is relevant at different time frames and scales. Explain why analyzing data in the context of appropriate time frames and scales is important, using an example from your investigation about carbon dioxide levels in the atmosphere.

6. It is critical to understand what makes a system stable or unstable and what controls rates of change in a system. Explain why it is important to determine whether a system is changing or is stable, using an example from your investigation about trends in average global temperatures.

Application Lab

Teacher Notes

Lab 19. Differences in Regional Climate: Why Do Two Cities Located at the Same Latitude and Near a Body of Water Have Such Different Climates?

Purpose

The purpose of this lab is for students to *apply* what they know about the disciplinary core idea (DCI) of Weather and Climate by having them develop a conceptual model that explains the temperature and precipitation patterns in different cities. In addition, students have an opportunity to learn about the crosscutting concepts (CCs) of (a) Patterns and (b) Systems and System Models. During the explicit and reflective discussion, students will also learn about (a) how models are used as tools for reasoning about natural phenomena and (b) how the culture of science, societal needs, and current events influence the work of scientists.

Important Earth and Space Science Content

Several factors affect the climate of a given region:

- *Latitude.* This factor is important because it determines changes in day length and sun angle throughout the year.

- *Elevation.* Generally, as the elevation of a region increases, the temperature of that region decreases.

- *Proximity to a large body of water.* Land heats and cools faster than water. Water can also store more heat energy than land. This makes the climate of a region that is located near a large body of water, such as an ocean, more moderate because the water absorbs extra heat energy during the summer and releases heat into the air during winter.

- *The nature of any nearby ocean currents.* Ocean currents move large amounts of water to different locations across the Earth (see Figure 19.1). These vast masses of water can be warm or cold. Regions that are located near a current that brings warm water to the area tend to be warmer than regions that are located near currents that transport cold water from the poles to the area.

- *The direction and strength of the prevailing winds in the region.* Winds can move air masses with specific properties from a source region to a different region. In the Northern Hemisphere, winds tend to blow from west to east (westerly winds) in the mid-latitudes and northeast to southwest between the Tropic of Cancer and the equator (see Figure 19.2).

- *Local topography.* The presence or absence of a mountain, for example, can affect precipitation patterns and therefore climate in a region.

FIGURE 19.1

Map of major ocean currents (blue lines represent cold-water currents and red lines represent warm-water currents)

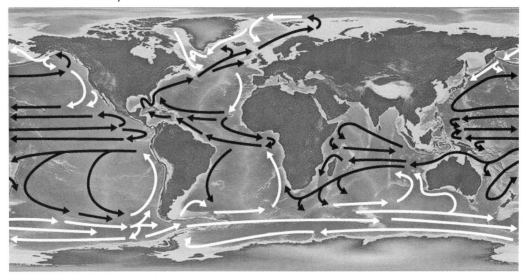

Note: A full-color version of this figure is available on the book's Extras page at *www.nsta.org/adi-ess*

FIGURE 19.2

Map of prevailing wind directions

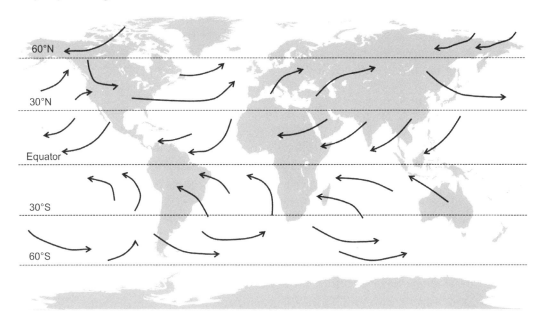

In this investigation, students must develop a conceptual model that they can use to explain why San Francisco, California, and Norfolk, Virginia, have different climates. San Francisco is located on the coast of the Pacific Ocean at 37.7° N latitude. It has mild summers and winters; the average high temperature between 1945 and 2017 has been 17.1°C (62.8°F) for July and 9.7°C (49.4°F) for January. San Francisco is very dry, averaging 19.7 inches of rain per year from 1945 to 2017. (Climate information from the National Oceanic and Atmospheric Administration [NOAA] National Centers for Environmental Information is available at *www.ncdc.noaa.gov.*) According to the Köppen Climate Classification System (the most widely used system for classifying the world's climates; see *www.britannica.com/ science/Koppen-climate-classification*), San Francisco has a temperate Mediterranean climate with warm summers (denoted Csb in the Köppen system). In contrast, Norfolk, which is located on the Atlantic Ocean at 36.9° N latitude, has hot summers and mild winters. The average high temperature between 1945 and 2017 has been 30.9°C (87.6°F) in July and 9.5°C (49.1°F) in January. Between 1945 and 2017, Norfolk has averaged 46.4 inches of rain a year. According to the Köppen classification system, Norfolk has a humid subtropical climate (Cfa in the Köppen system). In summary, these two cities have very different temperature and precipitation patterns throughout the year even though they are located at similar latitudes and elevations and are both located near an ocean.

The climate differences between San Francisco and Norfolk are caused by their respective locations on the North American continent. Both cities have westerly prevailing winds because they are in the mid-latitudes of the Northern Hemisphere. The winds that blow over San Francisco, however, form over the Pacific Ocean. Since the winds originate over a body of water, the winds will be relatively cool during warm months and warm during cold ones. In contrast, winds that blow over Norfolk form over continental regions, so an ocean does not moderate the wind's temperature. Norfolk receives warm prevailing winds in the summer months and cool prevailing winds during winter months. The California current in the Pacific Ocean and the Gulf Stream in the Atlantic Ocean also play a part in their climate differences. The California current transports cold water from the north to San Francisco, whereas the Gulf Stream transports warm water from the tropical and subtropical Atlantic (including some from the Gulf of Mexico) to Norfolk. The water off the coast of San Francisco, as a result, is much colder than the water off the coast of Norfolk. Warm ocean water off the coast of Norfolk results in more precipitation year-round and higher temperatures in the summer. In summary, despite the fact that San Francisco and Norfolk are located at similar latitudes and elevations and are both coastal cities, they have very different climates because of the different origins of prevailing winds and ocean currents in their respective regions.

Once the students have a conceptual model that they can use to explain why San Francisco and Norfolk have different season temperatures and rain patterns, they will need to test their model to see if it allows them to predict the climates of other pairs of cities. Examples of cities that are located at similar latitudes and elevations but on different coasts include

- San Diego, California, and Charleston, South Carolina;

- Portland, Oregon, and Bangor, Maine;

- Eureka, California, and New York City; and

- Santa Monica, California, and Wilmington, North Carolina.

In general, coastal cities on the West Coast of North America will have smaller average temperature ranges between the hottest months and coldest months than will cities on the East Coast of North America. For more climate data on specific cities, we recommend using the following website: *www.weatherbase.com/weather/state. php3?c=US&s=&countryname=United-States.*

Timeline

The instructional time needed to complete this lab investigation is 270–330 minutes. Appendix 3 (p. 573) provides options for implementing this lab investigation over several class periods. Option G (330 minutes) should be used if students are unfamiliar with scientific writing, because this option provides extra instructional time for scaffolding the writing process. You can scaffold the writing process by modeling, providing examples, and providing hints as students write each section of the report. Option H (270 minutes) should be used if students are familiar with scientific writing and have developed the skills needed to write an investigation report on their own. In option H, students complete stage 6 (writing the investigation report) and stage 8 (revising the investigation report) as homework. Both options give students time to test their models on day 3.

Materials and Preparation

The materials needed to implement this investigation are listed in Table 19.1 (p. 452). Students will need to access the following online resources:

- The U.S. Climate Data website (*www.usclimatedata.com/climate/united-states/us*)

- Wind rose charts showing wind speed and direction for 237 U.S. cities, available at the National Oceanic and Atmospheric Administration (NOAA) Climate.gov website (*www.climate.gov/maps-data/dataset/monthly-wind-rose-plots-charts*)

- Earth: A Global Map of Wind, Weather, and Ocean Conditions, an interactive animated map showing current wind speeds and direction for the entire planet (*https://earth.nullschool.net*)

- *My NASA Data.* You can access this database, which includes information about ocean surface temperatures and the average wind speed and direction by month over the entire year, at *https://mynasadata.larc.nasa.gov.*

- *State of the Ocean* (SOTO). You can access this visualization tool, which includes information about current and past ocean currents and changes in surface temperature, through the NASA Jet Propulsion Laboratory Physical

Oceanography Distributed Active Archive Center at *https://podaac.jpl.nasa.gov.* Click on "Data Access" and then "SOTO (State of the Ocean)."

All of these resources are free to use. You should access each website and learn how to find desired data before beginning the lab investigation. In addition, it is important to check if students can access and use the websites from a school computer or tablet, because some schools have set up firewalls and other restrictions on web browsing.

TABLE 19.1

Materials list for Lab 19

Item	Quantity
Computer or tablet with internet access	1 per group
Whiteboard, 2' × 3'*	1 per group
Lab Handout	1 per student
Peer-review guide and teacher scoring rubric	1 per student
Checkout Questions	1 per student

* As an alternative, students can use computer and presentation software such as Microsoft PowerPoint or Apple Keynote to create their arguments.

Safety Precautions

Remind students to follow all normal lab safety rules.

Topics for the Explicit and Reflective Discussion

Reflecting on the Use of Core Ideas and Crosscutting Concepts During the Investigation

Teachers should begin the explicit and reflective discussion by asking students to discuss what they know about the DCI they used during the investigation. The following are some important concepts related to the DCI of Weather and Climate that students need to develop a conceptual model that explains the temperature and precipitation patterns in different cities:

- *Climate* is the average weather for a particular location over a long period of time.

- *Weather* is the current atmospheric conditions.

- Interactions between sunlight, the atmosphere, landforms, and ocean temperatures affect weather and climate.

- The ocean exerts a major influence on weather and climate by absorbing energy from the sun, releasing it over time, and redistributing it around the globe.

- Latitude, elevation, geographic position, and ocean currents affect the temperature of a region.

- Wind patterns and local topography affect the precipitation patterns of a region.
- Winds in the middle latitudes (approximately 30° to 60° in both the Northern and Southern Hemispheres) blow from west to east and are called westerlies.

To help students reflect on what they know about these concepts, we recommend showing them two or three images using presentation software that help illustrate these important ideas. You can then ask the students the following questions to encourage students to share how they are thinking about these important concepts:

1. What do we see going on in this image?
2. Does anyone have anything else to add?
3. What might be going on that we can't see?
4. What are some things that we are not sure about here?

You can then encourage students to think about how CCs played a role in their investigation. There are at least two CCs that students need to develop a conceptual model that explains the temperature and precipitation patterns in different cities: (a) Patterns and (b) Systems and System Models (see Appendix 2 [p. 569] for a brief description of these CCs). To help students reflect on what they know about these CCs, we recommend asking them the following questions:

1. Why do scientists look for and attempt to explain patterns in nature?
2. What patterns did you identify and use during your investigation? Why was that useful?
3. Why do scientists often define a system and then develop a model of it as part of an investigation?
4. How did you use a model to understand the factors that affect regional climates? Why was that useful?

You can then encourage the students to think about how they used all these different concepts to help answer the guiding question and why it is important to use these ideas to help justify their evidence for their final arguments. Be sure to remind your students to explain why they included the evidence in their arguments and make the assumptions underlying their analysis and interpretation of the data explicit in order to provide an adequate justification of their evidence.

Reflecting on Ways to Design Better Investigations

It is important for students to reflect on the strengths and weaknesses of the investigation they designed during the explicit and reflective discussion. Students should therefore be encouraged to discuss ways to eliminate potential flaws, measurement errors, or sources of uncertainty in their investigations. To help students be more reflective about the design of their investigation and what they can do to make their investigations more rigorous in the future, you can ask the following questions:

1. What were some of the strengths of the way you planned and carried out your investigation? In other words, what made it scientific?

2. What were some of the weaknesses of the way you planned and carried out your investigation? In other words, what made it less scientific?

3. What rules can we make, as a class, to ensure that our next investigation is more scientific?

Reflecting on the Nature of Scientific Knowledge and Scientific Inquiry

This investigation can be used to illustrate two important concepts related to the nature of scientific knowledge and the nature of scientific inquiry: (a) how models are used as tools for reasoning about natural phenomena and (b) how the culture of science, societal needs, and current events influence the work of scientists (see Appendix 2 [p. 569] for a brief description of these two concepts). Be sure to review these concepts during and at the end of the explicit and reflective discussion. To help students think about these concepts in relation to what they did during the lab, you can ask the following questions:

- I asked you to develop a model to explain the climates of San Francisco and Norfolk as part of your investigation. Why is it useful to develop models in science?

- Can you work with your group to come up with a rule that you can use to decide what a model is and what a model is not in science? Be ready to share in a few minutes.

- People view some types of research as being more important than other types of research because of cultural values and current events. Can you come up with some examples of how cultural values and current events have influenced the work of climate scientists?

You can also use presentation software or other techniques to encourage your students to think about these concepts. You can show examples and non-examples of scientific models and ask students to classify each one and explain their thinking. You can also show examples of research projects that were influenced by cultural values and current events and ask students to think about what was going on in society when the research was conducted and why that research was viewed as being important for the greater good.

Remind your students that, to be proficient in science, it is important that they understand what counts as scientific knowledge and how that knowledge develops over time.

Hints for Implementing the Lab

- Learn how to use all the online resources before the lab begins. It is important for you to know how to use these resources so you can help students when they get stuck or confused.

- A group of three students per computer or tablet tends to work well.

- Allow the students to explore the online resources as part of the tool talk before they begin to design their investigation. This gives students a chance to see what they can and cannot do with these resources.

- Students often make mistakes when developing their conceptual models and/or initial arguments, but they should quickly realize these mistakes during the argumentation session. Be sure to allow students to revise their models and arguments at the end of the argumentation session. The explicit and reflective discussion will also give students an opportunity to reflect on and identify ways to improve how they develop and test models. This also offers an opportunity to discuss what scientists do when they realize a mistake is made.

- Be sure that students record actual values (e.g., wind speeds and direction) and are not just attempting to hand draw what they see on the computer screen.

Connections to Standards

Table 19.2 highlights how the investigation can be used to address specific (a) performance expectations from the *NGSS* and (b) *Common Core State Standards* in English language arts (*CCSS ELA*).

TABLE 19.2

Lab 19 alignment with standards

NGSS performance expectation	Weather and climate • MS-ESS2-6: Develop and use a model to describe how unequal heating and rotation of the Earth cause patterns of atmospheric and oceanic circulation that determine regional climates.
CCSS ELA—**Reading in Science and Technical Subjects**	Key ideas and details • CCSS.ELA-LITERACY.RST.6-8.1: Cite specific textual evidence to support analysis of science and technical texts. • CCSS.ELA-LITERACY.RST.6-8.2: Determine the central ideas or conclusions of a text; provide an accurate summary of the text distinct from prior knowledge or opinions.

Continued

TABLE 19.2 (*continued*)

CCSS ELA—**Reading in Science and Technical Subjects** (*continued*)	Craft and structure • CCSS.ELA-LITERACY.RST.6-8.4: Determine the meaning of symbols, key terms, and other domain-specific words and phrases as they are used in a specific scientific or technical context relevant to *grade 6–8 texts and topics*. • CCSS.ELA-LITERACY.RST.6-8.5: Analyze the structure an author uses to organize a text, including how the major sections contribute to the whole and to an understanding of the topic. • CCSS.ELA-LITERACY.RST.6-8.6: Analyze the author's purpose in providing an explanation, describing a procedure, or discussing an experiment in a text. Integration of knowledge and ideas • CCSS.ELA-LITERACY.RST.6-8.7: Integrate quantitative or technical information expressed in words in a text with a version of that information expressed visually (e.g., in a flowchart, diagram, model, graph, or table). • CCSS.ELA-LITERACY.RST.6-8.8: Distinguish among facts, reasoned judgment based on research findings, and speculation in a text. • CCSS.ELA-LITERACY.RST.6-8.9: Compare and contrast the information gained from experiments, simulations, video, or multimedia sources with that gained from reading a text on the same topic.
CCSS ELA—**Writing in Science and Technical Subjects**	Text types and purposes • CCSS.ELA-LITERACY.WHST.6-8.1: Write arguments focused on *discipline-specific content.* • CCSS.ELA-LITERACY.WHST.6-8.2: Write informative or explanatory texts, including the narration of historical events, scientific procedures/experiments, or technical processes. Production and distribution of writing • CCSS.ELA-LITERACY.WHST.6-8.4: Produce clear and coherent writing in which the development, organization, and style are appropriate to task, purpose, and audience. • CCSS.ELA-LITERACY.WHST.6-8.5: With some guidance and support from peers and adults, develop and strengthen writing as needed by planning, revising, editing, rewriting, or trying a new approach, focusing on how well purpose and audience have been addressed. • CCSS.ELA-LITERACY.WHST.6-8.6: Use technology, including the internet, to produce and publish writing and present the relationships between information and ideas clearly and efficiently.

Continued

TABLE 19.2 (*continued*)

CCSS ELA—Writing in Science and Technical Subjects (*continued*)	Range of writing • CCSS.ELA-LITERACY.WHST.6-8.10: Write routinely over extended time frames (time for reflection and revision) and shorter time frames (a single sitting or a day or two) for a range of discipline-specific tasks, purposes, and audiences.
CCSS ELA—Speaking and Listening	Comprehension and collaboration • CCSS.ELA-LITERACY.SL.6-8.1: Engage effectively in a range of collaborative discussions (one-on-one, in groups, and teacher-led) with diverse partners on grade 6–8 topics, texts, and issues, building on others' ideas and expressing their own clearly. • CCSS.ELA-LITERACY.SL.6-8.2:* Interpret information presented in diverse media and formats (e.g., visually, quantitatively, orally) and explain how it contributes to a topic, text, or issue under study. • CCSS.ELA-LITERACY.SL.6-8.3:* Delineate a speaker's argument and specific claims, distinguishing claims that are supported by reasons and evidence from claims that are not. Presentation of knowledge and ideas • CCSS.ELA-LITERACY.SL.6-8.4:* Present claims and findings, sequencing ideas logically and using pertinent descriptions, facts, and details to accentuate main ideas or themes; use appropriate eye contact, adequate volume, and clear pronunciation. • CCSS.ELA-LITERACY.SL.6-8.5:* Include multimedia components (e.g., graphics, images, music, sound) and visual displays in presentations to clarify information. • CCSS.ELA-LITERACY.SL.6-8.6: Adapt speech to a variety of contexts and tasks, demonstrating command of formal English when indicated or appropriate.

* Only the standard for grade 6 is provided because the standards for grades 7 and 8 are similar. Please see *www.corestandards.org/ELA-Literacy/SL* for the exact wording of the standards for grades 7 and 8.

LAB 19

Lab Handout

Lab 19. Differences in Regional Climate: Why Do Two Cities Located at the Same Latitude and Near a Body of Water Have Such Different Climates?

Introduction

Weather describes the current atmospheric conditions at a particular location. *Climate*, in contrast, is the aggregate or typical weather for a particular location over a long period of time. People often describe the climate of a region by reporting average temperatures and rainfall by month or by season. Cities located at higher latitudes (i.e., farther from the equator) experience greater changes in day length and Sun angle over the course of a year. These cities, as a result, have greater seasonal temperature differences than cities that are located closer to the equator. It is therefore not surprising that cities located at different latitudes often have very different climates. Cities located at the same latitude, however, can also have very different climates. A good example of this phenomenon is seen when we look at the climates of San Francisco, California, and Norfolk, Virginia (see Figure L19.1).

FIGURE L19.1
Locations of San Francisco and Norfolk

San Francisco is located on the coast of the Pacific Ocean at 37.7° N latitude. It has mild summers and winters; the average high temperature between 1945 and 2017 has been 17.1°C (62.8°F) for July and 9.7°C (49.4°F) for January. San Francisco is very dry, averaging 19.7 inches of rain per year from 1946 to 2017. (Climate information from the National Oceanic and Atmospheric Administration [NOAA] National Centers for Environmental Information is available at *www.ncdc.noaa.gov.*) According to the Köppen Climate Classification System (the most widely used system for classifying the world's climates; see *www.britannica.com/science/Koppen-climate-classification*), San Francisco has a temperate Mediterranean climate with warm summers (denoted Csb in the Köppen system). In contrast, Norfolk, which is located on the Atlantic Ocean at 36.9° N latitude, has hot summers and mild winters. The average high temperature between 1945 and 2017 has been 30.9°C (87.6°F) in July and 9.5°C (49.1°F) in January. Between 1945 and 2017, Norfolk has averaged 46.4 inches of rain a year. Norfolk has a humid subtropical climate (Cfa) based on the Köppen classification system. In summary, San Francisco and Norfolk have very different temperature and precipitation patterns throughout the year even though they are located at similar latitudes. To understand why these two cities have such different temperature and precipitation patterns, we must consider all the different factors than can affect the climate of a region.

There are at least six important factors to consider when someone attempts to explain a difference in two or more regional climates:

- *Latitude,* as noted earlier, determines changes in day length and sun angle throughout the year.

- *Elevation,* which is the height of an area above sea level. Generally, as elevation increases, temperature decreases.

- *Proximity to a large body of water.* Land heats up and cools down faster than water. Water can also store more heat energy than land. This makes the climate of a region that is located near a large body of water, such as an ocean, more moderate because the water absorbs extra heat energy during the summer and releases heat into the air during winter.

- *The nature of nearby ocean currents.* Ocean currents move large amounts of water with different properties to different locations across the Earth. Winds, tides, and differences in water temperature and salinity at different locations in the ocean affect the path an ocean current follows over time.

- *The direction and strength of prevailing winds.* Winds can move air masses with specific properties from a source region to a different region. In the Northern Hemisphere, winds tend to blow from west to east (westerly winds) in the mid-latitudes and northeast to southwest between the Tropic of Cancer and the equator.

- *Local topography.* The presence of absence of a mountain in a region, for example, can affect precipitation patterns and therefore climate.

LAB 19

These six factors can help us understand why there are different climates at different locations around the globe. Yet, some of them may or may not be useful when we need to explain the different temperature and precipitation patterns observed in cities such as San Francisco and Norfolk. The two cities, as noted earlier, have much in common. San Francisco and Norfolk are located at similar latitudes, at an elevation slightly above sea level, and are near an ocean but on different coasts. You will therefore need to learn more about the nature of nearby ocean currents, the nature of any prevailing winds in these regions, and the local topography around these cities to figure out why these two cities have such different climates. Next, you will put all these pieces of information together to develop a conceptual model that not only explains the different climates in San Francisco and Norfolk but can also explain differences in the climates of other cities that are located at similar latitudes.

Your Task

Develop a conceptual model that you can use to explain the temperature and precipitation patterns in San Francisco and Norfolk. Your conceptual model must reflect what we know about the various factors that can affect climate, patterns, and systems and system models. To be considered valid or acceptable, you should be able use your conceptual model to not only explain why San Francisco and Norfolk have different climates but also to predict the temperature and precipitation patterns of several other pairs of cities that are located at similar latitudes on Earth.

The guiding question of this investigation is, *Why do two cities located at the same latitude and near a body of water have such different climates?*

Materials

You can use the following online resources during your investigation:

- U.S. Climate Data. You can access this website, which includes detailed climate data and the location of most major U.S. Cities, at *www.usclimatedata.com/climate/united-states/us*.

- Wind rose data from the National Oceanic and Atmospheric Administration (NOAA). You can access this database, which includes monthly wind speed and direction information for 237 U.S. cities, at *www.climate.gov/maps-data/dataset/monthly-wind-rose-plots-charts*.

- *Earth: A Global Map of Wind, Weather, and Ocean Conditions.* You can access this interactive animated map that shows current wind speeds and direction for the entire planet at *https://earth.nullschool.net*.

- *My NASA Data.* You can access this database, which includes information about ocean surface temperatures and the average wind speed and direction by month over the entire year, at *https://mynasadata.larc.nasa.gov*.

- *State of the Ocean* (SOTO). You can access this visualization tool, which includes information about current and past ocean currents and changes in surface temperature, through the NASA Jet Propulsion Laboratory Physical Oceanography Distributed Active Archive Center at *https://podaac.jpl.nasa.gov*. Click on "Data Access" and "SOTO (State of the Ocean)" to open the visualization tool.

Safety Precautions

Follow all normal lab safety rules.

Investigation Proposal Required?   ☐ Yes   ☐ No

Getting Started

The first step in developing a conceptual model that explains differences in the temperature and precipitation patterns of San Francisco and Norfolk is to collect information about seasonal changes in temperature and precipitation in both cities. This information can be found at the U.S. Climate Data website. Be sure look for any patterns that you can use to help develop your conceptual model. Next, you can learn about the prevailing winds at each location using data provided by the NOAA wind rose website and the Earth: A Global Map of Wind, Weather, and Ocean Conditions website (which provides real-time data you can use). You can also access data about ocean temperatures and wind patterns by using the My NASA data. The final website, SOTO, will allow you to visualize a variety of ocean characteristics on a map of the world.

To learn more about how prevailing winds affect climate, you must first determine what type of data you need to collect, how you will collect it, and how you will analyze it. To determine *what type of data you need to collect,* think about the following questions:

- What are the boundaries and components of the system you are studying?
- How do the components of the system interact with each other?
- When is this system stable, and under which conditions does it change?
- What could be the underlying cause of this phenomenon?
- What type of measurements or observations will you need?
- What types of patterns could you look for in the available data?

To determine *how you will collect your data,* think about the following questions:

- What conditions need to be satisfied to establish a cause-and-effect relationship?
- How can you describe the components of the system quantitatively?
- What measurement scale or scales should you use to collect data?
- What type of comparisons will you need to make?
- How will you keep track of and organize the data you collect?

To determine *how you will analyze your data,* think about the following questions:

- What types of patterns might you look for as you analyze your data?
- How could you use mathematics to document a difference between conditions?
- What type of comparisons and calculations will you need to make?
- What type of graph could you create to help make sense of your data?

Once you feel you have gathered sufficient data and identified important patterns about how oceans and prevailing winds affect climate, your group can develop a conceptual model that can be used to explain why San Francisco and Norfolk have such different climates. To be valid or acceptable, your conceptual model must be able to explain

    a. why San Francisco and Norfolk have such different seasonal temperatures, and

    b. why San Francisco and Norfolk have such different rain patterns.

The last step in your investigation will be to generate the evidence that you need to convince others that your conceptual model is valid or acceptable. To accomplish this goal, you will use your model to predict the temperature and precipitation patterns in several additional cities. These cities should be ones that you have not looked up before but are located at similar latitudes on different coasts. Some good pairs of cities to compare are

- San Diego, California, and Charleston, South Carolina;
- Portland, Oregon, and Bangor, Maine;
- Eureka, California, and New York City; and
- Santa Monica, California, and Wilmington, North Carolina.

You can also attempt to show how using a different version of your model or making a specific change to a portion of your model will make your model inconsistent with data you have or the facts we know about climate. Scientists often make comparisons between different versions of a model in this manner to show that a model is valid or acceptable. If you are able to use your conceptual model to make accurate predictions about the climates of other cities or if you are able show how your conceptual model explains the climates of different cities better than other conceptual models, then you should be able to convince others that it is valid or acceptable.

Connections to the Nature of Scientific Knowledge and Scientific Inquiry

As you work through your investigation, be sure to think about

- how models are used as tools for reasoning about phenomena, and
- how the culture of science, societal needs, and current events influence the work of scientists.

Initial Argument

Once your group has finished collecting and analyzing your data, your group will need to develop an initial argument. Your initial argument needs to include a claim, evidence to support your claim, and a justification of the evidence. The *claim* is your group's answer to the guiding question. The *evidence* is an analysis and interpretation of your data. Finally, the *justification* of the evidence is why your group thinks the evidence matters. The justification of the evidence is important because scientists can use different kinds of evidence to support their claims. Your group will create your initial argument on a whiteboard. Your whiteboard should include all the information shown in Figure L19.2.

FIGURE L19.2 _____

Argument presentation on a whiteboard

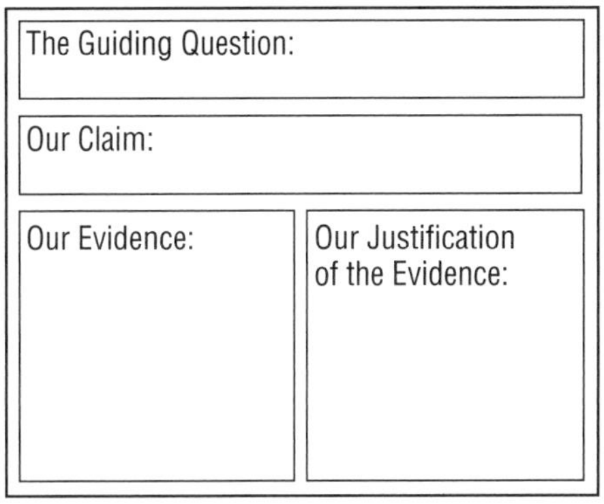

The Guiding Question:	
Our Claim:	
Our Evidence:	Our Justification of the Evidence:

Argumentation Session

The argumentation session allows all of the groups to share their arguments. One or two members of each group will stay at the lab station to share that group's argument, while the other members of the group go to the other lab stations to listen to and critique the other arguments. This is similar to what scientists do when they propose, support, evaluate, and refine new ideas during a poster session at a conference. If you are presenting your group's argument, your goal is to share your ideas and answer questions. You should also keep a record of the critiques and suggestions made by your classmates so you can use this feedback to make your initial argument stronger. You can keep track of specific critiques and suggestions for improvement that your classmates mention in the space below.

Critiques of our initial argument and suggestions for improvement:

If you are critiquing your classmates' arguments, your goal is to look for mistakes in their arguments and offer suggestions for improvement so these mistakes can be fixed. You should look for ways to make your initial argument stronger by looking for things that the other groups did well. You can keep track of interesting ideas that you see and hear during the argumentation in the space below. You can also use this space to keep track of any questions that you will need to discuss with your team.

Interesting ideas from other groups or questions to take back to my group:

Once the argumentation session is complete, you will have a chance to meet with your group and revise your initial argument. Your group might need to gather more data or design a way to test one or more alternative claims as part of this process. Remember, your goal at this stage of the investigation is to develop the best argument possible.

Report

Once you have completed your research, you will need to prepare an investigation report that consists of three sections. Each section should provide an answer for the following questions:

1. What question were you trying to answer and why?

2. What did you do to answer your question and why?

3. What is your argument?

Your report should answer these questions in two pages or less. You should write your report using a word processing application (such as Word, Pages, or Google Docs), if possible, to make it easier for you to edit and revise it later. You should embed any diagrams, figures, or tables into the document. Be sure to write in a persuasive style; you are trying to convince others that your claim is acceptable or valid.

Checkout Questions

Lab 19. Differences in Regional Climate: Why Do Two Cities Located at the Same Latitude and Near a Body of Water Have Such Different Climates?

1. What is climate?

2. How do prevailing winds and ocean currents affect the climates of different regions?

3. The map below shows the locations of three cities: Los Angeles, California; Oklahoma City, Oklahoma; and Myrtle Beach, South Carolina. The map also includes information about prevailing wind directions (dotted lines) and the location of cold water (solid lines) and warm water (dashed lines) ocean currents.

a. Rank the cities based on their summer high temperatures, with 1 being the coolest summer, and 3 being the warmest summer.

City	Rank
Los Angeles, CA	_____
Oklahoma City, OK	_____
Myrtle Beach, SC	_____

b. How do you know? Explain why you ranked the cities this way.

c. Rank the cities based on their winter low temperatures, with 1 being the warmest winter, and 3 being the coldest winter.

City	Rank
Los Angeles, CA	_____
Oklahoma City, OK	_____
Myrtle Beach, SC	_____

d. How do you know? Explain why you ranked the cities this way.

4. Scientists conduct investigation based on their own interests; societal needs and current events do not influence the work of scientists at all.

a. I agree with this statement.

b. I disagree with this statement.

Explain your answer, using an example from your investigation about differences in regional climate.

5. Scientists use models as tools for reasoning about natural phenomena.

    a. I agree with this statement.

    b. I disagree with this statement.

Explain your answer, using an example from your investigation about differences in regional climate.

6. Scientists often need to look for patterns that occur in the data they collect and analyze. Explain why identifying patterns are important, using an example from your investigation about differences in regional climate.

7. Defining a system under study and making a model of it are tools for developing a better understanding of natural phenomena in science. Give an example of how system models are used by scientists to investigate natural and designed systems.

SECTION 6
Human Impact

Introduction Labs

Teacher Notes

Lab 20. Predicting Hurricane Strength: How Can Someone Predict Changes in Hurricane Wind Speed Over Time?

Purpose

The purpose of this lab is to *introduce* students to the disciplinary core idea (DCI) of Natural Hazards by giving them an opportunity to develop a conceptual model that can be used to predict changes in wind speed inside a hurricane over time. In addition, students have an opportunity to learn about the crosscutting concepts (CCs) of (a) Cause and Effect: Mechanism and Explanation and (b) Energy and Matter: Flows, Cycles, and Conservation. During the explicit and reflective discussion, students will also learn about (a) the use of models as tools for reasoning about natural phenomena and (b) the types of questions that scientists can investigate.

Important Earth and Space Science Content

The strongest tropical storms are called hurricanes, typhoons, or cyclones. The different names are used to describe tropical storms that originate in different parts of the world (see Figure 20.1). If a huge storm forms off the west coast of Africa in the Atlantic, it is called a hurricane. Hurricanes have strong winds, a spiral shape, and a low-pressure center called an eye. Unlike other natural hazards, such as earthquakes or even tornadoes, scientists can observe the development of a hurricane over time and track how it moves across the ocean.

FIGURE 20.1 _____

Areas where tropical storms tend to form and the paths that they follow

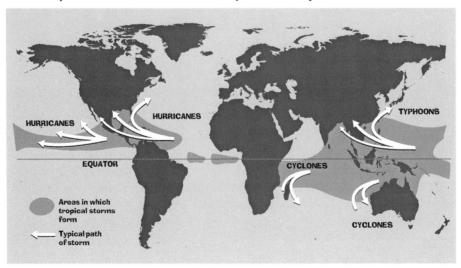

A hurricane begins as a tropical disturbance. A tropical disturbance forms in an area where the ocean surface temperature is at least 27°C (80°F). The warm humid air at that location rises and creates an area of low atmospheric pressure near the ocean surface. Cooler air in the region then rushes into the area of low pressure. This air picks up evaporated water from the surface, increases in temperature, and moves upward into the atmosphere. This process produces large water-filled thunderclouds around the area of low pressure. The trade winds (which blow from east to west) slowly push the disturbance to the west.

Over the next few days, more warm air will rise and the winds will begin to circulate around the center of the disturbance in counterclockwise (when viewed from above) direction. A layer of clouds called an outflow will also begin to form at the top of the storm. Winds inside the storm will increase in speed over time. When the winds within the storm are between 25 and 38 miles per hour (mph), the storm is called a tropical depression. When the wind speeds reach 39 mph, the storm is classified as a tropical storm rather than as a tropical depression. This is also the point in time when the storm gets a name. In a couple of days, as the system moves across the ocean, the clouds expand and the winds continue to speed up. When the wind speeds inside the storm reach 74 mph, it is classified as a hurricane (see Table 20.1).

TABLE 20.1

Types of tropical storms classified by wind speed according to the Saffir-Simpson Hurricane Scale and by colors used on historical hurricane track maps from the National Oceanic and Atmospheric Administration

Classification	Wind speed			Color used on historical hurricane track maps
	mph	kn	km/h	
Category 5 hurricane	≥ 157	≥ 137	≥ 252	Purple
Category 4 hurricane	130–156	113–136	209–251	Pink
Category 3 hurricane	111–129	96–112	178–208	Red
Category 2 hurricane	96–110	83–95	154–177	Orange
Category 1 hurricane	74–95	64–82	119–153	Yellow
Tropical storm	39–73	34–63	63–118	Green
Tropical depression	25–38	22–33	40–62	Blue

Hurricanes weaken whenever they (1) move over cooler ocean waters that cannot supply warm humid air, (2) move onto land, or (3) reach a location where there are high winds in the upper atmosphere. Whenever a hurricane moves onto land, its sustained wind speed will quickly begin to decrease. The most important reason for this rapid change in wind speed is the fact that less moisture is carried into the storm, cloud coverage lessens, and cools and sinks, which disrupts the circulation of the hurricane and the development of new thunderstorms. In addition, friction from the increased roughness of the land surface rapidly slows surface wind speeds. This factor causes the winds to move directly into the eye of the storm (instead of continuing to circulate around it), thus helping to eliminate the large differences in atmospheric pressure within the storm. Without a pressure difference in the atmosphere, there is no wind.

Timeline

The instructional time needed to complete this lab investigation is 270–330 minutes. Appendix 3 (p. 573) provides options for implementing this lab investigation over several class periods. Option G (330 minutes) should be used if students are unfamiliar with scientific writing, because this option provides extra instructional time for scaffolding the writing process. You can scaffold the writing process by modeling, providing examples, and providing hints as students write each section of the report. Option H (270 minutes) should be used if students are familiar with scientific writing and have developed the skills needed to write an investigation report on their own. In option H, students complete stage 6 (writing the investigation report) and stage 8 (revising the investigation report) as homework. Both options give students time to test their models on day 3.

Materials and Preparation

The materials needed to implement this investigation are listed in Table 20.2. The five handouts can be downloaded from the book's Extras page at *www.nsta.org/adi-ess*. Students will use the Hurricane Track A—Black and White handout and the Hurricane Track B—Black and White handout to test the accuracy of their models. These handouts must therefore be printed out in black and white so students can see the track of these hurricanes but not know the actual wind speed (the wind speed is color coded as shown in Table 20.1). Students will use the Hurricane Track A—Color handout and the Hurricane Track B—Color handout to check the accuracy of their models. These handouts must therefore printed out in color so students can determine hurricane wind speed. You should give the black-and-white hurricane track handouts to the students *after* they have developed their model and are ready to make their predictions. You should give the color hurricane track handouts to the students *after* they have made their prediction and are ready to check them for accuracy. This will enable the students to make a fair test of their conceptual models. Hurricane Track A is for Ivan 2004 and Hurricane Track B is for Jeanne 2004.

Students will need to access the following online resources:

- Historical hurricane track maps from the National Oceanic and Atmospheric Administration (NOAA) are available at *https://coast.noaa.gov/hurricanes.* The tracks on the maps are color-coded to reflect changes in the storm over time (see Table 20.1).
- Information about current sea surface temperatures in the North Atlantic can be found at *www.ospo.noaa.gov/Products/ocean/sst/contour/index.html.*

- Information about monthly sea surface temperatures for 1984–1998 is available at *www.ospo.noaa.gov/Products/ocean/sst/monthly_mean.html*.

- Information about land surface temperature during the daytime by month can be found at *http://earthobservatory.nasa.gov/GlobalMaps*.

All of these online resources are free to use. You should access each website and learn how to find the desired data before beginning the lab investigation. In addition, it is important to check if students can access and use the websites from a school computer or tablet, because some schools have set up restrictions on web browsing.

TABLE 20.2

Materials list for Lab 20

Item	Quantity
Some Major Hurricanes handout	1 per group
Hurricane Track A—Black and White handout	1 per group
Hurricane Track A—Color handout	1 per group
Hurricane Track B—Black and White handout	1 per group
Hurricane Track B—Color handout	1 per group
Computer or tablet with internet access	1 per group
Whiteboard, 2' × 3'*	1 per group
Lab Handout	1 per student
Peer-review guide and teacher scoring rubric	1 per student
Checkout Questions	1 per student

* As an alternative, students can use computer and presentation software such as Microsoft PowerPoint or Apple Keynote to create their arguments.

Safety Precautions

Remind students to follow all normal lab safety rules.

Topics for the Explicit and Reflective Discussion

Reflecting on the Use of Core Ideas and Crosscutting Concepts During the Investigation

Teachers should begin the explicit and reflective discussion by asking students to discuss what they know about the DCI they used during the investigation. The following are some important concepts related to the DCI of Natural Hazards that students need to develop a conceptual model that can be used to predict changes in wind speed inside a hurricane over time:

- Mapping the history of natural hazards in a region, combined with an understanding of related geologic forces, can help forecast the locations and likelihood of future events.
- Air flows from areas of high pressure to areas of low pressure
- Interactions between sunlight, the atmosphere, landforms, and ocean temperatures affect weather and climate.
- The ocean exerts a major influence on weather and climate by absorbing energy from the Sun, releasing it over time, and redistributing it around the globe.
- Hurricanes form in between the equator and the Tropic of Cancer where winds blow from east to west (easterly winds).
- The path of hurricanes can be observed and predicted.

To help students reflect on what they know about these concepts, we recommend showing them two or three images using presentation software that help illustrate these important ideas. You can then ask the students the following questions to encourage students to share how they are thinking about these important concepts:

1. What do we see going on in this image?

2. Does anyone have anything else to add?

3. What might be going on that we can't see?

4. What are some things that we are not sure about here?

You can then encourage students to think about how CCs played a role in their investigation. There are at least two CCs that students need to develop a conceptual model that can be used to predict changes in wind speed inside a hurricane over time: (a) Cause and Effect: Mechanism and Explanation and (b) Energy and Matter: Flows, Cycles, and Conservation (see Appendix 2 [p. 569] for a brief description of these CCs). To help students reflect on what they know about these CCs, we recommend asking them the following questions:

1. Why can cause-and-effect relationships be used to predict phenomena in natural systems?

2. What cause-and-effect relationships did you use during your investigation to make predictions? Why was that useful to do?

3. Why is it important to track how energy flows into, out of, or within a system during an investigation?

4. How did you track the flow of energy in the system you were studying? What did tracking the flow of energy allow you to do during your investigation?

You can then encourage the students to think about how they used all these different concepts to help answer the guiding question and why it is important to use these ideas to help justify their evidence for their final arguments. Be sure to remind your students to explain why they included the evidence in their arguments and make the assumptions underlying their analysis and interpretation of the data explicit in order to provide an adequate justification of their evidence.

Reflecting on Ways to Design Better Investigations

It is important for students to reflect on the strengths and weaknesses of the investigation they designed during the explicit and reflective discussion. Students should therefore be encouraged to discuss ways to eliminate potential flaws, measurement errors, or sources of uncertainty in their investigations. To help students be more reflective about the design of their investigation and what they can do to make their investigations more rigorous in the future, you can ask the following questions:

1. What were some of the strengths of the way you planned and carried out your investigation? In other words, what made it scientific?

2. What were some of the weaknesses of the way you planned and carried out your investigation? In other words, what made it less scientific?

3. What rules can we make, as a class, to ensure that our next investigation is more scientific?

Reflecting on the Nature of Scientific Knowledge and Scientific Inquiry

This investigation can be used to illustrate two important concepts related to the nature of scientific knowledge and the nature of scientific inquiry: (a) the use of models as tools for reasoning about natural phenomena and (b) the types of questions that scientists can investigate (see Appendix 2 [p. 569] for a brief description of these two concepts). Be sure to review these concepts during and at the end of the explicit and reflective discussion. To help students think about these concepts in relation to what they did during the lab, you can ask the following questions:

- I asked you to develop a model to that can be used to predict how hurricane wind strength will change over time as part of your investigation. Why is it useful to develop models in science?

- Can you work with your group to come up with a rule that you can use to decide what a model is and what a model is not in science? Be ready to share in a few minutes.

- Not all questions can be answered by science. Can you give me some examples of questions related to this investigation that can and cannot be answered by science?

- Can you work with your group to come up with a rule that you can use to decide if question can be answered by science or not? Be ready to share in a few minutes.

You can also use presentation software or other techniques to encourage the students to think about these concepts. You can show examples and non-examples of scientific models and then ask students to classify each one and then explain their thinking. You can also show one or more examples of questions that can be answered by science (e.g., What causes hurricanes? How do prevailing winds affect the path of a hurricane?) and cannot be answered by science (e.g., What is the best way to prevent property damage during a hurricane? Who should be required to have extra hurricane insurance?), and then ask students why each example is or is not a question that can be answered by science.

Remind your students that, to be proficient in science, it is important that they understand what counts as scientific knowledge and how that knowledge develops over time.

Hints for Implementing the Lab

- A group of three students per computer or tablet tends to work well.
- Learn how to use all the online resources before the lab begins. It is important for you to know how to use these resources so you can help students when they get stuck or confused.
- Allow the students to play with the online resources as part of the tool talk before they begin to design their investigation. This gives students a chance to see what data they can and cannot access with each resource.
- Be sure that students record actual values (e.g., wind speeds and direction) and are not just attempting to hand draw what they see on the computer screen.
- This is a good lab for students to make mistakes during the data collection or the development of the initial argument stage. Students will quickly figure out what they did wrong during the argumentation session, and it will only take them a short period of time to re-collect data. It will also create an opportunity for students to reflect on and identify ways to improve the way they design investigations.
- This lab also provides an excellent opportunity to discuss how scientists identify a signal (a pattern or trend) from the noise (measurement error) in their data and how to establish cause-and-effect relationships when experiments cannot be done. Be sure to use this activity as a concrete example during the explicit and reflective discussion.

Connections to Standards

Table 20.3 highlights how the investigation can be used to address specific (a) performance expectations from the *NGSS* and (b) *Common Core State Standards* in English language arts (*CCSS ELA*).

TABLE 20.3

Lab 20 alignment with standards

***NGSS* performance expectation**	Human impact • MS-ESS3-2: Analyze and interpret data on natural hazards to forecast future catastrophic events and inform the development of technologies to mitigate their effects.
***CCSS ELA*—Reading in Science and Technical Subjects**	Key ideas and details • CCSS.ELA-LITERACY.RST.6-8.1: Cite specific textual evidence to support analysis of science and technical texts. • CCSS.ELA-LITERACY.RST.6-8.2: Determine the central ideas or conclusions of a text; provide an accurate summary of the text distinct from prior knowledge or opinions. Craft and structure • CCSS.ELA-LITERACY.RST.6-8.4: Determine the meaning of symbols, key terms, and other domain-specific words and phrases as they are used in a specific scientific or technical context relevant to *grade 6–8 texts and topics.* • CCSS.ELA-LITERACY.RST.6-8.5: Analyze the structure an author uses to organize a text, including how the major sections contribute to the whole and to an understanding of the topic. • CCSS.ELA-LITERACY.RST.6-8.6: Analyze the author's purpose in providing an explanation, describing a procedure, or discussing an experiment in a text. Integration of knowledge and ideas • CCSS.ELA-LITERACY.RST.6-8.7: Integrate quantitative or technical information expressed in words in a text with a version of that information expressed visually (e.g., in a flowchart, diagram, model, graph, or table). • CCSS.ELA-LITERACY.RST.6-8.8: Distinguish among facts, reasoned judgment based on research findings, and speculation in a text. • CCSS.ELA-LITERACY.RST.6-8.9: Compare and contrast the information gained from experiments, simulations, video, or multimedia sources with that gained from reading a text on the same topic.
***CCSS ELA*—Writing in Science and Technical Subjects**	Text types and purposes • CCSS.ELA-LITERACY.WHST.6-8.1: Write arguments focused on *discipline-specific content.*

Continued

TABLE 20.3 (*continued*)

***CCSS ELA*—Writing in Science and Technical Subjects** (*continued*)	Text types and purposes (*continued*) • CCSS.ELA-LITERACY.WHST.6-8.2: Write informative or explanatory texts, including the narration of historical events, scientific procedures/experiments, or technical processes. Production and distribution of writing • CCSS.ELA-LITERACY.WHST.6-8.4: Produce clear and coherent writing in which the development, organization, and style are appropriate to task, purpose, and audience. • CCSS.ELA-LITERACY.WHST.6-8.5: With some guidance and support from peers and adults, develop and strengthen writing as needed by planning, revising, editing, rewriting, or trying a new approach, focusing on how well purpose and audience have been addressed. • CCSS.ELA-LITERACY.WHST.6-8.6: Use technology, including the internet, to produce and publish writing and present the relationships between information and ideas clearly and efficiently. Range of writing • CCSS.ELA-LITERACY.WHST.6-8.10: Write routinely over extended time frames (time for reflection and revision) and shorter time frames (a single sitting or a day or two) for a range of discipline-specific tasks, purposes, and audiences.
***CCSS ELA*—Speaking and Listening**	Comprehension and collaboration • CCSS.ELA-LITERACY.SL.6-8.1: Engage effectively in a range of collaborative discussions (one-on-one, in groups, and teacher-led) with diverse partners on grade 6-8 topics, texts, and issues, building on others' ideas and expressing their own clearly. • CCSS.ELA-LITERACY.SL.6-8.2:* Interpret information presented in diverse media and formats (e.g., visually, quantitatively, orally) and explain how it contributes to a topic, text, or issue under study. • CCSS.ELA-LITERACY.SL.6-8.3:* Delineate a speaker's argument and specific claims, distinguishing claims that are supported by reasons and evidence from claims that are not. Presentation of knowledge and ideas • CCSS.ELA-LITERACY.SL.6-8.4:* Present claims and findings, sequencing ideas logically and using pertinent descriptions, facts, and details to accentuate main ideas or themes; use appropriate eye contact, adequate volume, and clear pronunciation. • CCSS.ELA-LITERACY.SL.6-8.5:* Include multimedia components (e.g., graphics, images, music, sound) and visual displays in presentations to clarify information. • CCSS.ELA-LITERACY.SL.6-8.6: Adapt speech to a variety of contexts and tasks, demonstrating command of formal English when indicated or appropriate.

* Only the standard for grade 6 is provided because the standards for grades 7 and 8 are similar. Please see *www.corestandards.org/ELA-Literacy/SL* for the exact wording of the standards for grades 7 and 8.

Lab Handout

Lab 20. Predicting Hurricane Strength: How Can Someone Predict Changes in Hurricane Wind Speed Over Time?

Introduction

The strongest tropical storms are called hurricanes, typhoons, or cyclones. The different names all mean the same thing but are used to describe tropical storms that originate in different parts of the world. If a huge storm starts off the west coast of Africa in the Atlantic, it is called a hurricane. Hurricanes have strong winds, a spiral shape, and a low-pressure center called an eye. Unlike other natural hazards, such as earthquakes or even tornadoes, we can observe the development of a hurricane over time and track how it moves across the ocean.

A hurricane begins as a tropical disturbance in the ocean off the west coast of Africa. A tropical disturbance forms in an area where the ocean surface temperature is at least 27°C (80°F). The warm humid air at that location rises and creates an area of low atmospheric pressure near the ocean surface. Cooler air in the region then rushes into the area of low pressure. This air picks up evaporated water from the surface, increases in temperature, and moves upward into the atmosphere. This process produces large water-filled thunderclouds around the area of low pressure. The trade winds (which blow from east to west) slowly push the disturbance to the west.

Over the next few days, more warm air will rise and the winds will begin to circulate around the center of the disturbance in counterclockwise (when viewed from above) direction. A layer of clouds called an outflow will also begin to form at the

FIGURE L20.1

Image of Hurricane Isabel about 400 miles north of Puerto Rico on September 14, 2003, captured by the NASA Terra satellite; the sustained wind speed inside Hurricane Isabel at that time was 155 mph

Note: A full-color version of this figure is available on the book's Extras page at *www.nsta.org/adi-ess*.

top of the storm. Winds inside the storm will increase in speed over time. When the winds within the storm are between 25 and 38 miles per hour (mph), the storm is called a tropical depression. When the wind speeds reach 39 mph, the storm is classified as a tropical storm rather than as a tropical depression. This is also the point in time when the storm gets a name. In a couple of days, as the system moves across the ocean, the clouds expand and the winds continue to speed up. When the wind speeds inside the storm reach 74 mph, it is classified as a hurricane (see Figure L20.1).

In an average year, several different hurricanes will form over the Atlantic Ocean and head westward toward the Caribbean, the east coast of Central America, or the southeastern United States. Figure L20.2 shows the tracks of all North Atlantic Ocean hurricanes that developed between 1980 and 2005. The points on each track represent the location of that storm at six-hour intervals. Hurricanes will often last several weeks before they break down because they tend to move very slowly across the ocean. In fact, hurricanes usually travel across the ocean at only about 24 kilometers per hour (or 15 mph).

FIGURE L20.2

Map showing the tracks of all hurricanes in the North Atlantic Ocean from 1980 to 2005; the points show the locations of the storms at six-hour intervals

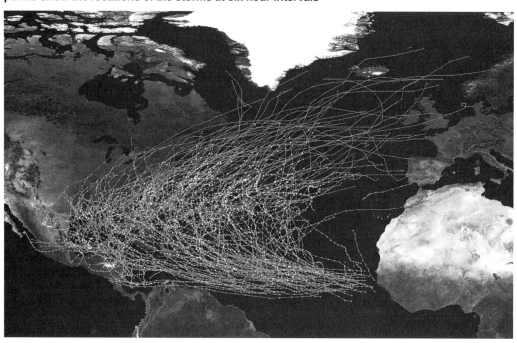

Note: A full-color version of this figure is available on the book's Extras page at *www.nsta.org/adi-ess*.

Scientists use the sustained wind speed inside a hurricane to classify it. The sustained wind speed inside a hurricane, however, can increase or decrease over time. It is therefore important for scientists to understand why the winds inside a hurricane change over time. This type of information is important because scientists are responsible for issuing evacuation warnings, and they need to know if the wind within a hurricane is likely to increase or decrease before it reaches landfall. It is also important to understand the factors that affect the wind speed of a hurricane over time as it moves over water or land; this information helps city planners to establish building codes for cities to ensure that new buildings will be able to withstand the winds of a typical hurricane for that area.

In this investigation, you will have an opportunity to learn more about the factors that affect wind speed within hurricanes. Your goal is to develop a conceptual model that you can use to not only explain why the wind speed within a hurricane changes over time as it moves over water or land but also predict how the strength of a hurricane will increase or decrease over time based on the path that it follows.

Your Task

Develop a conceptual model that can be used to explain why wind speed inside a hurricane changes over time as it moves over water or land. Your conceptual model must be based on what we know about natural hazards; weather; the importance of tracking how energy flows into, within, and out of a system; and cause-and-effect relationships. Once you have developed your model, you will need to test it to see if you can use it to make accurate predictions about how the strength of several hurricanes changed over time in the past.

The guiding question of this investigation is, **How can someone predict changes in hurricane wind speed over time?**

Materials

You may use any of the following materials during your investigation:

Equipment
- Computer or tablet with internet access

Other Resources
- Some Major Hurricanes handout
- Hurricane Track A—Black and White handout (use to test your model)
- Hurricane Track A—Color handout (use to check your predictions)
- Hurricane Track B—Black and White handout (use to test your model)
- Hurricane Track B—Color handout (use to check your predictions)

Safety Precautions

Follow all normal lab safety rules.

Investigation Proposal Required?   ☐ Yes   ☐ No

Getting Started

The first step in this investigation is to determine how the strength of a hurricane changes over time as it travels over water and land. To accomplish this goal, you will need to examine several different historical hurricane tracks and look for patterns that you can use to explain and predict changes in wind speed. You can access historical hurricane track maps from the National Oceanic and Atmospheric Administration (NOAA) at *https://coast.noaa.gov/hurricanes*. The tracks on the maps are color-coded using the Saffir-Simpson Hurricane Scale (see Table L20.1, p. 484) so you can keep track of how the strength of a storm changed over time.

LAB 20

Types of tropical storms classified by wind speed according to the Saffir-Simpson Hurricane Scale and by colors used on historical hurricane track maps from the National Oceanic and Atmospheric Administration

Classification	Wind speed			Color used on historical hurricane track maps
	mph	kn	km/h	
Category 5 hurricane	≥ 157	≥ 137	≥ 252	Purple
Category 4 hurricane	130–156	113–136	209–251	Pink
Category 3 hurricane	111–129	96–112	178–208	Red
Category 2 hurricane	96–110	83–95	154–177	Orange
Category 1 hurricane	74–95	64–-82	119–153	Yellow
Tropical storm	39–73	34–63	63–118	Green
Tropical depression	25–38	22–33	40–62	Blue

You can then examine how hurricane wind speed is related to the surface temperature of the ocean (also called sea surface temperature, or SST). Information about current SSTs in the North Atlantic can be found at *www.ospo.noaa.gov/Products/ocean/sst/contour/index.html*. You can also find information about monthly SSTs for 1984–1998 at *www.ospo.noaa.gov/Products/ocean/sst/monthly_mean.html*. Finally, you may want to compare land surface temperature to SSTs. Information about land surface temperature during the daytime by month can be found at *http://earthobservatory.nasa.gov/GlobalMaps*.

Once you finished analyzing these data, you can develop your conceptual model. A conceptual model is an idea or set of ideas that explains what causes a particular phenomenon in nature. People often use words, images, and arrows to describe a conceptual model. Your conceptual model needs to be able to explain why hurricanes wind speed changes over time. The model also needs to be consistent with what we know about natural hazards, weather, and how energy flows into, within, and out of systems.

The last step in this investigation is to test your model. To accomplish this goal, you can use your model to make predictions about how the wind speed of past hurricanes changed over time using the Hurricane Track A and Hurricane Track B handouts. The black-and-white track maps include letters that mark specific locations along these tracks. You goal is to predict the category of these hurricanes at these locations. Your teacher will then give you color versions of these hurricane track maps. The color versions include information about the strength of these hurricanes at each location. You can use these maps to determine if your predictions were accurate. If you are able to make accurate predictions about how the wind speed within these two hurricanes changed over time, then you will be able to generate the evidence you need to convince others that the conceptual model you developed is valid or acceptable.

Connections to the Nature of Scientific Knowledge and Scientific Inquiry

As you work through your investigation, be sure to think about

- the use of models as tools for reasoning about natural phenomena, and
- the types of questions that scientists can investigate.

Initial Argument

Once your group has finished collecting and analyzing your data, your group will need to develop an initial argument. Your initial argument needs to include a claim, evidence to support your claim, and a justification of the evidence. The *claim* is your group's answer to the guiding question. The *evidence* is an analysis and interpretation of your data. Finally, the *justification* of the evidence is why your group thinks the evidence matters. The justification of the evidence is important because scientists can use different kinds of evidence to support their claims. Your group will create your initial argument on a whiteboard. Your whiteboard should include all the information shown in Figure L20.3.

FIGURE L20.3

Argument presentation on a whiteboard

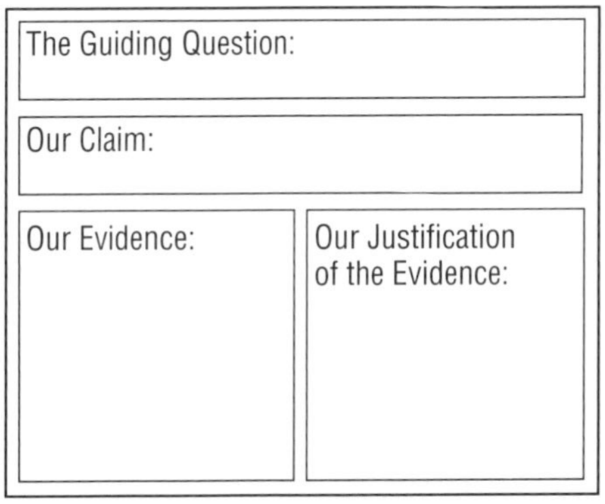

The Guiding Question:	
Our Claim:	
Our Evidence:	Our Justification of the Evidence:

Argumentation Session

The argumentation session allows all of the groups to share their arguments. One or two members of each group will stay at the lab station to share that group's argument, while the other members of the group go to the other lab stations to listen to and critique the other arguments. This is similar to what scientists do when they propose, support, evaluate, and refine new ideas during a poster session at a conference. If you are presenting your group's argument, your goal is to share your ideas and answer questions. You should also keep a record of the critiques and suggestions made by your classmates so you can use this feedback to make your initial argument stronger. You can keep track of specific critiques and suggestions for improvement that your classmates mention in the space below.

Critiques of our initial argument and suggestions for improvement:

If you are critiquing your classmates' arguments, your goal is to look for mistakes in their arguments and offer suggestions for improvement so these mistakes can be fixed. You should look for ways to make your initial argument stronger by looking for things that the other groups did well. You can keep track of interesting ideas that you see and hear during the argumentation in the space below. You can also use this space to keep track of any questions that you will need to discuss with your team.

Interesting ideas from other groups or questions to take back to my group:

Once the argumentation session is complete, you will have a chance to meet with your group and revise your initial argument. Your group might need to gather more data or design a way to test one or more alternative claims as part of this process. Remember, your goal at this stage of the investigation is to develop the best argument possible.

Report

Once you have completed your research, you will need to prepare an investigation report that consists of three sections. Each section should provide an answer for the following questions:

1. What question were you trying to answer and why?

2. What did you do to answer your question and why?

3. What is your argument?

Your report should answer these questions in two pages or less. You should write your report using a word processing application (such as Word, Pages, or Google Docs), if possible, to make it easier for you to edit and revise it later. You should embed any diagrams, figures, or tables into the document. Be sure to write in a persuasive style; you are trying to convince others that your claim is acceptable or valid.

Checkout Questions

Lab 20. Predicting Hurricane Strength: How Can Someone Predict Changes in Hurricane Wind Speed Over Time?

1. The maps below show the path of two different hurricanes. David was classified as a category 2 hurricane when it reached Florida. Andrew, in contrast, was classified as a category 4 hurricane when it passed over Florida 13 years later. Both hurricanes began as a tropical depression near Africa.

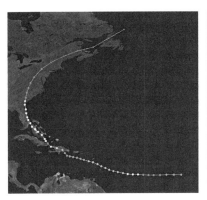

Path of Hurricane David (1979)   Path of Hurricane Andrew (1992)

   a. What are two factors that can affect the wind speed of a hurricane?

   b. Why was the sustained wind speed of Hurricane David less than then sustained wind speed of Hurricane Andrew when these two hurricanes made landfall in Florida?

2. Scientists create pictures of things to teach people about them. These pictures are models.

   a. I agree with this statement.

   b. I disagree with this statement.

   Explain your answer, using an example from your investigation about hurricanes.

3. All questions can be answered by science.

   a. I agree with this statement.

   b. I disagree with this statement.

   Explain your answer, using an example from your investigation about hurricanes.

4. Natural phenomena have causes, and uncovering causal relationships is a major activity of science. Explain why identifying cause-and-effect relationships is important, using an example from your investigation about hurricanes.

5. Tracking energy as it moves into, out of, and within systems is an important activity in science. Give an example for energy moving into, within, or out of a system from your investigation about hurricanes.

Teacher Notes

Lab 21. Forecasting Extreme Weather: When and Under What Atmospheric Conditions Are Tornadoes Likely to Develop in the Oklahoma City Area?

Purpose

The purpose of this lab is to *introduce* students to the disciplinary core ideas (DCIs) of (a) Weather and Climate and (b) Natural Hazards by having them determine when and under what conditions tornadoes tend to form in the Oklahoma City area. In addition, students have an opportunity to learn about the crosscutting concepts (CCs) of (a) Patterns and (b) Cause and Effect: Mechanism and Explanation. During the explicit and reflective discussion, students will also learn about (a) the difference between data and evidence in science and (b) how scientists use different methods to answer different types of questions.

Important Earth and Space Science Content

Tornadoes are short but violent windstorms. A characteristic feature of a tornado is a rotating column of air, which is called a *vortex,* that extends down from a cumulonimbus cloud (see Figure 21.1). Atmospheric pressure within some tornadoes can be as much as 10% lower than immediately outside the storm. This difference in pressure causes air near the ground to rush into the tornado from all directions. As the air moves inward, it is spiraled upward around the center of the vortex until it eventual merges with the airflow of the thunderstorm deep within the cumulonimbus cloud. Tornado winds can reach speeds of 480 km/h (almost 300 m/h).

FIGURE 21.1 _____

The vortex of the tornado is the thin tube reaching from a cloud to the ground; the wind of the tornado has a much wider radius than the vortex

Tornadoes can form in any situation that produces severe weather, including cold fronts and tropical cyclones. The most intense tornadoes usually form within huge thunderstorms called *supercells*. The appearance of a *mesocyclone* with the supercell often precedes the formations of a tornado by 30 minutes or so. A mesocyclone is a vertical cylinder of rotating air that is usually about 3–10 km (2–6 miles) across and develops in the updraft of a severe thunderstorm. Mesocyclones form when wind shear, which is a radical shift in wind speed and direction over a short distance, causes air within a thunderstorm to start spinning along a horizontal axis (see Figure 21.2a). A strong updraft with the thunderstorm then bends or tilts the spinning air upwards until it has nearly vertical alignment (see Figure 21.2b). At this point, the spinning air is a mesocyclone within the thunderstorm (see Figure 21.2c). A tornado, if it develops within the thunderstorm, will descend from a slowly rotating wall cloud in the lower portion of the mesocyclone. It is important to note, however, that the appearance of a mesocyclone within a thunderstorm does not necessarily mean that a tornado will develop as well, because only about half of all mesocyclones produce a tornado. Forecasters are still not sure what triggers the development of a tornado within a supercell.

FIGURE 21.2

The development of a mesocyclone and then a tornado within a thunderstorm

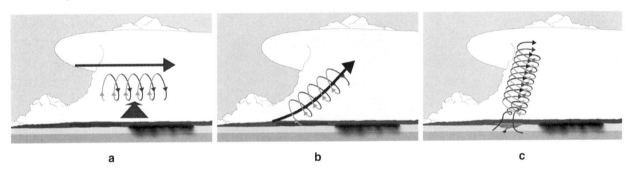

| a | b | c |

Severe thunderstorms, and therefore tornadoes, tend to form along the cold front of a *mid-latitude cyclone.* Mid-latitude cyclones are large centers of low pressure that are formed within the continental United States. These weather systems generally travel from west to east and have a counterclockwise circulation with inward airflow. Most mid-latitude cyclones also have a cold front and a warm front extending out from the central area of low pressure. Mid-latitude cyclones tend to last between 3 and 10 days.

The cold and warm air masses that are associated with the development of a mid-latitude cyclone tend to have very different characteristics during the spring. This is important because thunderstorms are more intense when two air masses with very different characteristics meet and interact with each other. Two contrasting air masses are most likely to meet in the central United States because there is no significant natural barrier separating the center of the country from the Arctic or the Gulf of Mexico. This region, as a result, generates more tornadoes than any other area of the country or, in fact, the world. Many people call this region tornado alley. Figure 21.3 (p. 492) shows the location of tornado alley and where the cold dry air masses moving in from the north tend to meet and interact with warm and humid air masses moving in from the south.

LAB 21

FIGURE 21.3

The location of tornado alley and associated air masses

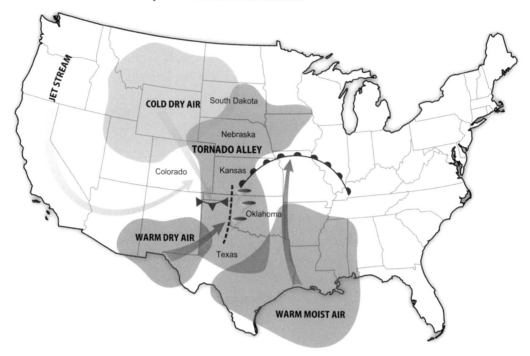

Not all tornadoes form along the cold front of a mid-latitude cyclone, but there are trends in tornado occurrences based on the time of year and the time of day. Tornadoes appear more frequently in the afternoon hours, when the atmosphere has absorbed a large amount of heat that can fuel thunderstorms and tornadoes. Tornado frequency also increases in the months of oncoming warmer seasons because of the nature of the warm and cold air masses that interact with each other during this time of year. The time of year that is associated with more tornadoes varies by region. Tornado season tends to start in March in southeastern states (see Figure 21.4a), in April for south central states (see 21.4b), and in May for the Great Plains and Great Lakes areas (21.4c). Tornado alley, consisting of northern Texas, Oklahoma, Kansas, Nebraska, and South Dakota, is especially active during April and May. There are fewer tornadoes observed throughout the continental United States in the winter. An animated map that shows how the daily tornado risk changes in different regions of the United States over the course of the year is available at *https://imgur. com/qxAwhDZ*. This map is based on tornado risk data from the National Oceanic and Atmospheric Administration (NOAA) National Centers for Environmental Information (see *www.ncdc.noaa.gov/climate-information/extreme-events/us-tornado-climatology*).

FIGURE 21.4

Average number of tornadoes in the United States between 1989 and 2013 in (a) March, (b) April, and (c) May

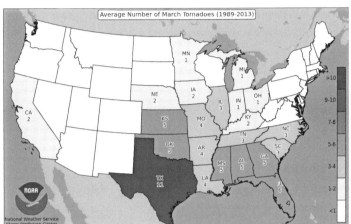

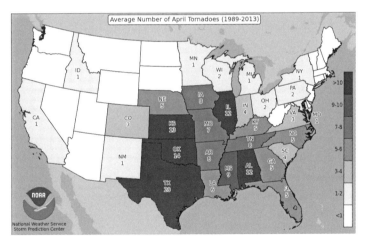

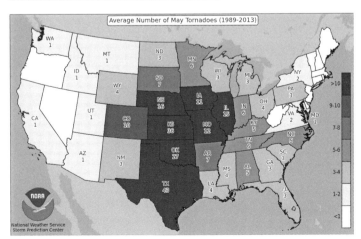

Note: A full-color version of this figure is available on the book's Extras page at *www.nsta.org/adi-ess*.

LAB 21

Timeline

The instructional time needed to complete this lab investigation is 200–280 minutes. Appendix 3 (p. 573) provides options for implementing this lab investigation over several class periods. Option E (280 minutes) should be used if students are unfamiliar with scientific writing, because this option provides extra instructional time for scaffolding the writing process. You can scaffold the writing process by modeling, providing examples, and providing hints as students write each section of the report. Option F (200 minutes) should be used if students are familiar with scientific writing and have developed the skills needed to write an investigation report on their own. In option F, students complete stage 6 (writing the investigation report) and stage 8 (revising the investigation report) as homework.

Materials and Preparation

The materials needed to implement this investigation are listed in Table 21.1. Students will need to access the following online resources:

- The NOAA National Weather Service provides information about every recorded tornado in the Oklahoma City area at *www.weather.gov/oun/tornadodata*.

- NOAA Central Library's Daily Weather Map Archive provides historical U.S. weather maps at *www.wpc.ncep.noaa.gov/dwm/dwm.shtml*.

- Weather Underground provides historical weather data at *www.wunderground.com/history*.

All of these online resources are free to use. You should access each website and learn how to find the desired data before beginning the lab investigation. In addition, it is important to check if students can access and use the websites from a school computer or tablet because some schools have set up restrictions on web browsing.

TABLE 21.1

Materials list for Lab 21

Item	Quantity
Computer or tablet with internet access	1 per group
Whiteboard, 2' × 3'*	1 per group
Lab Handout	1 per student
Peer-review guide and teacher scoring rubric	1 per student
Checkout Questions	1 per student

* As an alternative, students can use computer and presentation software such as Microsoft PowerPoint or Apple Keynote to create their arguments.

Safety Precautions

Remind students to follow all normal lab safety rules.

Topics for the Explicit and Reflective Discussion

Reflecting on the Use of Core Ideas and Crosscutting Concepts During the Investigation

Teachers should begin the explicit and reflective discussion by asking students to discuss what they know about the DCIs they used during the investigation. The following are some important concepts related to the DCIs of (a) Weather and Climate and (b) Natural Hazards that students need to determine what the atmospheric conditions tend to be like right before a severe tornado:

- *Weather* is the current atmospheric conditions.
- Interactions between sunlight, the atmosphere, landforms, and ocean temperatures affect weather and climate.
- The motion and interaction of air masses results in changes in weather conditions.
- People can use changes in atmospheric conditions over time to predict future weather.
- Mapping the history of natural weather hazards in a region, combined with an understanding of related atmospheric conditions, can help forecast the locations and likelihood of future events.

To help students reflect on what they know about these concepts, we recommend showing them two or three images using presentation software that help illustrate these important ideas. You can then ask the students the following questions to encourage students to share how they are thinking about these important concepts:

1. What do we see going on in this image?
2. Does anyone have anything else to add?
3. What might be going on that we can't see?
4. What are some things that we are not sure about here?

You can then encourage students to think about how CCs played a role in their investigation. There are at least two CCs that students need to determine what the atmospheric conditions tend to be like right before a severe tornado: (a) Patterns and (b) Cause and Effect: Mechanism and Explanation (see Appendix 2 [p. 569] for a brief description of these CCs). To help students reflect on what they know about these CCs, we recommend asking them the following questions:

1. Why do scientists look for and attempt to explain patterns in nature?

2. What patterns did you identify and use during your investigation? Why was that useful?

3. Why can cause-and-effect relationships be used to predict phenomena in natural systems?

4. What cause-and-effect relationships did you use during your investigation to make predictions? Why was that useful to do?

You can then encourage the students to think about how they used all these different concepts to help answer the guiding question and why it is important to use these ideas to help justify their evidence for their final arguments. Be sure to remind your students to explain why they included the evidence in their arguments and make the assumptions underlying their analysis and interpretation of the data explicit in order to provide an adequate justification of their evidence.

Reflecting on Ways to Design Better Investigations

It is important for students to reflect on the strengths and weaknesses of the investigation they designed during the explicit and reflective discussion. Students should therefore be encouraged to discuss ways to eliminate potential flaws, measurement errors, or sources of uncertainty in their investigations. To help students be more reflective about the design of their investigation and what they can do to make their investigations more rigorous in the future, you can ask them the following questions:

1. What were some of the strengths of the way you planned and carried out your investigation? In other words, what made it scientific?

2. What were some of the weaknesses of the way you planned and carried out your investigation? In other words, what made it less scientific?

3. What rules can we make, as a class, to ensure that our next investigation is more scientific?

Reflecting on the Nature of Scientific Knowledge and Scientific Inquiry

This investigation can be used to illustrate two important concepts related to the nature of scientific knowledge and the nature of scientific inquiry: (a) the difference between data and evidence in science and (b) how scientists use different methods to answer different types of questions (see Appendix 2 [p. 569] for a brief description of these two concepts). Be sure to review these concepts during and at the end of the explicit and reflective discussion. To help students think about these concepts in relation to what they did during the lab, you can ask the following questions:

1. You had to talk about data and evidence during your investigation. Can you give me some examples of data and evidence from your investigation?

2. Can you work with your group to come up with a rule that you can use to decide if a piece of information is data or evidence? Be ready to share in a few minutes.

3. There is no universal step-by-step scientific method that all scientists follow. Why do you think there is no universal scientific method?

4. Think about what you did during this investigation. How would you describe the method you used to determine the atmospheric conditions that are associated with the formation of tornadoes? Why would you call it that?

You can also use presentation software or other techniques to encourage your students to think about these concepts. You can show examples of information from the investigation that are either data or evidence and ask students to classify each example and explain their thinking. You can also show one or more images of a "universal scientific method" that misrepresent the nature of scientific inquiry (see, e.g., *https://commons.wikimedia.org/wiki/ File:The_Scientific_Method_as_an_Ongoing_Process.svg*) and ask students why each image is *not* a good representation of what scientists do to develop scientific knowledge. You can also ask students to suggest revisions to the image that would make it more consistent with the way scientists develop scientific knowledge.

Remind your students that, to be proficient in science, it is important that they understand what counts as scientific knowledge and how that knowledge develops over time.

Hints for Implementing the Lab

- A group of three students per computer or tablet tends to work well.

- Learn how to use all the online resources before the lab begins. It is important for you to know how to use these resources so you can help students when they get stuck or confused.

- Allow the students to play with the online resources as part of the tool talk before they begin to design their investigation. This gives students a chance to see what data they can and cannot access with each resource.

- Be sure that students record actual values (e.g., wind speeds and direction) and are not just attempting to hand draw what they see on the computer screen.

- This is a good lab for students to make mistakes during the data collection or the development of the initial argument stage. Students will quickly figure out what they did wrong during the argumentation session, and it will only take them a short period of time to re-collect data. It will also create an opportunity for students to reflect on and identify ways to improve the way they design investigations.

LAB 21

- This lab also provides an excellent opportunity to discuss how scientists identify a signal (a pattern or trend) from the noise (measurement error) in their data and how to establish cause-and-effect relationships when experiments cannot be done. Be sure to use this activity as a concrete example during the explicit and reflective discussion.

Connections to Standards

Table 21.2 highlights how the investigation can be used to address specific (a) performance expectations from the *NGSS* and (b) *Common Core State Standards* in English language arts (*CCSS ELA*).

TABLE 21.2

Lab 21 alignment with standards

NGSS performance expectation	Human impact • MS-ESS3-2: Analyze and interpret data on natural hazards to forecast future catastrophic events and inform the development of technologies to mitigate their effects.
CCSS ELA—Reading in Science and Technical Subjects	Key ideas and details • CCSS.ELA-LITERACY.RST.6-8.1: Cite specific textual evidence to support analysis of science and technical texts. • CCSS.ELA-LITERACY.RST.6-8.2: Determine the central ideas or conclusions of a text; provide an accurate summary of the text distinct from prior knowledge or opinions. Craft and structure • CCSS.ELA-LITERACY.RST.6-8.4: Determine the meaning of symbols, key terms, and other domain-specific words and phrases as they are used in a specific scientific or technical context relevant to *grade 6–8 texts and topics*. • CCSS.ELA-LITERACY.RST.6-8.5: Analyze the structure an author uses to organize a text, including how the major sections contribute to the whole and to an understanding of the topic. • CCSS.ELA-LITERACY.RST.6-8.6: Analyze the author's purpose in providing an explanation, describing a procedure, or discussing an experiment in a text. Integration of knowledge and ideas • CCSS.ELA-LITERACY.RST.6-8.7: Integrate quantitative or technical information expressed in words in a text with a version of that information expressed visually (e.g., in a flowchart, diagram, model, graph, or table).

Continued

TABLE 21.2 (*continued*)

CCSS ELA—**Reading in Science and Technical Subjects** (*continued*)	Integration of knowledge and ideas (*continued*) • CCSS.ELA-LITERACY.RST.6-8.8: Distinguish among facts, reasoned judgment based on research findings, and speculation in a text. • CCSS.ELA-LITERACY.RST.6-8.9: Compare and contrast the information gained from experiments, simulations, video, or multimedia sources with that gained from reading a text on the same topic.
CCSS ELA—**Writing in Science and Technical Subjects**	Text types and purposes • CCSS.ELA-LITERACY.WHST.6-8.1: Write arguments focused on *discipline-specific content.* • CCSS.ELA-LITERACY.WHST.6-8.2: Write informative or explanatory texts, including the narration of historical events, scientific procedures/experiments, or technical processes. Production and distribution of writing • CCSS.ELA-LITERACY.WHST.6-8.4: Produce clear and coherent writing in which the development, organization, and style are appropriate to task, purpose, and audience. • CCSS.ELA-LITERACY.WHST.6-8.5: With some guidance and support from peers and adults, develop and strengthen writing as needed by planning, revising, editing, rewriting, or trying a new approach, focusing on how well purpose and audience have been addressed. • CCSS.ELA-LITERACY.WHST.6-8.6: Use technology, including the internet, to produce and publish writing and present the relationships between information and ideas clearly and efficiently. Range of writing • CCSS.ELA-LITERACY.WHST.6-8.10: Write routinely over extended time frames (time for reflection and revision) and shorter time frames (a single sitting or a day or two) for a range of discipline-specific tasks, purposes, and audiences.
CCSS ELA—**Speaking and Listening**	Comprehension and collaboration • CCSS.ELA-LITERACY.SL.6-8.1: Engage effectively in a range of collaborative discussions (one-on-one, in groups, and teacher-led) with diverse partners on grade 6–8 topics, texts, and issues, building on others' ideas and expressing their own clearly. • CCSS.ELA-LITERACY.SL.6-8.2:* Interpret information presented in diverse media and formats (e.g., visually, quantitatively, orally) and explain how it contributes to a topic, text, or issue under study. • CCSS.ELA-LITERACY.SL.6-8.3:* Delineate a speaker's argument and specific claims, distinguishing claims that are supported by reasons and evidence from claims that are not.

Continued

TABLE 21.2 (*continued*)

CCSS ELA—**Speaking and Listening** (*continued*)	Presentation of knowledge and ideas • CCSS.ELA-LITERACY.SL.6-8.4:* Present claims and findings, sequencing ideas logically and using pertinent descriptions, facts, and details to accentuate main ideas or themes; use appropriate eye contact, adequate volume, and clear pronunciation. • CCSS.ELA-LITERACY.SL.6-8.5:* Include multimedia components (e.g., graphics, images, music, sound) and visual displays in presentations to clarify information. • CCSS.ELA-LITERACY.SL.6-8.6: Adapt speech to a variety of contexts and tasks, demonstrating command of formal English when indicated or appropriate.

* Only the standard for grade 6 is provided because the standards for grades 7 and 8 are similar. Please see *www.corestandards.org/ELA-Literacy/SL* for the exact wording of the standards for grades 7 and 8.

Lab Handout

Lab 21. Forecasting Extreme Weather: When and Under What Atmospheric Conditions Are Tornadoes Likely to Develop in the Oklahoma City Area?

Introduction

A tornado is a violent rotating column of air that extends from the clouds to the ground. The wind speeds in a tornado can reach as high as 480 kilometers per hour (almost 300 miles per hour). A tornado can destroy large buildings, uproot trees, and hurl vehicles hundreds of yards. The Oklahoma City metropolitan area is one spot in the United States that is known for frequent tornadoes. According to the National Weather Service, at least 162 tornadoes have touched down in the Oklahoma City metropolitan area between 1890 and 2013, an average of just over one per year (National Weather Service 2017). An example of a tornado in this area was the one that hit Moore, Oklahoma, on May 20th, 2013 at 2:46 p.m. (CDT) (National Weather Service n.d.; Thompson 2013). It stayed on the ground for 47 minutes, traveled 22.5 km (14 miles), and was 1.7 km (1.1 miles) wide at its peak (see Figure L21.1). The 2013 Moore tornado killed 24 people, injured another 212, destroyed 1,150 homes, and caused $2 billion in damage (see Figure L21.2).

FIGURE L21.1

The 2013 EF5 Moore tornado as it passed through south Oklahoma City

FIGURE L21.2

An overhead view of damage done by the 2013 Moore tornado

Tornadoes tend to develop within a supercell, which is a thunderstorm with a large rotating updraft. Thunderstorms tend to develop at locations where there is an interaction between a warm air mass and a cold air mass. Two air masses, however, can interact with each other in many different ways. For example, a warm air mass can move into an area that was formerly covered by cold air mass. A cold air mass can also move into an area that was occupied by a warm air mass. A warm air mass and a cold air mass can also move parallel to each other. These are just a few examples of how warm and cold air mass can interact with each other. Some of these interactions might lead to the development of a thunderstorm and some might not. Tornadoes also tend to develop in different regions during certain times of the year and at certain times of the day. People can make better forecasts about when tornadoes are likely to develop if they understand when tornadoes tend to happen and the atmospheric conditions that tend to be associated with them.

There are still many questions to be answered about what causes a tornado to develop and the factors that affect the wind speed of a tornado. One of the best ways to learn more about tornadoes is to keep a record of which months they happen, what time of day they happen, and what the atmospheric conditions are like in a region right before they develop and then look for patterns that might give us clues about when they are likely to develop and why. This information can then be used to help forecast future tornadoes and to inform the development of new technologies that can mitigate the damage that often results from this type of catastrophic event. In this investigation, you will have an opportunity to use historical weather records to determine when tornadoes tend to occur in the Oklahoma City area and what the atmospheric conditions were like in the region at that time. You can then use this information to identify a way to help predict when a tornado will likely appear in or around Oklahoma City.

Your Task

Use what you know about natural hazards, weather, patterns, and cause-and-effect relationships to analyze historical weather data from the Oklahoma City area to determine when tornadoes tend to happen in this area and what the atmospheric conditions tend to be like at that time. Your goal is to help people in the Oklahoma City area make better forecasts about these potentially catastrophic events.

The guiding question of this investigation is, *When and under what atmospheric conditions are tornadoes likely to develop in the Oklahoma City area?*

Materials

You can use the following online resources during your investigation:

- The National Oceanic and Atmospheric Administration (NOAA) National Weather Service provides information about every recorded tornado in the Oklahoma City area at *www.weather.gov/oun/tornadodata*.
- NOAA Central Library's Daily Weather Map Archive provides historical U.S. weather maps at *www.wpc.ncep.noaa.gov/dwm/dwm.shtml*.
- Weather Underground provides historical weather data at *www.wunderground.com/history*.

Safety Precautions

Follow all normal lab safety rules.

Investigation Proposal Required? ☐ Yes ☐ No

Getting Started

To answer the guiding question, you will need to analyze historical weather data. The first step in your analysis is to learn more about the tornadoes that have hit the Oklahoma City area. You will need to know when during the year these tornadoes developed, what time of the day they happened, how long they lasted, the path they followed, and their magnitude. NOAA provides this information about every recorded tornado in the Oklahoma City area since 1890. You can decide which ones and how many to study.

The next step in your investigation will be to learn more about what the atmospheric conditions were like in the Oklahoma City area before these different tornadoes developed. To accomplish this goal, you can use the NOAA Central Library's Daily Weather Map Archive or Weather Underground. The NOAA Central Library's Daily Weather Map Archive allows you to look up and download the weather maps for the entire United States for a specific date in the past. The Weather Underground website allows you to look up detailed information about the weather in any city dating back to 1945. You can use these two websites to investigate the changes in weather conditions near the dates of the tornadoes listed in the Oklahoma City Area Tornado Table. When you access these websites, you need to think about what data you need to collect, how you will collect it, and how you will analyze it.

To determine *what type of data you need to collect*, think about the following questions:

- Of all of the information you can access, which data are relevant and which data are irrelevant?
- How will you decide which tornadoes to include and which ones to exclude in your study?

To determine *how you will collect the data*, think about the following questions:

- How much data do you need to sufficiently answer your question?
- What scale or scales should you use?
- How will you keep track of and organize the data you collect?
- How will you organize your data?

To determine *how you will analyze the data*, think about the following questions:

- What types of patterns could you look for in your data?
- How could you use mathematics to describe a change over time?

- What type of table or chart could you create to help make sense of your data?
- How could you use mathematics to describe a relationship between variables?

Connections to the Nature of Scientific Knowledge and Scientific Inquiry

As you work through your investigation, be sure to think about

- the difference between data and evidence in science, and
- how scientists use different methods to answer different types of questions.

Initial Argument

Once your group has finished collecting and analyzing your data, your group will need to develop an initial argument. Your initial argument needs to include a claim, evidence to support your claim, and a justification of the evidence. The *claim* is your group's answer to the guiding question. The *evidence* is an analysis and interpretation of your data. Finally, the *justification* of the evidence is why your group thinks the evidence matters. The justification of the evidence is important because scientists can use different kinds of evidence to support their claims. Your group will create your initial argument on a whiteboard. Your whiteboard should include all the information shown in Figure L21.3.

FIGURE L21.3

Argument presentation on a whiteboard

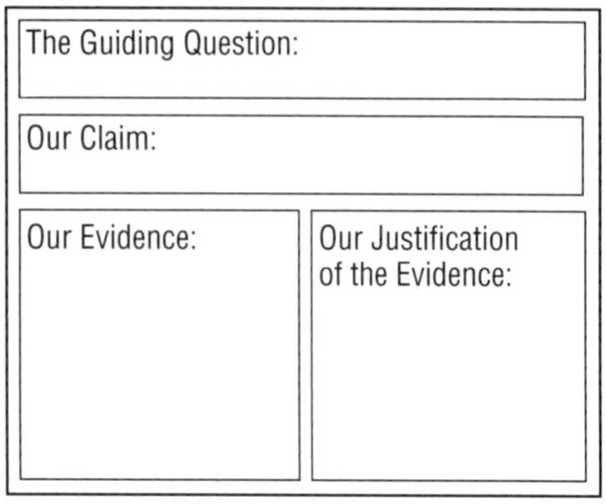

The Guiding Question:	
Our Claim:	
Our Evidence:	Our Justification of the Evidence:

Argumentation Session

The argumentation session allows all of the groups to share their arguments. One or two members of each group will stay at the lab station to share that group's argument, while the other members of the group go to the other lab stations to listen to and critique the other arguments. This is similar to what scientists do when they propose, support, evaluate, and refine new ideas during a poster session at a conference. If you are presenting your group's argument, your goal is to share your ideas and answer questions. You should also keep a record of the critiques and suggestions made by your classmates so you can use this feedback to make your initial argument stronger. You can keep track of specific critiques and suggestions for improvement that your classmates mention in the space provided.

Critiques of our initial argument and suggestions for improvement:

If you are critiquing your classmates' arguments, your goal is to look for mistakes in their arguments and offer suggestions for improvement so these mistakes can be fixed. You should look for ways to make your initial argument stronger by looking for things that the other groups did well. You can keep track of interesting ideas that you see and hear during the argumentation in the space below. You can also use this space to keep track of any questions that you will need to discuss with your team.

Interesting ideas from other groups or questions to take back to my group:

Once the argumentation session is complete, you will have a chance to meet with your group and revise your initial argument. Your group might need to gather more data or design a way to test one or more alternative claims as part of this process. Remember, your goal at this stage of the investigation is to develop the best argument possible.

Report

Once you have completed your research, you will need to prepare an *investigation report* that consists of three sections. Each section should provide an answer for the following questions:

1. What question were you trying to answer and why?

2. What did you do to answer your question and why?

3. What is your argument?

Your report should answer these questions in two pages or less. You should write your report using a word processing application (such as Word, Pages, or Google Docs), if possible, to make it easier for you to edit and revise it later. You should embed any diagrams, figures, or tables into the document. Be sure to write in a persuasive style; you are trying to convince others that your claim is acceptable or valid.

References

National Weather Service. n.d. Moore, Oklahoma tornadoes (1890-present). *www.weather.gov/oun/tornadodata-city-ok-moore.*

National Weather Service. 2017. Tornadoes in the Oklahoma City, Oklahoma area since 1890. *www.weather.gov/oun/tornadodata-okc.*

Thompson, A. 2013. New satellite image shows Moore tornado scar. *www.livescience.com/37176-moore-tornado-damage-satellite-image.html.*

Checkout Questions

Lab 21. Forecasting Extreme Weather: When and Under What Atmospheric Conditions Are Tornadoes Likely to Develop in the Oklahoma City Area?

1. What atmospheric conditions are typically present before tornado formation?

2. The table below shows data for four moments in Oklahoma City with different atmospheric conditions.

Moment	Date	Time	Atmospheric conditions
A	January 15	5:00–7:00 p.m.	Approaching warm front
B	May 3	5:00–7:00 p.m.	Approaching cold front
C	January 15	7:00–9:00 a.m.	Approaching cold front
D	May 3	7:00–9:00 a.m.	Stationary front

a. Rank the moments in the order of likelihood that a severe tornado will occur during this time, with 1 being most likely that a severe tornado will occur, and 4 being least likely that a severe tornado will occur.

b. How do you know?

3. A list of every date a tornado has struck the Oklahoma City area is an example of evidence.

    a. I agree with this statement.

    b. I disagree with this statement.

Explain your answer, using an example from your investigation about forecasting extreme weather.

4. Scientists do not always use lab experiments to further a scientific understanding of a natural phenomenon.

    a. I agree with this statement.

    b. I disagree with this statement.

Explain your answer, using an example from your investigation about forecasting extreme weather.

5. Scientists often need to look for patterns that occur in the data they collect and analyze. Explain why identifying patterns is important, using an example from your investigation about forecasting extreme weather.

6. Natural phenomena have causes, and uncovering causal relationships is a major activity of science. Explain why it is important to uncover causal relationships, using an example from your investigation about forecasting extreme weather.

Application Labs

LAB 22

Teacher Notes

Lab 22. Minimizing Carbon Emissions: What Type of Greenhouse Gas Emission Reduction Policy Will Different Regions of the World Need to Adopt to Prevent the Average Global Surface Temperature on Earth From Increasing by 2°C Between Now and the Year 2100?

Purpose

The purpose of this lab is for students to *apply* what they know about the disciplinary core ideas (DCIs) of (a) Natural Resources, (b) Human Impacts on Earth Systems, and (c) Global Climate Change by having them develop a plan to minimize human impact on climate. In addition, students have an opportunity to learn about the crosscutting concepts (CCs) of (a) Systems and System Models and (b) Stability and Change. During the explicit and reflective discussion, students will also learn about (a) the use of models as tools for reasoning about natural phenomena and (b) the types of questions that scientists can investigate.

Important Earth and Space Science Content

Carbon dioxide (CO_2) is an important gas in Earth's atmosphere. As one of the greenhouse gases that traps heat from the Sun, it makes the surface of Earth a habitable temperature. Carbon dioxide is added to the atmosphere in many natural ways like animal respiration, volcanic eruption, and ocean release. The amount of CO_2 in the atmosphere is naturally reduced when plants take in CO_2 to perform photosynthesis and when oceans absorb it as atmospheric CO_2 levels rise. These natural processes result in atmospheric CO_2 levels that oscillate around an equilibrium point.

Human activities have altered this equilibrium in two ways: by adding more CO_2 to the atmosphere and by reducing carbon sinks (a *carbon sink* is anything that absorbs more carbon than it releases as CO_2). Humans add CO_2 to the atmosphere through activities such as burning fossil fuels for energy or through industrial processes like creating cement, certain metals, and chemicals. Carbon dioxide is removed from the atmosphere by carbon sinks. The biggest carbon sinks on Earth are forests and oceans. Widespread deforestation has contributed to an increase of CO_2 in the atmosphere, and oceans are not an endless source of carbon sinks. Oceans can only absorb so much CO_2 before other adverse effects occur, like ocean acidification. Since the Industrial Revolution, human activities have added CO_2 and have removed carbon sinks, so an abundance of CO_2 has accumulated in the atmosphere. Such an abundance of CO_2 causes increased heat absorption by the greenhouse gas, and thus a rise in average global temperature.

Minimizing Carbon Emissions

What Type of Greenhouse Gas Emission Reduction Policy Will Different Regions of the World Need to Adopt to Prevent the Average Global Surface Temperature on Earth From Increasing by 2°C Between Now and the Year 2100?

The average global temperature has risen by 0.8°C (1.4°F) since 1880. This may seem like a small change, but small changes in average global temperatures can translate to large changes in weather patterns and local climates. A change in climate can change rain patterns, causing dangerous weather like severe rain, floods, or droughts. Scientists, policy makers, and citizens are therefore working to create a plan to reduce greenhouse gas emissions and the impact humans make on Earth's climate. Carbon dioxide can stay in the atmosphere for nearly a century, but an effort to reduce emissions can lessen the risk of further warming and even more severe climate changes.

The Intergovernmental Panel on Climate Change (IPCC), which is the leading international body for the assessment of change in global climate, expects the Earth's average temperature to increase by at least 4.5°C (8.1°F) by the year 2100 (IPCC 2013). This substantial increase in average global temperature will reduce ice and snow cover, change precipitation patterns across the globe, raise sea level, increase the acidity of the oceans, and shift the characteristics of many different natural habitats. These changes will have an impact on our food supply, water resources, infrastructure, and health. Many people are therefore examining different ways to reduce the amount of CO_2 that is released into the atmosphere or to find ways to remove more CO_2 from the atmosphere to help prevent this substantial increase in average global temperature. The current goal is for various regions of the world to identify and enact CO_2 emission reduction policies that will prevent the average global surface temperature from increasing by more than 2°C (3.6°F) before the year 2100.

The *C-ROADS* (Climate Rapid Overview and Decision Support) simulator, available from Climate Interactive, is designed to test strategies for tackling climate change (see *www.climateinteractive.org*). The online simulation allows users to visualize the impact of adopting different emission reduction policies in different regions of the world. This model is useful because it allows users to see and understand the gap between a proposed policy (such as reducing deforestation in a region by 10% a year) and what actually needs to happen to stabilize the concentration of greenhouse gases in the atmosphere and prevent an increase in average global surface temperature. C-ROADS therefore offers students a way to assess the impact of different policies and see what works to address climate change and related issues such as energy, water, food, and the reduction in risks of disasters.

Timeline

The instructional time needed to complete this lab investigation is 200–280 minutes. Appendix 3 (p. 573) provides options for implementing this lab investigation over several class periods. Option E (280 minutes) should be used if students are unfamiliar with scientific writing, because this option provides extra instructional time for scaffolding the writing process. You can scaffold the writing process by modeling, providing examples, and providing hints as students write each section of the report. Option F (200 minutes) should be used if students are familiar with scientific writing and have developed the skills needed to write an investigation report on their own. In option F, students complete stage 6 (writing the investigation report) and stage 8 (revising the investigation report) as homework.

Materials and Preparation

The materials needed to implement this investigation are listed in Table 22.1. The *C-ROADS* simulator is available at *www.climateinteractive.org/tools/c-roads*. It is free to use and can be run online using an internet browser. You should access the website and learn how the simulation works before beginning the lab investigation. In addition, it is important to check if students can access and use the simulation from a school computer or tablet, because some schools have set up firewalls and other restrictions on web browsing.

TABLE 22.1

Materials list for Lab 22

Item	Quantity
Computer or tablet with internet access	1 per group
Whiteboard, 2' × 3'*	1 per group
Lab Handout	1 per student
Peer-review guide and teacher scoring rubric	1 per student
Checkout Questions	1 per student

* As an alternative, students can use computer and presentation software such as Microsoft PowerPoint or Apple Keynote to create their arguments.

Safety Precautions

Remind students to follow all normal lab safety rules.

Topics for the Explicit and Reflective Discussion

Reflecting on the Use of Core Ideas and Crosscutting Concepts During the Investigation

Teachers should begin the explicit and reflective discussion by asking students to discuss what they know about the DCIs they used during the investigation. The following are some important concepts related to the DCIs of (a) Natural Resources, (b) Human Impacts on Earth Systems, and (c) Global Climate Change that students need to develop a plan to minimize human impact on climate:

- An abundance of atmospheric carbon dioxide leads to increased average global temperatures.

- An increase in average global temperature has the potential to cause dangerous changes in climate and weather patterns.

- Human activities, such as the release of greenhouse gases from burning fossil fuels, are major factors in the current rise in Earth's average surface temperature.

- Reducing the level of climate change and reducing human vulnerability to whatever climate changes do occur depend on the understanding of climate science, engineering abilities, and other kinds of knowledge such as understanding human behavior.

- As human populations increase, so do the negative impacts on Earth, unless the activities and technologies involved are engineered to minimize those impacts.

To help students reflect on what they know about these concepts, we recommend showing them two or three images using presentation software that help illustrate these important ideas. You can then ask the students the following questions to encourage students to share how they are thinking about these important concepts:

1. What do we see going on in this image?

2. Does anyone have anything else to add?

3. What might be going on that we can't see?

4. What are some things that we are not sure about here?

You can then encourage students to think about how CCs played a role in their investigation. There are at least two CCs that students need to develop a plan to minimize human impact on climate: (a) Systems and System Models and (b) Stability and Change (see Appendix 2 [p. 569] for a brief description of these CCs). To help students reflect on what they know about these CCs, we recommend asking them the following questions:

1. Why do scientists often define a system and then develop a model of it as part of an investigation?

2. How did you use a model to understand the factors that affect climate change? Why was that useful?

3. Why is it important to think about what controls or affects the rate of change in a system? How can we measure a rate of change?

4. Which factor(s) might have controlled the rate of change in atmospheric carbon dioxide or average global temperature in your investigation? What did exploring these factors allow you to do?

You can then encourage the students to think about how they used all these different concepts to help answer the guiding question and why it is important to use these ideas to help justify their evidence for their final arguments. Be sure to remind your students to explain why they included the evidence in their arguments and make the assumptions underlying their analysis and interpretation of the data explicit in order to provide an adequate justification of their evidence.

Reflecting on Ways to Design Better Investigations

It is important for students to reflect on the strengths and weaknesses of the investigation they designed during the explicit and reflective discussion. Students should therefore be encouraged to discuss ways to eliminate potential flaws, measurement errors, or sources of uncertainty in their investigations. To help students be more reflective about the design of their investigation and what they can do to make their investigations more rigorous in the future, you can ask the following questions:

1. What were some of the strengths of the way you planned and carried out your investigation? In other words, what made it scientific?

2. What were some of the weaknesses of the way you planned and carried out your investigation? In other words, what made it less scientific?

3. What rules can we make, as a class, to ensure that our next investigation is more scientific?

Reflecting on the Nature of Scientific Knowledge and Scientific Inquiry

This investigation can be used to illustrate two important concepts related to the nature of scientific knowledge and the nature of scientific inquiry: (a) how models are used as tools for reasoning about natural phenomena and (b) the types of questions that scientists can investigate (see Appendix 2 [p. 569] for a brief description of these two concepts). Be sure to review these concepts during and at the end of the explicit and reflective discussion. To help students think about these concepts in relation to what they did during the lab, you can ask the following questions:

- I asked you to use a model to develop a plan to minimize human impact on climate as part of your investigation. Why is it useful to develop models in science?

- Can you work with your group to come up with a rule that you can use to decide what a model is and what a model is not in science? Be ready to share in a few minutes.

- Not all questions can be answered by science. Can you give me some examples of questions related to this investigation that can and cannot be answered by science?

- Can you work with your group to come up with a rule that you can use to decide if a question can be answered by science or not? Be ready to share in a few minutes.

Minimizing Carbon Emissions

What Type of Greenhouse Gas Emission Reduction Policy Will Different Regions of the World Need to Adopt to Prevent the Average Global Surface Temperature on Earth From Increasing by 2°C Between Now and the Year 2100?

You can also use presentation software to encourage the students to think about these concepts. You can show examples and non-examples of scientific models and then ask students to classify each one and explain their thinking. You can also show one or more examples of questions that can be answered by science (e.g., What are the sources of greenhouse gases emissions? How do increased greenhouse gases in the atmosphere affect average global surface temperature?) and cannot be answered by science (e.g., What is the best way to reduce carbon dioxide emissions? Who should be required to cut their greenhouse gas emissions?) and then ask students why each example is or is not a question that can be answered by science.

Remind your students that, to be proficient in science, it is important for them to understand what counts as scientific knowledge and how that knowledge develops over time.

Hints for Implementing the Lab

- A group of three students per computer or tablet tends to work well.
- Learn how to use the simulation before the lab begins. It is important for you to know how to use the simulation so you can help students when they get stuck or confused.
- Allow the students to play with the simulation as part of the tool talk before they begin to design their investigation. This gives students a chance to see what they can and cannot do with the simulation.
- Be sure that students record actual values (e.g., fossil fuel emissions by year, temperature increase by year) and are not just attempting to hand draw what they see on the computer screen.

Connections to Standards

Table 22.2 (p. 518) highlights how the investigation can be used to address specific (a) performance expectations from the *NGSS* and (b) *Common Core State Standards* in English language arts (*CCSS ELA*).

LAB 22

TABLE 22.2

Lab 22 alignment with standards

***NGSS* performance expectations**	Weather and climate; Human impact • MS-ESS3-3: Apply scientific principles to design a method for monitoring and minimizing a human impact on the environment. • MS-ESS3-4: Construct an argument supported by evidence for how increases in human population and per-capita consumption of natural resources impact Earth's systems. • HS-ESS3-1: Construct an explanation based on evidence for how the availability of natural resources, occurrence of natural hazards, and changes in climate have influenced human activity. • HS-ESS3-3: Create a computational simulation to illustrate the relationships among the management of natural resources, the sustainability of human populations, and biodiversity. • HS-ESS3-5: Analyze geoscience data and the results from global climate models to make an evidence-based forecast of the current rate of global or regional climate change and associated future impacts to Earth systems.
***CCSS ELA*—Reading in Science and Technical Subjects**	Key ideas and details • CCSS.ELA-LITERACY.RST.6-8.1: Cite specific textual evidence to support analysis of science and technical texts. • CCSS.ELA-LITERACY.RST.6-8.2: Determine the central ideas or conclusions of a text; provide an accurate summary of the text distinct from prior knowledge or opinions. • CCSS.ELA-LITERACY.RST.9-10.1: Cite specific textual evidence to support analysis of science and technical texts, attending to the precise details of explanations or descriptions. • CCSS.ELA-LITERACY.RST.9-10.2: Determine the central ideas or conclusions of a text; trace the text's explanation or depiction of a complex process, phenomenon, or concept; provide an accurate summary of the text. • CCSS.ELA-LITERACY.RST.9-10.3: Follow precisely a complex multistep procedure when carrying out experiments, taking measurements, or performing technical tasks, attending to special cases or exceptions defined in the text. Craft and structure • CCSS.ELA-LITERACY.RST.6-8.4: Determine the meaning of symbols, key terms, and other domain-specific words and phrases as they are used in a specific scientific or technical context relevant to *grade 6–8 texts and topics*.

Continued

Minimizing Carbon Emissions

What Type of Greenhouse Gas Emission Reduction Policy Will Different Regions of the World Need to Adopt to Prevent the Average Global Surface Temperature on Earth From Increasing by 2°C Between Now and the Year 2100?

TABLE 22.2 (*continued*)

CCSS ELA—**Reading in Science and Technical Subjects** (*continued*)	Craft and structure (*continued*) • CCSS.ELA-LITERACY.RST.6-8.5: Analyze the structure an author uses to organize a text, including how the major sections contribute to the whole and to an understanding of the topic. • CCSS.ELA-LITERACY.RST.6-8.6: Analyze the author's purpose in providing an explanation, describing a procedure, or discussing an experiment in a text. • CCSS.ELA-LITERACY.RST.9-10.4: Determine the meaning of symbols, key terms, and other domain-specific words and phrases as they are used in a specific scientific or technical context relevant to *grade 9–10 texts and topics*. • CCSS.ELA-LITERACY.RST.9-10.5: Analyze the structure of the relationships among concepts in a text, including relationships among key terms (e.g., *force, friction, reaction force, energy*). • CCSS.ELA-LITERACY.RST.9-10.6: Analyze the author's purpose in providing an explanation, describing a procedure, or discussing an experiment in a text, defining the question the author seeks to address. Integration of knowledge and ideas • CCSS.ELA-LITERACY.RST.6-8.7: Integrate quantitative or technical information expressed in words in a text with a version of that information expressed visually (e.g., in a flowchart, diagram, model, graph, or table). • CCSS.ELA-LITERACY.RST.6-8.8: Distinguish among facts, reasoned judgment based on research findings, and speculation in a text. • CCSS.ELA-LITERACY.RST.6-8.9: Compare and contrast the information gained from experiments, simulations, video, or multimedia sources with that gained from reading a text on the same topic. • CCSS.ELA-LITERACY.RST.9-10.7: Translate quantitative or technical information expressed in words in a text into visual form (e.g., a table or chart) and translate information expressed visually or mathematically (e.g., in an equation) into words. • CCSS.ELA-LITERACY.RST.9-10.8: Assess the extent to which the reasoning and evidence in a text support the author's claim or a recommendation for solving a scientific or technical problem. • CCSS.ELA-LITERACY.RST.9-10.9: Compare and contrast findings presented in a text to those from other sources (including their own experiments), noting when the findings support or contradict previous explanations or accounts.

Continued

TABLE 22.2 (*continued*)

CCSS ELA—Writing in Science and Technical Subjects	Text types and purposes • CCSS.ELA-LITERACY.WHST.6-10.1: Write arguments focused on *discipline-specific content.* • CCSS.ELA-LITERACY.WHST.6-10.2: Write informative or explanatory texts, including the narration of historical events, scientific procedures/experiments, or technical processes. Production and distribution of writing • CCSS.ELA-LITERACY.WHST.6-8.4: Produce clear and coherent writing in which the development, organization, and style are appropriate to task, purpose, and audience. • CCSS.ELA-LITERACY.WHST.6-8.5: With some guidance and support from peers and adults, develop and strengthen writing as needed by planning, revising, editing, rewriting, or trying a new approach, focusing on how well purpose and audience have been addressed. • CCSS.ELA-LITERACY.WHST.6-8.6: Use technology, including the internet, to produce and publish writing and present the relationships between information and ideas clearly and efficiently. • CCSS.ELA-LITERACY.WHST.9-10.5: Develop and strengthen writing as needed by planning, revising, editing, rewriting, or trying a new approach, focusing on addressing what is most significant for a specific purpose and audience. • CCSS.ELA-LITERACY.WHST.9-10.6: Use technology, including the internet, to produce, publish, and update individual or shared writing products, taking advantage of technology's capacity to link to other information and to display information flexibly and dynamically. Range of writing • CCSS.ELA-LITERACY.WHST.6-10.10: Write routinely over extended time frames (time for reflection and revision) and shorter time frames (a single sitting or a day or two) for a range of discipline-specific tasks, purposes, and audiences.
CCSS ELA—Speaking and Listening	Comprehension and collaboration • CCSS.ELA-LITERACY.SL.6-8.1: Engage effectively in a range of collaborative discussions (one-on-one, in groups, and teacher-led) with diverse partners on grade 6–8 topics, texts, and issues, building on others' ideas and expressing their own clearly. • CCSS.ELA-LITERACY.SL.6-8.2:* Interpret information presented in diverse media and formats (e.g., visually, quantitatively, orally) and explain how it contributes to a topic, text, or issue under study. • CCSS.ELA-LITERACY.SL.6-8.3:* Delineate a speaker's argument and specific claims, distinguishing claims that are supported by reasons and evidence from claims that are not.

Continued

Minimizing Carbon Emissions

What Type of Greenhouse Gas Emission Reduction Policy Will Different Regions of the World Need to Adopt to Prevent the Average Global Surface Temperature on Earth From Increasing by 2°C Between Now and the Year 2100?

TABLE 22.2 (*continued*)

CCSS ELA—Speaking and Listening (*continued*)	Comprehension and collaboration (*continued*) • CCSS.ELA-LITERACY.SL.9-10.1: Initiate and participate effectively in a range of collaborative discussions (one-on-one, in groups, and teacher-led) with diverse partners on grade 6-10 topics, texts, and issues, building on others' ideas and expressing their own clearly and persuasively. • CCSS.ELA-LITERACY.SL.9-10.2: Integrate multiple sources of information presented in diverse media or formats (e.g., visually, quantitatively, orally) evaluating the credibility and accuracy of each source. • CCSS.ELA-LITERACY.SL.9-10.3: Evaluate a speaker's point of view, reasoning, and use of evidence and rhetoric, identifying any fallacious reasoning or exaggerated or distorted evidence. Presentation of knowledge and ideas • CCSS.ELA-LITERACY.SL.6-8.4:* Present claims and findings, sequencing ideas logically and using pertinent descriptions, facts, and details to accentuate main ideas or themes; use appropriate eye contact, adequate volume, and clear pronunciation. • CCSS.ELA-LITERACY.SL.6-8.5:* Include multimedia components (e.g., graphics, images, music, sound) and visual displays in presentations to clarify information. • CCSS.ELA-LITERACY.SL.6-8.6: Adapt speech to a variety of contexts and tasks, demonstrating command of formal English when indicated or appropriate. • CCSS.ELA-LITERACY.SL.9-10.4: Present information, findings, and supporting evidence clearly, concisely, and logically such that listeners can follow the line of reasoning and the organization, development, substance, and style are appropriate to purpose, audience, and task. • CCSS.ELA-LITERACY.SL.9-10.5: Make strategic use of digital media (e.g., textual, graphical, audio, visual, and interactive elements) in presentations to enhance understanding of findings, reasoning, and evidence and to add interest. • CCSS.ELA-LITERACY.SL.9-10.6: Adapt speech to a variety of contexts and tasks, demonstrating command of formal English when indicated or appropriate.

* Only the standard for grade 6 is provided because the standards for grades 7 and 8 are similar. Please see *www. corestandards.org/ELA-Literacy/SL* for the exact wording of the standards for grades 7 and 8.

Reference

Intergovernmental Panel on Climate Change (IPCC). 2013. *Climate change 2013: The physical science basis. Working Group I contribution to the Fifth Assessment Report of the Intergovernmental Panel on Climate Change.* [Stocker, T.F., D. Qin, G.-K. Plattner, M. Tignor, S.K. Allen, J. Boschung, A. Nauels, Y. Xia, V. Bex and P.M. Midgley (eds.)]. Cambridge, England: Cambridge University Press.

LAB 22

Lab Handout

Lab 22. Minimizing Carbon Emissions: What Type of Greenhouse Gas Emission Reduction Policy Will Different Regions of the World Need to Adopt to Prevent the Average Global Surface Temperature on Earth From Increasing by 2°C Between Now and the Year 2100?

Introduction

When sunlight first reaches Earth, some of it is either reflected back out into space or absorbed by the atmosphere. The rest of the sunlight travels to Earth's surface. The sunlight that travels through the atmosphere hits Earth's surface, and the energy from the sunlight increases the temperature of Earth's surface.

Earth's surface emits infrared radiation, and the amount of infrared radiation emitted by the surface increases at higher temperatures. The atmosphere traps some of this infrared radiation before it can escape into space. The trapped infrared radiation in the atmosphere helps keep the temperature of Earth warmer than it would be without the atmosphere. Scientists call the warming of the atmosphere that is caused by trapped infrared radiation the greenhouse effect. Many gases that are found naturally in Earth's atmosphere, including water vapor, carbon dioxide (CO_2), methane, nitrous oxide, and ozone, are called greenhouse gases because these gases trap infrared energy in the atmosphere (see Figure L22.1).

The amount of energy that enters and leaves the Earth system is directly related to the average global temperature of the Earth. The Earth system, which includes the surface and the atmosphere, currently absorbs an average of about 240 watts of solar power per square meter over the course of the year. The Earth system also emits about the same amount of infrared energy into space. The average global surface temperature, as a result, tends to be stable over time. However, if something was to change the amount of energy that enters or leaves this system, then the flow of energy would be unbalanced and the average global temperature would change in response. Therefore, any change to the Earth system that affects how much energy enters or leaves the system can cause a significant change in Earth's average global temperature.

Climate scientists agree that human activities have led to an increase in the concentration of greenhouse gases found in the atmosphere (IPCC 2013). The greenhouse gas that has increased the most is carbon dioxide. Plants and animals release CO_2 into the atmosphere as a waste product of respiration. Plants also absorb CO_2 from the atmosphere for

Minimizing Carbon Emissions

What Type of Greenhouse Gas Emission Reduction Policy Will Different Regions of the World Need to Adopt to Prevent the Average Global Surface Temperature on Earth From Increasing by 2°C Between Now and the Year 2100?

FIGURE L22.1

The greenhouse effect

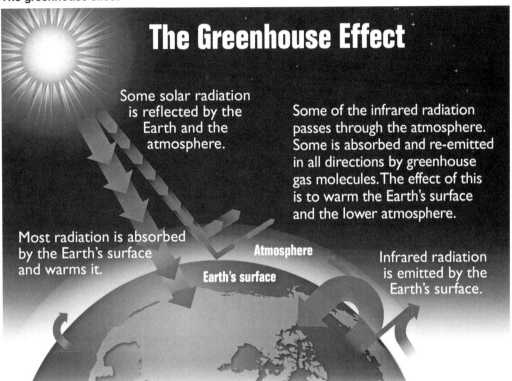

photosynthesis. People, however, also add CO_2 to the atmosphere when they burn fossil fuels for energy. This energy is used for electricity; as fuel for cars, planes, or boats; for heating and cooling homes; and for industrial manufacturing. The concentration of CO_2 in the atmosphere increased from about 320 parts per million (ppm) to almost 390 ppm between 1960 and 2010 (see Figure L22.2). The increase in greenhouse gas emissions such as CO_2 has led to an increase in the greenhouse effect of the atmosphere. The increased greenhouse effect has caused the average global surface temperature to increase approximately 0.8°C (1.4°F) since 1880.

FIGURE L22.2

Concentration of carbon dioxide in the atmosphere between 1960 and 2010; measurements recorded at Mauna Loa, Hawaii (the levels at this site are considered representative of global carbon dioxide levels)

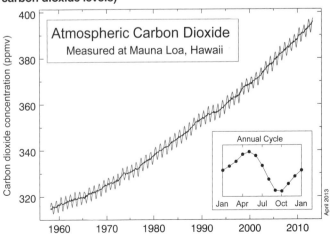

LAB 22

Greenhouse gas concentrations in the atmosphere will continue to increase unless the billions of tons of greenhouse gas that people add to the atmosphere each year decreases substantially. Even if greenhouse emissions were to stop completely, atmospheric greenhouse gas concentrations will continue to remain elevated for hundreds of years because many greenhouse gases stay in the atmosphere for long periods of time. Moreover, if the concentration of greenhouse gases found in the current atmosphere remained steady (which would require a dramatic reduction in current greenhouse gas emissions), surface air temperatures will continue to increase. This is because the oceans, which store heat, take many decades to fully respond to changes in greenhouse gas concentrations. The oceans will therefore continue to have an impact on global climate over the next several decades to hundreds of years.

The Intergovernmental Panel on Climate Change (IPCC), which is the leading international body for the assessment of change in global climate, expects the Earth's average temperature to increase by at least 4.5°C (8.1°F) by the year 2100 (IPCC 2013). This substantial increase in average global temperature will reduce ice and snow cover, change precipitation patterns across the globe, raise sea level, increase the acidity of the oceans, and shift the characteristics of many different natural habitats. These changes will have an impact on our food supply, water resources, infrastructure, and health. Many people are therefore examining different ways to reduce the amount of CO_2 that is released into the atmosphere or to find ways to remove more CO_2 from the atmosphere to help prevent this substantial increase in average global temperature. The current goal is for various regions of the world to identify and enact CO_2 emission reduction policies that will prevent the average global surface temperature from increasing by more than 2°C (3.6°F) before the year 2100.

Your Task

Use what you know about the greenhouse effect, systems and system models, and stability and change to develop a plan to minimize human impact on climate. Your goal is to identify a greenhouse gas emission reduction plan for six different regions of the world that will, when enacted together, prevent the average global surface temperature of Earth from increasing by more than 2°C until the year 2100. The plan you will develop for each region must include recommendations for when the greenhouse gas emissions in the region will need to stop increasing, when the greenhouse gas emissions will need to start decreasing in that region, and the annual greenhouse gas emission reduction rate for that the region. Each plan will also need to include recommendations for a policy that will dictate how much deforestation will need to be reduced in that region by 2050 and how much effort should go into creating new forests in that region starting in 2016.

The guiding question of this investigation is, *What type of greenhouse gas emission reduction policy will different regions of the world need to adopt to prevent the average global surface temperature on Earth from increasing by 2°C between now and the year 2100?*

Minimizing Carbon Emissions

What Type of Greenhouse Gas Emission Reduction Policy Will Different Regions of the World Need to Adopt to Prevent the Average Global Surface Temperature on Earth From Increasing by 2°C Between Now and the Year 2100?

Materials

You will use an online simulation called C-ROADS (Climate Rapid Overview and Decision Support) that helps people understand the long-term climate impacts of actions that reduce greenhouse gas emissions; the simulation is available at *www.climateinteractive.org/tools/c-roads*.

Safety Precautions

Follow all normal lab safety rules.

Investigation Proposal Required?   ☐ Yes   ☐ No

Getting Started

You can use C-ROADS (see Figure L22.3) to test strategies for tackling climate change. The online simulation allows users to visualize the impact of adopting different emission reduction policies in different regions of the world. This model is useful because it allows users to see and understand the gap between a proposed policy (such as reducing deforestation in a region by 10% a year) and what actually needs to happen to stabilize the concentration of greenhouse gases in the atmosphere and prevent an increase in average global surface temperature. C-ROADS therefore offers you a way to test different policies.

FIGURE L22.3

Screenshot of C-ROADS simulation

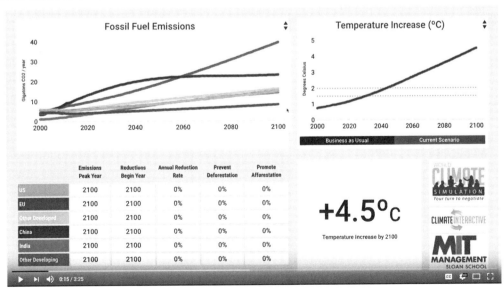

The first step in this investigation is to develop a greenhouse gas emission reduction policy for different regions of the world and then see how these policies, when enacted together, will affect the average global surface temperature over time. *C-ROADS* allows you to create greenhouse gas emission reduction policies for (1) the United States, (2) the European Union, (3) Other Developed Countries (e.g., Japan, Russia, Australia, Korea), (4) China, (5) India, and (6) Other Developing Countries (e.g., nations in Africa, South America, Central America, and the Middle East). The policy you create for each region will need to include

- the year that greenhouse gas emissions will *stop increasing* in the region (emission peak year),
- the year that greenhouse gas emission will *start decreasing* in the region (reduction begin year),
- the rate at which each region *decreases* emission each year (annual reduction rate),
- the rate at which deforestation will *decrease* in the region each year (reduce deforestation), and
- the rate at which new forests will *added* in the region each year (promote afforestation).

You can set values for these five aspects of a greenhouse gas emission reduction policy for the six different regions in *C-ROADS* using the table in the lower-left corner of the simulation. You can then examine how your choices will affect the fossil fuel emissions in each region over time (graph in the upper-left corner) and the average global temperature over time (graph in the upper-right corner). The box in the lower-right corner also tells you how much the average global temperature will change between now and 2100.

One you have developed an initial policy for each region and tested how these policies, when enacted together, will affect fossil fuel emissions and average global temperature over time, you can then begin to modify each of your proposed policies to make your over-all plan more effective. Your goal for this step of the investigation is to determine how your proposed policies for each region will need to be changed to close the gap between what you are proposing and what actually needs to happen to prevent a 2°C increase in average global surface temperature between now and 2100. As you modify your greenhouse gas emission policies, think about the following questions to help guide the changes you will make:

- What are the components of the system and how do they interact with each other?
- When is this system stable and under which conditions does it change?
- Which factor(s) might control the rate of change in this system?
- What scale or scales should you use to when you take measurements?

Minimizing Carbon Emissions

What Type of Greenhouse Gas Emission Reduction Policy Will Different Regions of the World Need to Adopt to Prevent the Average Global Surface Temperature on Earth From Increasing by 2°C Between Now and the Year 2100?

The last step in your investigation will be to identify the potential challenges and consequences that are associated with putting a proposed greenhouse gas emission reduction policy into action. This is important because policies affect the lives of people. For example, what are the consequences of a policy that requires an immediate 20% reduction in fossil fuel emissions for the people in a region that relies on coal or oil as an energy source? Will the people in that region be able to get to work? Heat their homes? Will they lose their jobs? You must also consider what is ethical and what is fair in terms of any proposed policy. For example, is it ethical or fair for people who live in a country in South America to have to cut greenhouse gas emissions at the same rate as people who live in the United States? You will therefore need to take into account these issues to ensure that your proposed greenhouse gas emission reduction policy is valid and acceptable. You can include information about any political, economic, or social issues you considered as you developed and evaluated your policy as part of your justification of your evidence.

Connections to the Nature of Scientific Knowledge and Scientific Inquiry

As you work through your investigation, be sure to think about

- how models are used as tools for reasoning about natural phenomena, and
- the types of questions that scientists can investigate.

Initial Argument

Once your group has finished collecting and analyzing your data, your group will need to develop an initial argument. Your initial argument needs to include a claim, evidence to support your claim, and a justification of the evidence. The *claim* is your group's answer to the guiding question. The *evidence* is an analysis and interpretation of your data. Finally, the *justification* of the evidence is why your group thinks the evidence matters. The justification of the evidence is important because scientists can use different kinds of evidence to support their claims. Your group will create your initial argument on a whiteboard. Your whiteboard should include all the information shown in Figure L22.4.

FIGURE L22.4

Argument presentation on a whiteboard

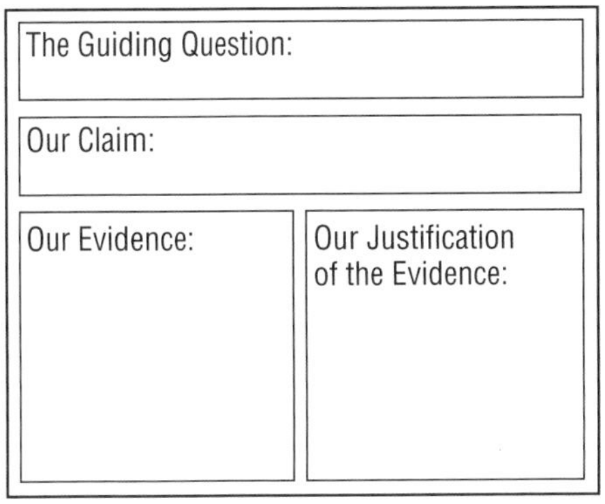

The Guiding Question:	
Our Claim:	
Our Evidence:	Our Justification of the Evidence:

Argumentation Session

The argumentation session allows all of the groups to share their arguments. One or two members of each group will stay at the lab station to share that group's argument, while the other members of the group go to the other lab stations to listen to and critique the other arguments. This is similar to what scientists do when they propose, support, evaluate, and refine new ideas during a poster session at a conference. If you are presenting your group's argument, your goal is to share your ideas and answer questions. You should also keep a record of the critiques and suggestions made by your classmates so you can use this feedback to make your initial argument stronger. You can keep track of specific critiques and suggestions for improvement that your classmates mention in the space below.

Critiques of our initial argument and suggestions for improvement:

If you are critiquing your classmates' arguments, your goal is to look for mistakes in their arguments and offer suggestions for improvement so these mistakes can be fixed. You should look for ways to make your initial argument stronger by looking for things that the other groups did well. You can keep track of interesting ideas that you see and hear during the argumentation in the space below. You can also use this space to keep track of any questions that you will need to discuss with your team.

Interesting ideas from other groups or questions to take back to my group:

Minimizing Carbon Emissions

What Type of Greenhouse Gas Emission Reduction Policy Will Different Regions of the World Need to Adopt to Prevent the Average Global Surface Temperature on Earth From Increasing by 2°C Between Now and the Year 2100?

Once the argumentation session is complete, you will have a chance to meet with your group and revise your initial argument. Your group might need to gather more data or design a way to test one or more alternative claims as part of this process. Remember, your goal at this stage of the investigation is to develop the best argument possible.

Report

Once you have completed your research, you will need to prepare an *investigation report* that consists of three sections. Each section should provide an answer for the following questions:

1. What question were you trying to answer and why?

2. What did you do to answer your question and why?

3. What is your argument?

Your report should answer these questions in two pages or less. You should write your report using a word processing application (such as Word, Pages, or Google Docs), if possible, to make it easier for you to edit and revise it later. You should embed any diagrams, figures, or tables into the document. Be sure to write in a persuasive style; you are trying to convince others that your claim is acceptable or valid.

Reference

Intergovernmental Panel on Climate Change (IPCC). 2013. *Climate change 2013: The physical science basis. Working Group I contribution to the Fifth Assessment Report of the Intergovernmental Panel on Climate Change.* [Stocker, T.F., D. Qin, G.-K. Plattner, M. Tignor, S.K. Allen, J. Boschung, A. Nauels, Y. Xia, V. Bex and P.M. Midgley (eds.)]. Cambridge, England: Cambridge University Press.

Checkout Questions

Lab 22. Minimizing Carbon Emissions: What Type of Greenhouse Gas Emission Reduction Policy Will Different Regions of the World Need to Adopt to Prevent the Average Global Surface Temperature on Earth From Increasing by 2°C Between Now and the Year 2100?

1. In the past century, how have humans increased the amount of carbon dioxide in the atmosphere?

2. What methods can produce usable energy without carbon dioxide emissions?

Minimizing Carbon Emissions

What Type of Greenhouse Gas Emission Reduction Policy Will Different Regions of the World Need to Adopt to Prevent the Average Global Surface Temperature on Earth From Increasing by 2°C Between Now and the Year 2100?

3. A student has submitted the following plan to reduce carbon dioxide emissions to 20% of the levels in 1990.

Action area	Policy	Result
Biofuel production	Government subsidies on land used for biofuel	Use 8,000 square miles of land to grow biofuels. Additionally, import the same amount of biofuels.
Oil, gas, and coal power	Some oil refineries and coal plants shut down to reduce availability of fossil fuels	Reduce the amount of fossil fuels used by 50%
Nuclear power	Tax reductions on nuclear power plants	Build 13 large nuclear power plants around the country.
Wind turbines	Tax credits for businesses that install wind turbines; state funding allocated for building wind turbines	Build 13,000 wind turbines on land and 17,000 wind turbines offshore.
Manufacturing growth	Strict regulations on new and existing manufacturing companies	Manufacturing declines, becoming roughly one-third smaller than existing manufacturing.
Home efficiency	Tax breaks and refunds for adding insulation to new and existing homes	Additional installation is installed in 75% of homes.
Home temperature	Radio, television, and internet advertisements that encourage thermostat changes	The average home temperature in the winter decreases from 17.5°C to 17°C.
Heating fuel	Increased taxes on coal- or gas-burning furnaces	About 20% of domestic heat is powered by electricity.
How we travel	Decreased prices on public transportation, as well as an increase in number of stops	People use public transportation for about one-quarter of their journeys.
Transport fuel	Government regulation that three out of five of the cars sold must be powered by electricity	Three out of five of cars driven are powered by electricity.

a. Which actions seem feasible for a country to carry out by 2050?

b. Why?

c. Which actions do not seem feasible for a country to carry out by 2050?

d. Why not?

4. Models are pictures created by scientists to teach something to others.

   a. I agree with this statement.

   b. I disagree with this statement.

   Explain your answer, using an example from your investigation about minimizing carbon emissions.

Minimizing Carbon Emissions

What Type of Greenhouse Gas Emission Reduction Policy Will Different Regions of the World Need to Adopt to Prevent the Average Global Surface Temperature on Earth From Increasing by 2°C Between Now and the Year 2100?

5. Science can answer any question.

   a. I agree with this statement.

   b. I disagree with this statement.

   Explain your answer, using an example from your investigation about minimizing carbon emissions.

6. Defining a system under study and making a model of it are tools for developing a better understanding of natural phenomena in science. Explain why it is important to make models of natural phenomena, using an example from your investigation about minimizing carbon emissions.

7. When studying a system, one of the main objectives is to determine how the system is changing over time and which factors are causing the system to become unstable. Explain why determining how a system changes over time is important, using an example from your investigation about minimizing carbon emissions.

LAB 23

Teacher Notes

Lab 23. Human Use of Natural Resources: Which Combination of Water Use Policies Will Ensure That the Phoenix Metropolitan Area Water Supply Is Sustainable?

Purpose

The purpose of this lab is for students to *apply* what they know about the disciplinary core ideas (DCIs) of (a) Natural Resources and (b) Human Impacts on Earth Systems by having them develop a plan to ensure that there is enough water available to meet the current and near-future needs of the people living in the Phoenix metropolitan area. In addition, students have an opportunity to learn about the crosscutting concepts (CCs) of (a) Cause and Effect: Mechanism and Explanation and (b) Stability and Change. During the explicit and reflective discussion, students will also learn about (a) how scientific knowledge changes over time and (b) the types of questions that scientists can investigate.

Important Earth and Space Science Content

In modern society, most communities draw the water they use for drinking, cooking, and other life activities from groundwater stored in underground reservoirs. This water is also used for farming and for industrial activities. The demand for freshwater is increasing as human populations grow, but the amount of groundwater in most reservoirs is dwindling. By formal definition, groundwater is a renewable resource because it can be replenished over relatively short periods of time. Water moves through the atmosphere and Earth's surface and recharges underground reservoirs in the hydrologic cycle. However, the rate of groundwater recharge is much slower than that of current human consumption. Additionally, urban development typically redirects water from natural recharge areas. Large human populations have an impact on the groundwater supply by both increasing demand for water and reducing the amount of groundwater recharge that would normally increase supply. Therefore, many regions are running out of freshwater.

Water supply is influenced by many factors. Some factors are controllable by policies, like whether to redirect or dam a river or how many golf courses are allowed in a region. Other factors, like home water use, are not controllable or predictable because these water decisions are made by individuals. Scientists have created mathematical models that can describe the effects of many variables on groundwater supply and can predict the ability to sustain water supply for future populations. Students will get an opportunity to interact with and use these mathematical models that are built into the *WaterSim* visualization tool to assist them in developing a plan that would ensure that there is enough water available for the current and future needs of the people living in the Phoenix Metropolitan area.

Timeline

The instructional time needed to complete this lab investigation is 200–280 minutes. Appendix 3 (p. 573) provides options for implementing this lab investigation over several class periods. Option E (280 minutes) should be used if students are unfamiliar with scientific writing, because this option provides extra instructional time for scaffolding the writing process. You can scaffold the writing process by modeling, providing examples, and providing hints as students write each section of the report. Option F (200 minutes) should be used if students are familiar with scientific writing and have developed the skills needed to write an investigation report on their own. In option F, students complete stage 6 (writing the investigation report) and stage 8 (revising the investigation report) as homework. Both options give students time to test their plans on day 2.

Materials and Preparation

The materials needed to implement this investigation are listed in Table 23.1. The *WaterSim* visualization tool was developed by researchers at Arizona State University and is available at *https://sustainability.asu.edu/dcdc/watersim*. It is free to use and can be run online using an internet browser. You should access the website and learn how the visualization tool works before beginning the lab investigation. In addition, it is important to check if students can access and use the visualization tool from a school computer or tablet, because some schools have set up firewalls and other restrictions on web browsing.

TABLE 23.1 _____

Materials list for Lab 23

Item	Quantity
Computer or tablet with internet access	1 per group
Whiteboard, 2' × 3'*	1 per group
Lab Handout	1 per student
Peer-review guide and teacher scoring rubric	1 per student
Checkout Questions	1 per student

* As an alternative, students can use computer and presentation software such as Microsoft PowerPoint or Apple Keynote to create their arguments.

Safety Precautions

Remind students to follow all normal lab safety rules.

Topics for the Explicit and Reflective Discussion

Reflecting on the Use of Core Ideas and Crosscutting Concepts During the Investigation

Teachers should begin the explicit and reflective discussion by asking students to discuss what they know about the DCIs they used during the investigation. The following are some important concepts related to the DCIs of (a) Natural Resources and (b) Human Impacts on Earth Systems that students need to develop a plan to ensure that there is enough water available to meet the current and near-future needs of the people living in the Phoenix metropolitan area:

- People depend on Earth's land, oceans, atmosphere, and biosphere for many different natural resources.

- Natural resources are limited, and many are not renewable or replaceable over human lifetimes.

- Many natural resources are distributed unevenly around the planet as a result of past geologic processes.

- Water reservoirs include natural or human-made lakes and the groundwater that is found in large aquifers. Scientists use the term *recharge* to describe how a reservoir fills with water over time.

- Human activities have significantly altered the biosphere, sometimes damaging or destroying natural habitats and causing the extinction of other species.

- As human populations increase, so do the negative impacts on Earth, unless the activities and technologies involved are engineered otherwise.

- Natural resources can be used in a sustainable manner. One way that a group of people can ensure that their use of resource is sustainable over time is for them to establish and then follow policies that will always keep the amount or rate of use (*demand*) at a level that is equal to or below the amount available (*supply*).

To help students reflect on what they know about these concepts, we recommend showing them two or three images using presentation software that help illustrate these important ideas. You can then ask the students the following questions to encourage students to share how they are thinking about these important concepts:

1. What do we see going on in this image?

2. Does anyone have anything else to add?

3. What might be going on that we can't see?

4. What are some things that we are not sure about here?

You can then encourage students to think about how CCs played a role in their investigation. There are at least two CCs that students need to develop a plan to ensure that there is enough water available to meet the current and near-future needs of the people living in the Phoenix metropolitan area: (a) Cause and Effect: Mechanism and Explanation and (b) Stability and Change (see Appendix 2 [p. 569] for a brief description of these CCs). To help students reflect on what they know about these CCs, we recommend asking them the following questions:

1. Why can cause-and-effect relationships be used to predict phenomena in natural systems?

2. What cause-and-effect relationships did you use during your investigation to make predictions? Why was that useful to do?

3. Why is it important to think about what controls or affects the rate of change in a system? How can we measure a rate of change?

4. Which factor(s) might have controlled the rate of change in the Phoenix metropolitan area water supply in your investigation? What did exploring these factors allow you to do?

You can then encourage the students to think about how they used all these different concepts to help answer the guiding question and why it is important to use these ideas to help justify their evidence for their final arguments. Be sure to remind your students to explain why they included the evidence in their arguments and make the assumptions underlying their analysis and interpretation of the data explicit in order to provide an adequate justification of their evidence.

Reflecting on Ways to Design Better Investigations

It is important for students to reflect on the strengths and weaknesses of the investigation they designed during the explicit and reflective discussion. Students should therefore be encouraged to discuss ways to eliminate potential flaws, measurement errors, or sources of uncertainty in their investigations. To help students be more reflective about the design of their investigation and what they can do to make their investigations more rigorous in the future, you can ask the following questions:

1. What were some of the strengths of the way you planned and carried out your investigation? In other words, what made it scientific?

2. What were some of the weaknesses of the way you planned and carried out your investigation? In other words, what made it less scientific?

3. What rules can we make, as a class, to ensure that our next investigation is more scientific?

Reflecting on the Nature of Scientific Knowledge and Scientific Inquiry

This investigation can be used to illustrate two important concepts related to the nature of scientific knowledge and the nature of scientific inquiry: (a) how scientific knowledge changes over time and (b) the types of questions that scientists can investigate (see Appendix 2 [p. 569] for a brief description of these two concepts). Be sure to review these concepts during and at the end of the explicit and reflective discussion. To encourage students to think about these concepts in relation to what they did during the lab, you can ask the following questions:

- Scientific knowledge can and does change over time. Can you tell me why it changes?

- Can you work with your group to come up with some examples of how scientific knowledge related to our understanding of water use and sustainability has changed over time? Be ready to share in a few minutes.

- Not all questions can be answered by science. Can you give me some examples of questions related to this investigation that can and cannot be answered by science?

- Can you work with your group to come up with a rule that you can use to decide if question can be answered by science or not? Be ready to share in a few minutes.

You can use presentation software or other techniques to encourage your students to think about these concepts. You can show examples of how our thinking about water use and sustainability has changed over time and ask students to discuss what they think led to those changes. You can also show one or more examples of questions that can be answered by science (e.g., How does the addition of a human-made water reservoir affect biodiversity in the area? How does increased per capita water affect the sustainability of a water supply over time) and cannot be answered by science (e.g., What is the best way to reduce water use? Who should be required to cut their water use?) and then ask students why each example is or is not a question that can be answered by science.

Remind your students that, to be proficient in science, it is important that they understand what counts as scientific knowledge and how that knowledge develops over time.

Hints for Implementing the Lab

- A group of three students per computer tends to work well.

- Learn how to use the WaterSim visualization tool before the lab begins. It is important for you to know how to use the visualization tool so you can help students when they get stuck or confused.

- Allow the students to play with the WaterSim visualization tool as part of the tool talk before they begin to design their investigation. This gives students a chance to see what they can and cannot do with the visualization tool. Be sure to encourage them to click on the question marks so they can see definitions for each variable.

- Be sure that students record actual values (e.g., amount of groundwater that needs to be extracted from aquifers, per capita water use, how many years the water supply can support the current population) and are not just attempting to hand draw or summarize what they see on the computer screen.

Connections to Standards

Table 23.2 highlights how the investigation can be used to address specific (a) performance expectations from the *NGSS* and (b) *Common Core State Standards* in English language arts (*CCSS ELA*).

TABLE 23.2 _____

Lab 23 alignment with standards

NGSS performance expectations	Human impact • MS-ESS3-3: Apply scientific principles to design a method for monitoring and minimizing a human impact on the environment. • MS-ESS3-4: Construct an argument supported by evidence for how increases in human population and per-capita consumption of natural resources impact Earth's systems. • HS-ESS3-1: Construct an explanation based on evidence for how the availability of natural resources, occurrence of natural hazards, and changes in climate have influenced human activity. • HS-ESS3-2: Evaluate competing design solutions for developing, managing, and utilizing energy and mineral resources based on cost-benefit ratios. • HS-ESS3-3: Create a computational simulation to illustrate the relationships among the management of natural resources, the sustainability of human populations, and biodiversity.
CCSS ELA—Reading in Science and Technical Subjects	Key ideas and details • CCSS.ELA-LITERACY.RST.6-8.1: Cite specific textual evidence to support analysis of science and technical texts. • CCSS.ELA-LITERACY.RST.6-8.2: Determine the central ideas or conclusions of a text; provide an accurate summary of the text distinct from prior knowledge or opinions. • CCSS.ELA-LITERACY.RST.9-10.1: Cite specific textual evidence to support analysis of science and technical texts, attending to the precise details of explanations or descriptions. • CCSS.ELA-LITERACY.RST.9-10.2: Determine the central ideas or conclusions of a text; trace the text's explanation or depiction of a complex process, phenomenon, or concept; provide an accurate summary of the text.

Continued

TABLE 23.2 (*continued*)

***CCSS ELA—*Reading in Science and Technical Subjects** (*continued*)	Key ideas and details (*continued*) • CCSS.ELA-LITERACY.RST.9-10.3: Follow precisely a complex multistep procedure when carrying out experiments, taking measurements, or performing technical tasks, attending to special cases or exceptions defined in the text. Craft and structure • CCSS.ELA-LITERACY.RST.6-8.4: Determine the meaning of symbols, key terms, and other domain-specific words and phrases as they are used in a specific scientific or technical context relevant to *grade 6–8 texts and topics*. • CCSS.ELA-LITERACY.RST.6-8.5: Analyze the structure an author uses to organize a text, including how the major sections contribute to the whole and to an understanding of the topic. • CCSS.ELA-LITERACY.RST.6-8.6: Analyze the author's purpose in providing an explanation, describing a procedure, or discussing an experiment in a text. • CCSS.ELA-LITERACY.RST.9-10.4: Determine the meaning of symbols, key terms, and other domain-specific words and phrases as they are used in a specific scientific or technical context relevant to *grade 9–10 texts and topics*. • CCSS.ELA-LITERACY.RST.9-10.5: Analyze the structure of the relationships among concepts in a text, including relationships among key terms (e.g., *force, friction, reaction force, energy*). • CCSS.ELA-LITERACY.RST.9-10.6: Analyze the author's purpose in providing an explanation, describing a procedure, or discussing an experiment in a text, defining the question the author seeks to address. Integration of knowledge and ideas • CCSS.ELA-LITERACY.RST.6-8.7: Integrate quantitative or technical information expressed in words in a text with a version of that information expressed visually (e.g., in a flowchart, diagram, model, graph, or table). • CCSS.ELA-LITERACY.RST.6-8.8: Distinguish among facts, reasoned judgment based on research findings, and speculation in a text. • CCSS.ELA-LITERACY.RST.6-8.9: Compare and contrast the information gained from experiments, simulations, video, or multimedia sources with that gained from reading a text on the same topic. • CCSS.ELA-LITERACY.RST.9-10.7: Translate quantitative or technical information expressed in words in a text into visual form (e.g., a table or chart) and translate information expressed visually or mathematically (e.g., in an equation) into words.

Continued

TABLE 23.2 (*continued*)

***CCSS ELA*—Reading in Science and Technical Subjects** (*continued*)	Integration of knowledge and ideas (*continued*) • CCSS.ELA-LITERACY.RST.9-10.8: Assess the extent to which the reasoning and evidence in a text support the author's claim or a recommendation for solving a scientific or technical problem. • CCSS.ELA-LITERACY.RST.9-10.9: Compare and contrast findings presented in a text to those from other sources (including their own experiments), noting when the findings support or contradict previous explanations or accounts.
***CCSS ELA*—Writing in Science and Technical Subjects**	Text types and purposes • CCSS.ELA-LITERACY.WHST.6-10.1: Write arguments focused on *discipline-specific content.* • CCSS.ELA-LITERACY.WHST.6-10.2: Write informative or explanatory texts, including the narration of historical events, scientific procedures/experiments, or technical processes. Production and distribution of writing • CCSS.ELA-LITERACY.WHST.6-10.4: Produce clear and coherent writing in which the development, organization, and style are appropriate to task, purpose, and audience. • CCSS.ELA-LITERACY.WHST.6-8.5: With some guidance and support from peers and adults, develop and strengthen writing as needed by planning, revising, editing, rewriting, or trying a new approach, focusing on how well purpose and audience have been addressed. • CCSS.ELA-LITERACY.WHST.6-8.6: Use technology, including the internet, to produce and publish writing and present the relationships between information and ideas clearly and efficiently. • CCSS.ELA-LITERACY.WHST.9-10.5: Develop and strengthen writing as needed by planning, revising, editing, rewriting, or trying a new approach, focusing on addressing what is most significant for a specific purpose and audience. • CCSS.ELA-LITERACY.WHST.9-10.6: Use technology, including the internet, to produce, publish, and update individual or shared writing products, taking advantage of technology's capacity to link to other information and to display information flexibly and dynamically. Range of writing • CCSS.ELA-LITERACY.WHST.6-10.10: Write routinely over extended time frames (time for reflection and revision) and shorter time frames (a single sitting or a day or two) for a range of discipline-specific tasks, purposes, and audiences.

Continued

TABLE 23.2 (*continued*)

***CCSS ELA*—Speaking and Listening**	Comprehension and collaboration • CCSS.ELA-LITERACY.SL.6-8.1: Engage effectively in a range of collaborative discussions (one-on-one, in groups, and teacher-led) with diverse partners on grade 6–8 topics, texts, and issues, building on others' ideas and expressing their own clearly. • CCSS.ELA-LITERACY.SL.6-8.2*: Interpret information presented in diverse media and formats (e.g., visually, quantitatively, orally) and explain how it contributes to a topic, text, or issue under study. • CCSS.ELA-LITERACY.SL.6-8.3:* Delineate a speaker's argument and specific claims, distinguishing claims that are supported by reasons and evidence from claims that are not. • CCSS.ELA-LITERACY.SL.9-10.1: Initiate and participate effectively in a range of collaborative discussions (one-on-one, in groups, and teacher-led) with diverse partners on grade 9–10 topics, texts, and issues, building on others' ideas and expressing their own clearly and persuasively. • CCSS.ELA-LITERACY.SL.9-10.2: Integrate multiple sources of information presented in diverse media or formats (e.g., visually, quantitatively, orally) evaluating the credibility and accuracy of each source. • CCSS.ELA-LITERACY.SL.9-10.3: Evaluate a speaker's point of view, reasoning, and use of evidence and rhetoric, identifying any fallacious reasoning or exaggerated or distorted evidence. Presentation of knowledge and ideas • CCSS.ELA-LITERACY.SL.6-8.4:* Present claims and findings, sequencing ideas logically and using pertinent descriptions, facts, and details to accentuate main ideas or themes; use appropriate eye contact, adequate volume, and clear pronunciation. • CCSS.ELA-LITERACY.SL.6-8.5:* Include multimedia components (e.g., graphics, images, music, sound) and visual displays in presentations to clarify information. • CCSS.ELA-LITERACY.SL.6-8.6: Adapt speech to a variety of contexts and tasks, demonstrating command of formal English when indicated or appropriate. • CCSS.ELA-LITERACY.SL.9-10.4: Present information, findings, and supporting evidence clearly, concisely, and logically such that listeners can follow the line of reasoning and the organization, development, substance, and style are appropriate to purpose, audience, and task. • CCSS.ELA-LITERACY.SL.9-10.5: Make strategic use of digital media (e.g., textual, graphical, audio, visual, and interactive elements) in presentations to enhance understanding of findings, reasoning, and evidence and to add interest. • CCSS.ELA-LITERACY.SL.9-10.6: Adapt speech to a variety of contexts and tasks, demonstrating command of formal English when indicated or appropriate.

* Only the standard for grade 6 is provided because the standards for grades 7 and 8 are similar. Please see *www. corestandards.org/ELA-Literacy/SL* for the exact wording of the standards for grades 7 and 8.

Lab Handout

Lab 23. Human Use of Natural Resources: Which Combination of Water Use Policies Will Ensure That the Phoenix Metropolitan Area Water Supply Is Sustainable?

Introduction

Water is an essential resource for us. We must drink water to survive. People also use water to cook, to clean, and for recreational purposes. We also need water to grow the food that we need and to produce many of the products that we use on a daily basis. Many people think water is an unlimited resource because oceans cover about 70% of Earth's surface. The water found in oceans, however, is high in salt and not fit for human consumption. In fact, 97% of the water found on Earth is classified as salt water (see Figure L23.1), so only 3% of all the water found on Earth is fresh. Of all this freshwater, about 69% of it is frozen in glaciers and the ice caps, 30% is groundwater, and the remaining 1% is located on the surface in lakes, rivers, marshes,

FIGURE L23.1

Distribution of water on Earth

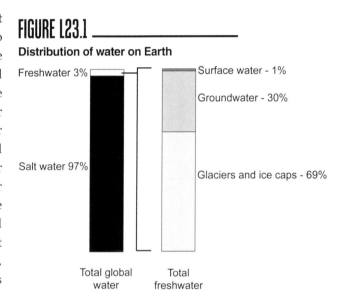

and swamps. All our drinking water and the water we use for agriculture, manufacturing, and sanitation comes from this relatively small amount of surface water and groundwater.

Water reservoirs include natural or human-made lakes (see Figure L23.2, p. 544) and the groundwater that is found in large aquifers, which are underground stores of freshwater (see Figure L23.3, p. 544). Scientists use the term *recharge* to describe how a reservoir fills with water over time. The rate of recharge depends on the amount of precipitation that happens in an area. Many people depend on an aboveground and/or belowground water reservoir to supply all the water they use on a daily basis. Unfortunately, a water reservoir can be depleted of water faster than it can recharge when people consume too much water and there is not much precipitation in an area for an extended amount of time. When a water reservoir runs dry, people who live in that area are forced to do without the water they need until it fills again. High consumption of water can also have a negative impact on the local environment. Typically, as human population and per capita (per person) consumption of water in a region increases, so does the likelihood that the people and the environment in that region will experience a negative impact. It is therefore important for people in a region to find a way to use water in a sustainable manner.

FIGURE L23.2

The Lake Mead Reservoir in Nevada and Arizona

FIGURE L23.3

Illustration of an aquifer

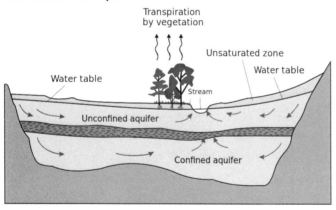

High hydraulic-conductivity aquifer

Low hydraulic-conductivity confining unit

Very low hydraulic-conductivity bedrock

Direction of ground-water flow

One way that a group of people can ensure that their use of water over time is sustainable is for them to establish and then follow policies that will always keep the amount or rate of water use (*demand*) at a level that is equal to or below the amount of water that is or will be available in the local water reservoir (*supply*). The state of Arizona is a good example of how people can work together to ensure that their consumption of water is sustainable over time despite having a limited supply

Arizona is one of the driest states in the United States. In fact, it only receives a statewide average of 12.5 inches of rain per year. Arizona is also one of the fastest-growing states in terms of population. Arizona's population in 2010 was 6.4 million and is projected to increase to over 9.5 million people by 2025 (U.S. Census Bureau 2012). Most of the people who live in Arizona reside within the Phoenix metropolitan area (PMA). The population in the PMA in 2010 was about 4.2 million (U.S. Census Bureau 2012). This many people living in the same area can use a lot of water over the course of a year. In fact, people living in the PMA used 3,667 acre-feet (1 acre-foot = 325,851 gallons) of water in 2008 (Arizona

Department of Water Resources [2009]). The dry climate presents numerous challenges for the people living in Arizona because they can quickly deplete their limited water supply. Therefore, the people of Arizona must find sustainable ways to use their limited supply of water to ensure that they can maintain their quality of life and grow their economy without damaging the local environment.

The people of Arizona get the water they need for agriculture, industry, and municipal use from three major sources. *The first major source is surface water*, and the largest portion of surface water comes from two main reservoirs on the Colorado River (Lake Mead and Lake Powell). The Colorado River starts in the central Rocky Mountains and drains into the Gulf of California (see Figure L23.4). The Colorado River supplies water to people in Arizona, California, Nevada, New Mexico, Utah, Colorado, Wyoming, and Mexico.

FIGURE L23.4

A map of the Colorado River watershed

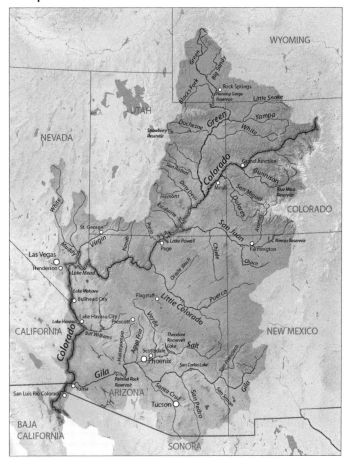

The people of Arizona are only allowed to take 2.8 million acre-feet of water from the Colorado River reservoirs annually (Arizona Department of Water Resources n.d.).

Smaller rivers and lakes in Arizona also provide some water for the people of Arizona. However, the amount of water available from these sites varies from year to year, season to season, and place to place because of the desert climate. The Arizona government has therefore built reservoir storage systems in most of the major rivers within the state, such as the Salt, Verde, Gila, and Agua Fria.

The second major source of water is groundwater. Aquifers supply about 43% of the state's water (Arizona Department of Water Resources n.d.). Throughout the 20th and 21st centuries, however, groundwater has been pumped out of the aquifers faster than it could recharge, which has left them depleted. Though a large amount of water remains stored underground, its availability is limited.

The third major source of water in Arizona is effluent. Effluent, or reclaimed water, is wastewater that has been collected and treated. The people of Arizona use reclaimed water for agriculture, golf courses, industrial cooling, and to maintain parks and other wildlife areas.

To help the people of the PMA identify and enact policies that will ensure that the available water supply is used in a sustainable manner, scientists at Arizona State University created a visualization tool called *WaterSim*. *WaterSim* uses a mathematical model to estimate water supply and demand for the PMA. People can use the *WaterSim* visualization tool to explore how various regional population growth, drought, and climate change scenarios and water management policies affect water sustainability. This visualization tool is valuable because it takes a lot of data that are usually collected separately (including water supply, water demand, climate, population, and policy data) and puts them together to give the user a way to see how all these different variables interact with each other. It also allows a user to change one variable in the system at a time and see how that change affects the other components of the system. In this investigation, you will have an opportunity to use the *WaterSim* visualization tool to explore different water use policies to determine how these different policies, if enacted, will affect the sustainability of the PMA water supply.

Your Task

Use what you know about natural resources, human impacts on Earth systems, stability and change, and cause-and-effect relationships to determine how different water use policies will affect the current and near-future needs of the people living in the PMA. To accomplish this goal, you will need to use the *WaterSim* visualization tool to examine how different water use choices will affect the water supply between now and the year 2050, assuming best- and worst-case scenarios for population growth, drought, and climate change. Your analysis, at a minimum, will need to examine per capita water use, the percentage of wastewater to be reclaimed, and the percentage of farm water to be used by cities in the PMA. It should also include an analysis of how much of the water in the Colorado River can be used to help restore the Colorado River delta of northern Mexico to prevent future habitat and biodiversity loss.

The guiding question of this investigation is, **Which combination of water use policies will ensure that the Phoenix Metropolitan Area water supply is sustainable?**

Materials

You will use a visualization tool called WaterSim to explore how water policy decisions influence water supply and sustainability; the tool is available at *https://sustainability.asu.edu/dcdc/watersim*.

Safety Precautions

Follow all normal lab safety rules.

Investigation Proposal Required?   ☐ Yes   ☐ No

Getting Started

You can use the *WaterSim* visualization tool (see Figure L23.5, p. 548) to examine how different water use choices affect the sustainability of water in the PMA over time, assuming different conditions in Arizona. This tool is useful because it allows users to see and understand the gap between a proposed policy choice (such as the percentage of wastewater to be reclaimed or how much farm water should be diverted to cities) and what actually needs to happen to ensure the sustainability of a water supply. *WaterSim* therefore offers you a way to test different policies.

The first step in this investigation is to develop an overall water use plan for the PMA and then see how this plan will affect the sustainability of water over time. *WaterSim* allows you to create a water use plan for the PMA that includes the following policy choices:

- the amount of total wastewater from residential, commercial, and industrial water users that is diverted to the reclaimed wastewater treatment plant ("% of Wastewater Reclaimed");

- the agriculture (farming) water made available for urban water use ("Farm Water Used by Cities");

- the portion of the Colorado River water flow intentionally left in the river to be diverted for use by the Colorado River delta ("Environmental Flows"); and

- the amount of water that people use at home, which is measured in units of *gallons per capita per day* (GPCD) per person ("Per Capita Water Use")

You can set values for these four water use policy choices using the sliders in the lower-left corner of the simulation.

Next, you can examine how these choices will affect the sustainability of the PMA water system using five different indicators at the top of the simulation. Each indicator represents a different aspect of water sustainability:

- *Groundwater.* How much groundwater needs to be extracted from the aquifers to meet the demands of the community as a percentage of the total amount of water use.

- *Environment.* The percentage contribution of Arizona's commitment to leave water in the Colorado River to restore the Colorado River delta

- *Ag to Urban.* The amount of agriculture water credits being diverted from farming and agriculture for municipal water use. Agriculture is a major part of the economy in Arizona, so less water available for agriculture means less economic growth.

- *Personal.* The amount of water that can be used for people to cook, clean, bathe, dispose of waste, and irrigate yards. Water is needed for health, comfort, and overall well-being, so less water available for personal use means more inconveniences for people.

- *Population.* How many years the water supply can support the current population before the population would experience water deficits or, alternatively, have to find more water at a higher cost.

The value presented inside the box for each indicator represents the final value (end of 2050) over the simulation period. The values for each indicator from the previous simulation are retained at the bottom of each indicator (in parentheses).

FIGURE L23.5

Screenshot of the *WaterSim* visualization tool

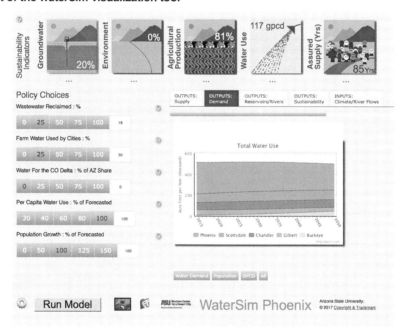

Once you have identified an overall water use plan for the PMA and determined how it will affect water sustainability over time, you can then begin to modify each water use policy choice to make the overall plan more effective. You can also change how fast the population will grow over time, how often and how long droughts will occur, and how the climate may change over time in the *WaterSim* visualization tool. Your goal for this step of the investigation is to determine how different policies will need to be modified to close the gap between what is proposed and what actually needs to happen to ensure the sustainability of the PMA water supply. As you modify the different water use policies and your assumptions about population growth, droughts, and climate change in the *WaterSim* visualization tool, think about the following questions to help guide the changes you will make:

- What are the components of the system and how do they interact with each other?
- When is this system stable and under which conditions does it change?
- Which factor(s) might control the rate of change in this system?
- What scale or scales should you use when you take measurements?

The last step in your investigation will be to identify the potential challenges and consequences that are associated with putting a proposed water use plan into action. This is important because policies affect the lives of people. For example, what are the consequences to farmers of a policy that requires more water to be diverted from farms to cities? Will farmers lose their jobs or their land? What will this do to the overall economy? You must also consider what is ethical and what is fair in terms of any proposed policy. For example, does a policy impact the people who live in cities and in rural areas the same way? You will need to take these issues into account to ensure that your proposed water use plan is valid and acceptable. You can include information about any political, economic, or social issues you considered as you developed and evaluated your policy as part of your justification of your evidence.

Connections to the Nature of Scientific Knowledge and Scientific Inquiry

As you work through your investigation, be sure to think about

- how scientific knowledge can change over time, and
- the types of questions that scientists can investigate.

Initial Argument

Once your group has finished collecting and analyzing your data, your group will need to develop an initial argument. Your initial argument needs to include a claim, evidence to support your claim, and a justification of the evidence. The *claim* is your group's answer to the guiding question. The *evidence* is an analysis and interpretation of your data. Finally, the *justification* of the evidence is why your group thinks the evidence matters. The justification of the evidence is important because scientists can use different kinds of evidence to support their claims. Your group will create your initial argument on a whiteboard. Your whiteboard should include all the information shown in Figure L23.6.

FIGURE L23.6
Argument presentation on a whiteboard

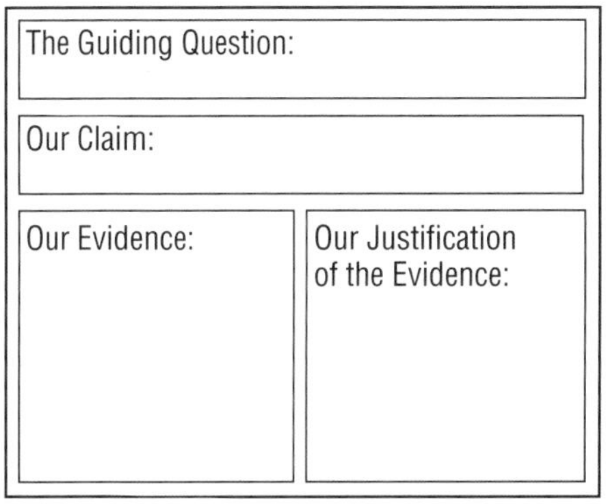

The Guiding Question:	
Our Claim:	
Our Evidence:	Our Justification of the Evidence:

Argumentation Session

The argumentation session allows all of the groups to share their arguments. One or two members of each group will stay at the lab station to share that group's argument, while the other members of the group go to the other lab stations to listen to and critique the other arguments. This is similar to what scientists do when they propose, support, evaluate, and refine new ideas during a poster session at a conference. If you are presenting your group's argument, your goal is to share your ideas and answer questions. You should also keep a record of the critiques and suggestions made by your classmates so you can use this feedback to make your initial argument stronger. You can keep track of specific critiques and suggestions for improvement that your classmates mention in the space below.

Critiques of our initial argument and suggestions for improvement:

If you are critiquing your classmates' arguments, your goal is to look for mistakes in their arguments and offer suggestions for improvement so these mistakes can be fixed. You should look for ways to make your initial argument stronger by looking for things that the other groups did well. You can keep track of interesting ideas that you see and hear during the argumentation in the space below. You can also use this space to keep track of any questions that you will need to discuss with your team.

Interesting ideas from other groups or questions to take back to my group:

Once the argumentation session is complete, you will have a chance to meet with your group and revise your initial argument. Your group might need to gather more data or design a way to test one or more alternative claims as part of this process. Remember, your goal at this stage of the investigation is to develop the best argument possible.

Report

Once you have completed your research, you will need to prepare an *investigation report* that consists of three sections. Each section should provide an answer for the following questions:

1. What question were you trying to answer and why?

2. What did you do to answer your question and why?

3. What is your argument?

Your report should answer these questions in two pages or less. You should write your report using a word processing application (such as Word, Pages, or Google Docs), if possible, to make it easier for you to edit and revise it later. You should embed any diagrams, figures, or tables into the document. Be sure to write in a persuasive style; you are trying to convince others that your claim is acceptable or valid.

References

U.S. Census Bureau. 2012. Arizona: 2010. Population and housing unit counts. 2010 Census of population and housing. CPH-2-4. Washington, DC: U.S. Government Printing Office. Also available online at *www.census.gov/prod/cen2010/cph-2-4.pdf*.

Arizona Department of Water Resources. n.d. Securing Arizona's water future. *www.azwater.gov/AzDWR/PublicInformationOfficer/documents/supplydemand.pdf*.

Arizona Department of Water Resources Drought Program. [2009]. Community water systems 2008 annual water use reporting summary. *www.azwater.gov/AzDWR/StatewidePlanning/drought/2008AnnualWaterUse.htm*.

Checkout Questions

Lab 23. Human Use of Natural Resources: Which Combination of Water Use Policies Will Ensure That the Phoenix Metropolitan Area Water Supply Is Sustainable?

1. List two ways increasing human population affects the supply and sustainability of groundwater.

2. Three city plans for water use are listed in the table below. Assume that the population size and per capita use are 100% as forecasted.

City	Percent wastewater reclaimed	Percent farm water used by cities	Percent environmental flows (water that remains unused in river)
A	82	58	46
B	47	22	62
C	84	18	41

a. Which city plan will result in the most sustainable groundwater stores?

b. How do you know?

c. Which city plan will result in the least sustainable groundwater stores?

d. How do you know?

3. Scientific explanations are subject to revision and improvement in light of new evidence.

    a. I agree with this statement.

    b. I disagree with this statement.

    Explain your answer, using an example from your investigation about human use of natural resources.

4. Science can answer all questions.

    a. I agree with this statement.

    b. I disagree with this statement.

    Explain your answer, using an example from your investigation about human use of natural resources.

5. One of the main objectives of science is to identify and establish relationships between a cause and an effect. Explain why identifying cause-and-effect relationships is important, using an example from your investigation about human use of natural resources.

6. Explain why it is important to determine how a system is changing over time and which factors are causing the system to become unstable, using an example from your investigation about human use of natural resources.

SECTION 7
Appendixes

APPENDIX 1
Standards Alignment Matrixes

Standards Matrix A: Alignment of the Argument-Driven Inquiry Lab Investigations With the Scientific Practices, Crosscutting Concepts, and Core Ideas in *A Framework for K–12 Science Education* (NRC 2012)

Aspect of the NRC *Framework*	Lab 1. Moon Phases	Lab 2. Seasons	Lab 3. Gravity and Orbits	Lab 4. Habitable Worlds	Lab 5. Geologic Time and the Fossil Record	Lab 6. Plate Interactions	Lab 7. Formation of Geologic Features	Lab 8. Surface Erosion by Wind	Lab 9. Sediment Transport by Water	Lab 10. Deposition of Sediments	Lab 11. Soil Texture and Soil Water Permeability	Lab 12. Cycling of Water on Earth	Lab 13. Characteristics of Minerals	Lab 14. Distribution of Natural Resources	Lab 15. Air Masses and Weather Conditions	Lab 16. Surface Materials and Temperature Change	Lab 17. Factors That Affect Global Temperature	Lab 18. Atmospheric Carbon Dioxide Levels	Lab 19. Differences in Regional Climate	Lab 20. Predicting Hurricane Strength	Lab 21. Forecasting Extreme Weather	Lab 22. Minimizing Carbon Emissions	Lab 23. Human Use of Natural Resources
Scientific practices																							
Asking questions	■	■	■	■	■	■	■	■	■	■	■	■	■	■	■	■	■	■	■	■	■	■	■
Developing and using models	■	■	■				■	■	■	■		■				■		■		■	■	■	
Planning and carrying out investigations	■	■	■	■	■	■	■	■	■	■	■	■	■	■	■	■	■	■	■	■	■	■	■
Analyzing and interpreting data	■	■	■	■	■	■	■	■	■	■	■	■	■	■	■	■	■	■	■	■	■	■	■
Using mathematics and computational thinking	■	■	■	■	■	■	■	■	■	■	■	■	■	■	■	■	■	■	■	■	■	■	■
Constructing explanations	■	■	■	■	■	■	■	■	■	■	■	■	■	■	■	■	■	■	■	■	■	■	■
Engaging in argument from evidence	■	■	■	■	■	■	■	■	■	■	■	■	■	■	■	■	■	■	■	■	■	■	■
Obtaining, evaluating, and communicating information	■	■	■	■	■	■	■	■	■	■	■	■	■	■	■	■	■	■	■	■	■	■	■

Key: ■ = strong alignment; □ = moderate alignment.

Aspect of the NRC *Framework*	Lab 1. Moon Phases	Lab 2. Seasons	Lab 3. Gravity and Orbits	Lab 4. Habitable Worlds	Lab 5. Geologic Time and the Fossil Record	Lab 6. Plate Interactions	Lab 7. Formation of Geologic Features	Lab 8. Surface Erosion by Wind	Lab 9. Sediment Transport by Water	Lab 10. Deposition of Sediments	Lab 11. Soil Texture and Soil Water Permeability	Lab 12. Cycling of Water on Earth	Lab 13. Characteristics of Minerals	Lab 14. Distribution of Natural Resources	Lab 15. Air Masses and Weather Conditions	Lab 16. Surface Materials and Temperature Change	Lab 17. Factors That Affect Global Temperature	Lab 18. Atmospheric Carbon Dioxide Levels	Lab 19. Differences in Regional Climate	Lab 20. Predicting Hurricane Strength	Lab 21. Forecasting Extreme Weather	Lab 22. Minimizing Carbon Emissions	Lab 23. Human Use of Natural Resources
Crosscutting concepts																							
Patterns	■	■		■	■	■	■							■	■	■			■		■		
Cause and effect: Mechanism and explanation											■				■	■	■			■	■		■
Scale, proportion, and quantity			■	■	■	■		■	■		■						■						
Systems and system models	■	■	■				■												■			■	
Energy and matter: Flows, cycles, and conservation									■		■	■					■			■			
Structure and function										■			■										
Stability and change								■				■			■	■	■					■	■

Key: ■ = strong alignment; □ = moderate alignment.

Aspect of the NRC *Framework*	Lab 1. Moon Phases	Lab 2. Seasons	Lab 3. Gravity and Orbits	Lab 4. Habitable Worlds	Lab 5. Geologic Time and the Fossil Record	Lab 6. Plate Interactions	Lab 7. Formation of Geologic Features	Lab 8. Surface Erosion by Wind	Lab 9. Sediment Transport by Water	Lab 10. Deposition of Sediments	Lab 11. Soil Texture and Soil Water Permeability	Lab 12. Cycling of Water on Earth	Lab 13. Characteristics of Minerals	Lab 14. Distribution of Natural Resources	Lab 15. Air Masses and Weather Conditions	Lab 16. Surface Materials and Temperature Change	Lab 17. Factors That Affect Global Temperature	Lab 18. Atmospheric Carbon Dioxide Levels	Lab 19. Differences in Regional Climate	Lab 20. Predicting Hurricane Strength	Lab 21. Forecasting Extreme Weather	Lab 22. Minimizing Carbon Emissions	Lab 23. Human Use of Natural Resources
Core ideas																							
The Universe and Its Stars	■	■	■	■																			
Earth and the Solar System	■	■	■																				
The History of Planet Earth					■		■																
Earth Materials and Systems								■	■	■	■	■	■	■	■								
Plate Tectonics and Large-Scale System Interactions						■	■																
The Roles of Water in Earth's Surface Processes									■	■	■												
Weather and Climate															■	■	■		■		■		
Biogeology																							
Natural Resources														■								■	■
Natural Hazards																				■	■		
Human Impacts on Earth Systems														■				■				■	■
Global Climate Change																		■				■	

Key: ■ = strong alignment; □ = moderate alignment.

Standards Matrix B: Alignment of the Argument-Driven Inquiry Lab Investigations With the Nature of Scientific Knowledge (NOSK) and the Nature of Scientific Inquiry (NOSI) Concepts*

NOSK and NOSI concepts	Lab 1. Moon Phases	Lab 2. Seasons	Lab 3. Gravity and Orbits	Lab 4. Habitable Worlds	Lab 5. Geologic Time and the Fossil Record	Lab 6. Plate Interactions	Lab 7. Formation of Geologic Features	Lab 8. Surface Erosion by Wind	Lab 9. Sediment Transport by Water	Lab 10. Deposition of Sediments	Lab 11. Soil Texture and Soil Water Permeability	Lab 12. Cycling of Water on Earth	Lab 13. Characteristics of Minerals	Lab 14. Distribution of Natural Resources	Lab 15. Air Masses and Weather Conditions	Lab 16. Surface Materials and Temperature Change	Lab 17. Factors That Affect Global Temperature	Lab 18. Atmospheric Carbon Dioxide Levels	Lab 19. Differences in Regional Climate	Lab 20. Predicting Hurricane Strength	Lab 21. Forecasting Extreme Weather	Lab 22. Minimizing Carbon Emissions	Lab 23. Human Use of Natural Resources
NOSK																							
How scientific knowledge changes over time					■									■			■	■					■
The difference between laws and theories in science			■													■							
The use of models as tools for reasoning about natural phenomena	■	■					■		■	■									■	■		■	
The difference between data and evidence in science				■				■					■								■		
The difference between observations and inferences in science						■					■	■			■								
NOSI																							
How the culture of science, societal needs, and current events influence the work of scientists						■					■								■				
The types of questions that scientists can investigate														■		■	■			■		■	■
How scientists use different methods to answer different types of questions	■	■			■								■		■						■		
The nature and role of experiments in science								■	■	■		■											
The assumptions made by scientists about order and consistency in nature			■	■			■											■					

Key: ■ = strong alignment; □ = moderate alignment.

*The NOSK/NOSI concepts listed in this matrix are based on the work of Abd-El-Khalick and Lederman 2000; Akerson, Abd-El-Khalick, and Lederman 2000; Lederman et al. 2002, 2014; Schwartz, Lederman, and Crawford 2004; and NGSS Lead States, 2013

Standards Matrix C: Alignment of the Argument-Driven Inquiry Lab Investigations With the *NGSS* Performance Expectations for Middle School and High School Earth and Space Sciences

NGSS performance expectations	Lab 1. Moon Phases	Lab 2. Seasons	Lab 3. Gravity and Orbits	Lab 4. Habitable Worlds	Lab 5. Geologic Time and the Fossil Record	Lab 6. Plate Interactions	Lab 7. Formation of Geologic Features	Lab 8. Surface Erosion by Wind	Lab 9. Sediment Transport by Water	Lab 10. Deposition of Sediments	Lab 11. Soil Texture and Soil Water Permeability	Lab 12. Cycling of Water on Earth	Lab 13. Characteristics of Minerals	Lab 14. Distribution of Natural Resources	Lab 15. Air Masses and Weather Conditions	Lab 16. Surface Materials and Temperature Change	Lab 17. Factors That Affect Global Temperature	Lab 18. Atmospheric Carbon Dioxide Levels	Lab 19. Differences in Regional Climate	Lab 20. Predicting Hurricane Strength	Lab 21. Forecasting Extreme Weather	Lab 22. Minimizing Carbon Emissions	Lab 23. Human Use of Natural Resources
Space Systems																							
MS-ESS1-1: Develop and use a model of the Earth-Sun-Moon system to describe the cyclic patterns of lunar phases, eclipses of the Sun and Moon, and seasons.	■	■																					
MS-ESS1-2: Develop and use a model to describe the role of gravity in the motions within galaxies and the solar system.			■																				
MS-ESS1-3: Analyze and interpret data to determine scale properties of objects in the solar system.				■																			
HS-ESS1-1: Develop a model based on evidence to illustrate the life span of the Sun and the role of nuclear fusion in the Sun's core to release energy that eventually reaches Earth in the form of radiation.																							
HS-ESS1-2: Construct an explanation of the Big Bang theory based on astronomical evidence of light spectra, motion of distant galaxies, and composition of matter in the universe.																							
HS-ESS1-3: Communicate scientific ideas about the way stars, over their life cycle, produce elements.																							
HS-ESS1-4: Use mathematical or computational representations to predict the motion of orbiting objects in the solar system.			■																				

Key: ■ = strong alignment; □ = moderate alignment.

NGSS performance expectations	Lab investigation																						
	Lab 1. Moon Phases	Lab 2. Seasons	Lab 3. Gravity and Orbits	Lab 4. Habitable Worlds	Lab 5. Geologic Time and the Fossil Record	Lab 6. Plate Interactions	Lab 7. Formation of Geologic Features	Lab 8. Surface Erosion by Wind	Lab 9. Sediment Transport by Water	Lab 10. Deposition of Sediments	Lab 11. Soil Texture and Soil Water Permeability	Lab 12. Cycling of Water on Earth	Lab 13. Characteristics of Minerals	Lab 14. Distribution of Natural Resources	Lab 15. Air Masses and Weather Conditions	Lab 16. Surface Materials and Temperature Change	Lab 17. Factors That Affect Global Temperature	Lab 18. Atmospheric Carbon Dioxide Levels	Lab 19. Differences in Regional Climate	Lab 20. Predicting Hurricane Strength	Lab 21. Forecasting Extreme Weather	Lab 22. Minimizing Carbon Emissions	Lab 23. Human Use of Natural Resources
History of Earth																							
MS-ESS1-4: Construct a scientific explanation based on evidence from rock strata for how the geologic time scale is used to organize Earth's 4.6-billion-year-old history.					■																		
MS-ESS2-2: Construct an explanation based on evidence for how geoscience processes have changed Earth's surface at varying time and spatial scales.						■	■	■	■	■													
MS-ESS2-3: Analyze and interpret data on the distribution of fossils and rocks, continental shapes, and seafloor structures to provide evidence of the past plate motions.						■	■																
HS-ESS1-5: Evaluate evidence of the past and current movements of continental and oceanic crust and the theory of plate tectonics to explain the ages of crustal rocks.							■																
HS-ESS1-6: Apply scientific reasoning and evidence from ancient Earth materials, meteorites, and other planetary surfaces to construct an account of Earth's formation and early history.																							

Key: ■ = strong alignment; □ = moderate alignment.

NGSS performance expectations	Lab 1. Moon Phases	Lab 2. Seasons	Lab 3. Gravity and Orbits	Lab 4. Habitable Worlds	Lab 5. Geologic Time and the Fossil Record	Lab 6. Plate Interactions	Lab 7. Formation of Geologic Features	Lab 8. Surface Erosion by Wind	Lab 9. Sediment Transport by Water	Lab 10. Deposition of Sediments	Lab 11. Soil Texture and Soil Water Permeability	Lab 12. Cycling of Water on Earth	Lab 13. Characteristics of Minerals	Lab 14. Distribution of Natural Resources	Lab 15. Air Masses and Weather Conditions	Lab 16. Surface Materials and Temperature Change	Lab 17. Factors That Affect Global Temperature	Lab 18. Atmospheric Carbon Dioxide Levels	Lab 19. Differences in Regional Climate	Lab 20. Predicting Hurricane Strength	Lab 21. Forecasting Extreme Weather	Lab 22. Minimizing Carbon Emissions	Lab 23. Human Use of Natural Resources
HS-ESS2-1: Develop a model to illustrate how Earth's internal and surface processes operate at different spatial and temporal scales to form continental and ocean-floor features.							□																

Earth's Systems

NGSS performance expectations	Lab 1	Lab 2	Lab 3	Lab 4	Lab 5	Lab 6	Lab 7	Lab 8	Lab 9	Lab 10	Lab 11	Lab 12	Lab 13	Lab 14	Lab 15	Lab 16	Lab 17	Lab 18	Lab 19	Lab 20	Lab 21	Lab 22	Lab 23
MS-ESS2-1: Develop a model to describe the cycling of Earth's materials and the flow of energy that drives this process.								■	■	■	□	■											
MS-ESS2-4: Develop a model to describe the cycling of water through Earth's systems driven by energy from the Sun and the force of gravity.									□	□	□	■											
MS-ESS3-1: Construct a scientific explanation based on evidence for how the uneven distributions of Earth's mineral, energy, and groundwater resources are the result of past and current geoscience processes.													□	■									
HS-ESS2-2: Analyze geoscience data to make the claim that one change to Earth's surface can create feedbacks that cause changes to other Earth systems.																	□						
HS-ESS2-3: Develop a model based on evidence of Earth's interior to describe the cycling of matter by thermal convection.																							
HS-ESS2-5: Plan and conduct an investigation of the properties of water and its effects on Earth materials and surface processes.									■	■	■	■											

Key: ■ = strong alignment; □ = moderate alignment.

Argument-Driven Inquiry in Earth and Space Science: Lab Investigations for Grades 6–10

NGSS performance expectations	Lab 1. Moon Phases	Lab 2. Seasons	Lab 3. Gravity and Orbits	Lab 4. Habitable Worlds	Lab 5. Geologic Time and the Fossil Record	Lab 6. Plate Interactions	Lab 7. Formation of Geologic Features	Lab 8. Surface Erosion by Wind	Lab 9. Sediment Transport by Water	Lab 10. Deposition of Sediments	Lab 11. Soil Texture and Soil Water Permeability	Lab 12. Cycling of Water on Earth	Lab 13. Characteristics of Minerals	Lab 14. Distribution of Natural Resources	Lab 15. Air Masses and Weather Conditions	Lab 16. Surface Materials and Temperature Change	Lab 17. Factors That Affect Global Temperature	Lab 18. Atmospheric Carbon Dioxide Levels	Lab 19. Differences in Regional Climate	Lab 20. Predicting Hurricane Strength	Lab 21. Forecasting Extreme Weather	Lab 22. Minimizing Carbon Emissions	Lab 23. Human Use of Natural Resources
HS-ESS2-6: Develop a quantitative model to describe the cycling of carbon among the hydrosphere, atmosphere, geosphere, and biosphere.																							
HS-ESS2-7: Construct an argument based on evidence about the simultaneous coevolution of Earth's systems and life on Earth.																							
Weather and Climate																							
MS-ESS2-5: Collect data to provide evidence for how the motions and complex interactions of air masses results in changes in weather conditions.															■								
MS-ESS2-6: Develop and use a model to describe how unequal heating and rotation of the Earth cause patterns of atmospheric and oceanic circulation that determine regional climates.																■			■				
MS-ESS3-5: Ask questions to clarify evidence of the factors that have caused the rise in global temperatures over the past century.																	■	■					
HS-ESS2-4: Use a model to describe how variations in the flow of energy into and out of Earth's systems result in changes in climate.																	■						
HS-ESS3-5: Analyze geoscience data and the results from global climate models to make an evidence-based forecast of the current rate of global or regional climate change and associated future impacts to Earth systems.																		□				■	

Key: ■ = strong alignment; □ = moderate alignment.

564

NGSS performance expectations	Lab 1. Moon Phases	Lab 2. Seasons	Lab 3. Gravity and Orbits	Lab 4. Habitable Worlds	Lab 5. Geologic Time and the Fossil Record	Lab 6. Plate Interactions	Lab 7. Formation of Geologic Features	Lab 8. Surface Erosion by Wind	Lab 9. Sediment Transport by Water	Lab 10. Deposition of Sediments	Lab 11. Soil Texture and Soil Water Permeability	Lab 12. Cycling of Water on Earth	Lab 13. Characteristics of Minerals	Lab 14. Distribution of Natural Resources	Lab 15. Air Masses and Weather Conditions	Lab 16. Surface Materials and Temperature Change	Lab 17. Factors That Affect Global Temperature	Lab 18. Atmospheric Carbon Dioxide Levels	Lab 19. Differences in Regional Climate	Lab 20. Predicting Hurricane Strength	Lab 21. Forecasting Extreme Weather	Lab 22. Minimizing Carbon Emissions	Lab 23. Human Use of Natural Resources
Human Impact																							
MS-ESS3-2: Analyze and interpret data on natural hazards to forecast future catastrophic events and inform the development of technologies to mitigate their effects.																				■	■		
MS-ESS3-3: Apply scientific principles to design a method for monitoring and minimizing a human impact on the environment.																						■	■
MS-ESS3-4: Construct an argument supported by evidence for how increases in human population and per-capita consumption of natural resources impact Earth's systems.																						■	■
HS-ESS3-1: Construct an explanation based on evidence for how the availability of natural resources, occurrence of natural hazards, and changes in climate have influenced human activity.														■								■	■
HS-ESS3-2: Evaluate competing design solutions for developing, managing, and utilizing energy and mineral resources based on cost-benefit ratios														■									■
HS-ESS3-3: Create a computational simulation to illustrate the relationships among the management of natural resources, the sustainability of human populations, and biodiversity.																						■	■
HS-ESS3-4: Evaluate or refine a technological solution that reduces impacts of human activities on natural systems																							

Key: ■ = strong alignment; □ = moderate alignment.

	Lab investigation																						
NGSS performance expectations	Lab 1. Moon Phases	Lab 2. Seasons	Lab 3. Gravity and Orbits	Lab 4. Habitable Worlds	Lab 5. Geologic Time and the Fossil Record	Lab 6. Plate Interactions	Lab 7. Formation of Geologic Features	Lab 8. Surface Erosion by Wind	Lab 9. Sediment Transport by Water	Lab 10. Deposition of Sediments	Lab 11. Soil Texture and Soil Water Permeability	Lab 12. Cycling of Water on Earth	Lab 13. Characteristics of Minerals	Lab 14. Distribution of Natural Resources	Lab 15. Air Masses and Weather Conditions	Lab 16. Surface Materials and Temperature Change	Lab 17. Factors That Affect Global Temperature	Lab 18. Atmospheric Carbon Dioxide Levels	Lab 19. Differences in Regional Climate	Lab 20. Predicting Hurricane Strength	Lab 21. Forecasting Extreme Weather	Lab 22. Minimizing Carbon Emissions	Lab 23. Human Use of Natural Resources
HS-ESS3-6: Use a computational representation to illustrate the relationships among Earth systems and how those relationships are being modified due to human activity																							

Key: ■ = strong alignment; □ = moderate alignment.

National Science Teachers Association

Standards Matrix D. Alignment of the Argument-Driven Inquiry Lab Investigations With the *Common Core State Standards for English Language Arts* (*CCSS ELA;* NGAC and CCSSO 2010)

Grades 6–12 literacy in science and technical subjects	Lab 1. Moon Phases	Lab 2. Seasons	Lab 3. Gravity and Orbits	Lab 4. Habitable Worlds	Lab 5. Geologic Time and the Fossil Record	Lab 6. Plate Interactions	Lab 7. Formation of Geologic Features	Lab 8. Surface Erosion by Wind	Lab 9. Sediment Transport by Water	Lab 10. Deposition of Sediments	Lab 11. Soil Texture and Soil Water Permeability	Lab 12. Cycling of Water on Earth	Lab 13. Characteristics of Minerals	Lab 14. Distribution of Natural Resources	Lab 15. Air Masses and Weather Conditions	Lab 16. Surface Materials and Temperature Change	Lab 17. Factors That Affect Global Temperature	Lab 18. Atmospheric Carbon Dioxide Levels	Lab 19. Atmospheric Movement and Climate	Lab 20. Predicting Hurricane Strength	Lab 21. Forecasting Extreme Weather	Lab 22. Minimizing Carbon Emissions	Lab 23. Human Use of Natural Resources
Reading																							
Key ideas and details	■	■	■	■	■	■	■	■	■	■	■	■	■	■	■	■	■	■	■	■	■	■	■
Craft and structure	■	■	■	■	■	■	■	■	■	■	■	■	■	■	■	■	■	■	■	■	■	■	■
Integration of knowledge and ideas	■	■	■	■	■	■	■	■	■	■	■	■	■	■	■	■	■	■	■	■	■	■	■
Writing																							
Text types and purposes	■	■	■	■	■	■	■	■	■	■	■	■	■	■	■	■	■	■	■	■	■	■	■
Production and distribution of writing	■	■	■	■	■	■	■	■	■	■	■	■	■	■	■	■	■	■	■	■	■	■	■
Research to build and present knowledge	□	□	□	□	□	□	□	□	□	□	□	□	□	□	□	□	□	□	□	□	□	□	□
Range of writing	■	■	■	■	■	■	■	■	■	■	■	■	■	■	■	■	■	■	■	■	■	■	■
Speaking and Listening																							
Comprehension and collaboration	■	■	■	■	■	■	■	■	■	■	■	■	■	■	■	■	■	■	■	■	■	■	■
Presentation of knowledge and ideas	■	■	■	■	■	■	■	■	■	■	■	■	■	■	■	■	■	■	■	■	■	■	■

Key: ■ = strong alignment; □ = moderate alignment.

References

Abd-El-Khalick, F., and N. G. Lederman. 2000. Improving science teachers' conceptions of nature of science: A critical review of the literature. *International Journal of Science Education* 22: 665–701.

Akerson, V., F. Abd-El-Khalick, and N. Lederman. 2000. Influence of a reflective explicit activity-based approach on elementary teachers' conception of nature of science. *Journal of Research in Science Teaching* 37 (4): 295–317.

Lederman, N. G., F. Abd-El-Khalick, R. L. Bell, and R. S. Schwartz. 2002. Views of nature of science questionnaire: Toward a valid and meaningful assessment of learners' conceptions of nature of science. *Journal of Research in Science Teaching* 39 (6): 497–521.

Lederman, J., N. Lederman, S. Bartos, S. Bartels, A. Meyer, and R. Schwartz. 2014. Meaningful assessment of learners' understanding about scientific inquiry: The Views About Scientific Inquiry (VASI) questionnaire. *Journal of Research in Science Teaching* 51 (1): 65–83.

National Governors Association Center for Best Practices and Council of Chief State School Officers (NGAC and CCSSO). 2010. *Common core state standards.* Washington, DC: NGAC and CCSSO.

NGSS Lead States. 2013. *Next Generation Science Standards: For states, by states.* Washington, DC: National Academies Press. *www.nextgenscience.org/next-generation-science-standards.*

National Research Council (NRC). 2012. *A framework for K–12 science education: Practices, crosscutting concepts, and core ideas.* Washington, DC: National Academies Press.

Schwartz, R. S., N. Lederman, and B. Crawford. 2004. Developing views of nature of science in an authentic context: An explicit approach to bridging the gap between nature of science and scientific inquiry. *Science Education* 88: 610–645.

APPENDIX 2

OVERVIEW OF THE *NGSS* CROSSCUTTING CONCEPTS

Patterns

Scientists look for patterns in nature and attempt to understand the underlying cause of these patterns. For example, scientists often collect data and then look for patterns to identify a relationship between two variables, a trend over time, or a difference between groups.

Cause and Effect: Mechanism and Explanation

Natural phenomena have causes, and uncovering causal relationships (e.g., how changes in x affect y) is a major activity of science. Scientists also need to understand that correlation does not imply causation, some effects can have more than one cause, and some cause-and-effect relationships in systems can only be described using probability.

Scale, Proportion, and Quantity

It is critical for scientists to be able to recognize what is relevant at different sizes, times, and scales. An understanding of scale involves not only understanding how systems and processes vary in size, time span, and energy, but also how different mechanisms operate at different scales. Scientists must also be able to recognize proportional relationships between categories, groups, or quantities.

Systems and System Models

Scientists often need to define a system under study, and making a model of the system is a tool for developing a better understanding of natural phenomena in science. Scientists also need to understand that a system may interact with other systems and a system might include several different subsystems. Scientists often describe a system in terms of inputs and outputs or processes and interactions. All models of a system have limitations because they only represent certain aspects of the system under study.

Energy and Matter: Flows, Cycles, and Conservation

It is important to track how energy and matter move into, out of, and within systems during investigations. Scientists understand that the total amount of energy and matter remains the same in a closed system and that energy cannot be created or destroyed; it only moves between objects and/or fields, between one place and another place, or between systems. Energy drives the cycling and transformation of matter within and between systems.

Structure and Function

The way an object or a material is structured or shaped determines how it functions and places limits on what it can and cannot do. Scientists can make inferences about the function of an object or system by making observations about the structure or shape of its component parts and how these components interact with each other.

Stability and Change

It is critical to understand what makes a system stable or unstable and what controls rates of change in systems. Scientists understand that changes in one part of a system might cause large changes in another part. They also understand that systems in dynamic equilibrium are stable due to a balance of feedback mechanisms, but the stability of these systems can be disturbed by a sudden change in the system or a series of gradual changes that accumulate over time.

OVERVIEW OF NATURE OF SCIENTIFIC KNOWLEDGE AND SCIENTIFIC INQUIRY CONCEPTS

Nature of Scientific Knowledge Concepts

How scientific knowledge changes over time

A person can have confidence in the validity of scientific knowledge but must also accept that scientific knowledge may be abandoned or modified in light of new evidence or because existing evidence has been reconceptualized by scientists. There are many examples in the history of science of both *evolutionary changes* (i.e., the slow or gradual refinement of ideas) and *revolutionary changes* (i.e., the rapid abandonment of a well-established idea) in scientific knowledge.

The difference between laws and theories in science

A *scientific law* describes the behavior of a natural phenomenon or a generalized relationship under certain conditions; a *scientific theory* is a well-substantiated explanation of some aspect of the natural world. Theories do not become laws even with additional evidence; they explain laws. However, not all scientific laws have an accompanying explanatory theory. It is also important for students to understand that scientists do not discover laws or theories; the scientific community develops them over time.

The use of models as tools for reasoning about natural phenomena

Scientists use conceptual models as tools to understand natural phenomena and to make predictions. A *conceptual model* is a representation of a set of ideas about how something works or why something happens. Models can take the form of diagrams, mathematical relationships, analogies, or simulations. Scientists often develop, use, test, and refine models as part of an investigation. All models are based on a set of assumptions and include approximations that limit how a model can be used and its overall predictive power.

The difference between data and evidence in science

Data are measurements, observations, and findings from other studies that are collected as part of an investigation. *Evidence,* in contrast, is analyzed data and an interpretation of the analysis. Scientists do not collect evidence; they collect data and then transform the data they collect into evidence through a process of analysis and interpretation.

The difference between observations and inferences in science

An *observation* is a descriptive statement about a natural phenomenon, whereas an *inference* is an interpretation of an observation. Students should also understand that current scientific knowledge and the perspectives of individual scientists guide both observations and inferences. Thus, different scientists can have different but equally valid interpretations of the same observations due to differences in their perspectives and background knowledge.

Nature of Scientific Inquiry Concepts

How the culture of science, societal needs, and current events influence the work of scientists

Scientists share a set of values, norms, and commitments that shape what counts as knowing, how to represent or communicate information, and how to interact with other scientists. The culture of science affects who gets to do science, what scientists choose to investigate, how investigations are conducted, how research findings are interpreted, and what people see as implications. People also view some research as being more important than other research because of cultural values and current events.

The types of questions that scientists can investigate

Scientists answer questions about the natural or material world, but not all questions can be answered by science. Science and technology may raise ethical issues for which science, by itself, does not provide answers and solutions. Scientists attempt to answer questions about what can happen in natural systems, why things happen, or how things happen. Scientists do not attempt to answer questions about what should happen. To answer questions about what should happen requires consideration of issues related to ethics, morals, values, politics, and economics.

How scientists use different methods to answer different types of questions

Examples of methods include experiments, systematic observations of a phenomenon, literature reviews, and analysis of existing data sets; the choice of method depends on the objectives of the research. There is no universal step-by step scientific method that all scientists follow; rather, different scientific disciplines (e.g., geoscience vs. chemistry) and fields within a discipline (e.g., geophysics vs. paleontology) use different types of methods, use different core theories, and rely on different standards to develop scientific knowledge.

The nature and role of experiments in science

Scientists use experiments to test the validity of a hypothesis (i.e., a tentative explanation) for an observed phenomenon. Experiments include a test and the formulation of predictions (expected results) if the test is conducted and the hypothesis is valid. The experiment is then carried out and the predictions are compared with the actual results of the experiment. If the predictions match the actual results, then the hypothesis is supported. If the actual results do not match the predicted results, then the hypothesis is not supported. A signature feature of an experiment is the control of variables to help eliminate alternative explanations for the results.

The assumptions made by scientists about order and consistency in nature

Scientific investigations are designed based on the assumptions that natural laws operate today as they did in the past and that they will continue to do so in the future. Scientists also assume that the universe is a vast single system in which basic laws are consistent.

APPENDIX 3
Timeline Options for Implementing ADI Lab Investigations

Option A: 6 days (280 minutes), no homework

Day	Stage	Time
1	1: Introduce the task and the guiding question	20 minutes
	2: Design a method	30 minutes
2	2: Collect data	50 minutes
3	3: Develop an initial argument	20 minutes
	4: Argumentation session (and revise initial argument)	30 minutes
4	5: Explicit and reflective discussion	20 minutes
	6: Write investigation report (draft)	30 minutes
5	7: Double-blind peer review	50 minutes
6	8: Revise and submit the investigation report	30 minutes

Option B: 5 days (220 minutes), writing done as homework

Day	Stage	Time
1	1: Introduce the task and the guiding question	20 minutes
	2: Design a method	30 minutes
2	2: Collect data	50 minutes
3	3: Develop an initial argument	20 minutes
	4: Argumentation session (and revise initial argument)	30 minutes
4	5: Explicit and reflective discussion	20 minutes
	6: Write investigation report (draft)	Homework
5	7: Double-blind peer review	50 minutes
	8: Revise and submit the investigation report	Homework

Option C: 5 days (230 minutes), no homework

Day	Stage	Time
1	1: Introduce the task and the guiding question	20 minutes
	2: Design a method and collect data	30 minutes
2	3: Develop an initial argument	20 minutes
	4: Argumentation session (and revise initial argument)	30 minutes
3	5: Explicit and reflective discussion	20 minutes
	6: Write investigation report (draft)	30 minutes
4	7: Double-blind peer review	50 minutes
5	8: Revise and submit the investigation report	30 minutes

Option D: 4 days (170 minutes), writing done as homework

Day	Stage	Time
1	1: Introduce the task and the guiding question	20 minutes
	2: Design a method and collect data	30 minutes
2	3: Develop an initial argument	20 minutes
	4: Argumentation session (and revise initial argument)	30 minutes
3	5: Explicit and reflective discussion	20 minutes
	6: Write investigation report (draft)	Homework
4	7: Double-blind peer review	50 minutes
	8: Revise and submit the investigation report	Homework

Option E: 6 days (280 minutes), no homework

Day	Stage	Time
1	1: Introduce the task and the guiding question	20 minutes
	2: Design a method	30 minutes
2	2: Collect data	30 minutes
	3: Develop an initial argument	20 minutes
3	4: Argumentation session (and revise initial argument)	30 minutes
	5: Explicit and reflective discussion	20 minutes
4	6: Write investigation report (draft)	50 minutes
5	7: Double-blind peer review	50 minutes
6	8: Revise and submit the investigation report	30 minutes

Option F: 4 days (200 minutes), writing done as homework

Day	Stage	Time
1	1: Introduce the task and the guiding question	20 minutes
	2: Design a method	30 minutes
2	2: Collect data	30 minutes
	3: Develop an initial argument	20 minutes
3	4: Argumentation session (and revise initial argument)	30 minutes
	5: Explicit and reflective discussion	20 minutes
	6: Write investigation report (draft)	Homework
4	7: Double-blind peer review	50 minutes
	8: Revise and submit the investigation report	Homework

Option G: 7 Days (330 minutes), no homework

Day	Stage	Time
1	1. Introduce the task and the guiding question	20 minutes
	2. Design a method	30 minutes
2	2. Collect data	50 minutes
3	2. Collect data (develop and test model)	50 minutes
4	3. Develop an initial argument	20 minutes
	4. Argumentation session (and revise initial argument)	30 minutes
5	5. Explicit and reflective discussion	20 minutes
	6. Write investigation report (draft)	30 minutes
6	7. Double-blind peer review	50 minutes
7	8. Revise and submit the investigation report	30 minutes

Option H: 6 Days (270 minutes), writing done as homework

Day	Stage	Time
1	1. Introduce the task and the guiding question	20 Minutes
	2. Design a method	30 Minutes
2	2. Collect data	50 Minutes
3	2. Collect data (develop and test model)	50 Minutes
4	3. Develop an initial argument	20 Minutes
	4. Argumentation session (and revise initial argument)	30 Minutes
5	5. Explicit and reflective discussion	20 Minutes
	6. Write investigation report (draft)	Homework
6	7. Double-blind peer review	50 Minutes
	8. Revise and submit the investigation report	Homework

APPENDIX 4
Investigation Proposal Options

This appendix presents six investigation proposals (long and short versions of three different types of proposals) that may be used in most labs. Investigation Proposal A is appropriate for descriptive studies, whereas Investigation Proposal B and Investigation Proposal C are appropriate for comparative or experimental studies. The development of these proposals was supported by the Institute of Education Sciences, U.S. Department of Education, through grant R305A100909 to Florida State University.

The format of investigation proposals B and C is modeled after a hypothetical deductive-reasoning guide described in *Exploring the Living World* (Lawson 1995) and modified from an investigation guide described in an article by Maguire, Myerowitz, and Sampson (2010).

References

Lawson, A. E. 1995. Exploring the living world: A laboratory manual for biology. McGraw-Hill College.

Maguire, L., L. Myerowitz, and V. Sampson. 2010. Diffusion and osmosis in cells: A guided inquiry activity. *The Science Teacher* 77 (8): 55–60.

Investigation Proposal A: Descriptive Study

The Guiding Question...	

↓

What data will you collect?	

↓

How will you collect your data?	Your Procedure What safety precautions will you follow?

↓

How will you analyze your data?	

I approve of this investigation. _____   _____

Your actual data

Your analysis of the data

↓

The claim you will make

↓

Investigation Proposal A: Descriptive Study (Short Form)

The Guiding Question...	

↓

What data will you collect?	

↓

How will you collect your data?	Your Procedure What safety precautions will you follow?

↓

How will you analyze your data?	

↓

Your actual data	

I approve of this investigation. _____ _____

Instructor's Signature   Date

Investigation Proposal B: Comparative or Experimental Study

The Guiding Question...

Hypothesis 1

IF...

Hypothesis 2

IF...

The Test

AND...
Procedure

What data will you collect?

How will you analyze the data?

What safety precautions will you follow?

Predicted Result if hypothesis 1 is valid

THEN...

Predicted Result if hypothesis 2 is valid

THEN...

I approve of this investigation. _____   _____
                                          Instructor's Signature   Date

AND...

Your actual data

Your analysis of
the data

The claim you
will make

Investigation Proposal B: Comparative or Experimental Study (Short Form)

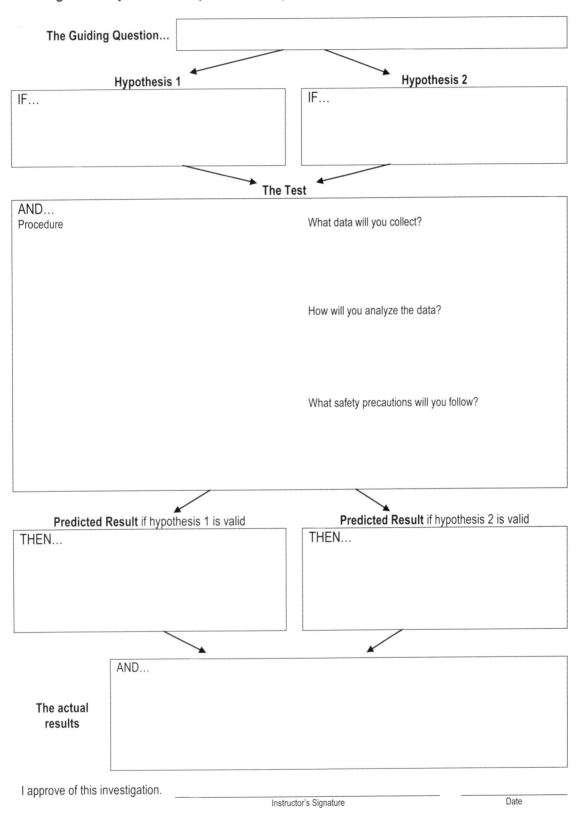

The Guiding Question...

Hypothesis 1

IF...

Hypothesis 2

IF...

The Test

AND...
Procedure

What data will you collect?

How will you analyze the data?

What safety precautions will you follow?

Predicted Result if hypothesis 1 is valid

THEN...

Predicted Result if hypothesis 2 is valid

THEN...

AND...

The actual
results

I approve of this investigation.

Instructor's Signature

Date

Investigation Proposal C: Comparative or Experimental Study

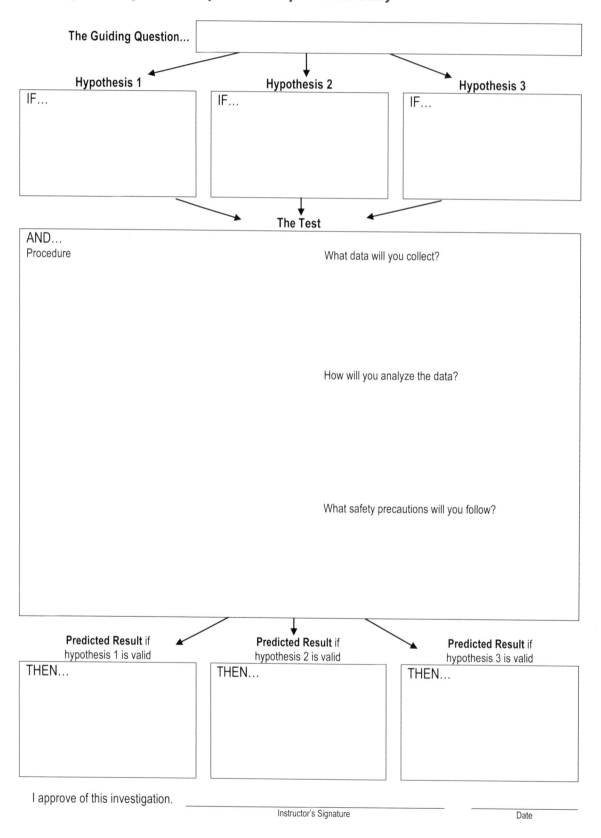

The Guiding Question...

Hypothesis 1
IF...

Hypothesis 2
IF...

Hypothesis 3
IF...

The Test

AND...
Procedure

What data will you collect?

How will you analyze the data?

What safety precautions will you follow?

Predicted Result if
hypothesis 1 is valid
THEN...

Predicted Result if
hypothesis 2 is valid
THEN...

Predicted Result if
hypothesis 3 is valid
THEN...

I approve of this investigation. _____   _____
Instructor's Signature   Date

AND...

Your actual data

Your analysis of the data

The claim you will make

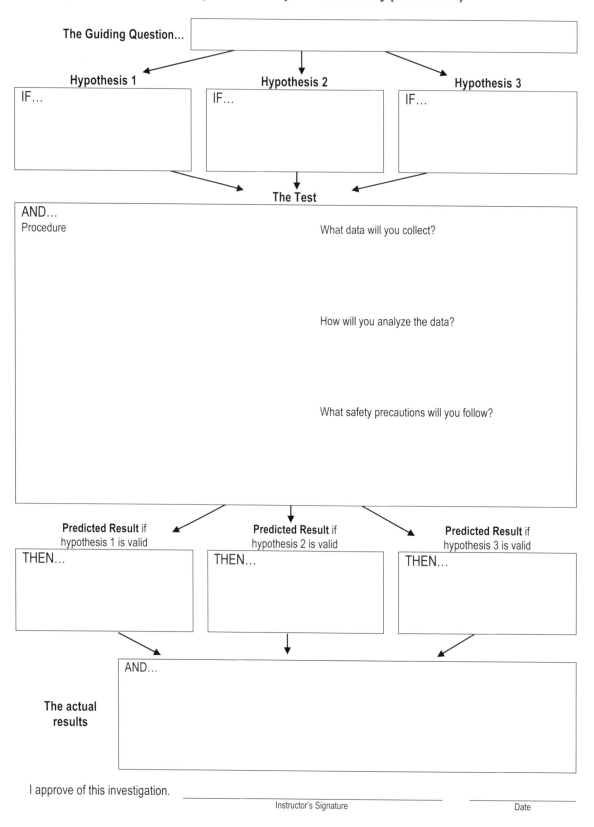

Investigation Proposal C: Comparative or Experimental Study (Short Form)

The Guiding Question...

Hypothesis 1

IF...

Hypothesis 2

IF...

Hypothesis 3

IF...

The Test

AND...
Procedure

What data will you collect?

How will you analyze the data?

What safety precautions will you follow?

Predicted Result if
hypothesis 1 is valid

THEN...

Predicted Result if
hypothesis 2 is valid

THEN...

Predicted Result if
hypothesis 3 is valid

THEN...

The actual results

AND...

I approve of this investigation. _____   _____

Instructor's Signature   Date

ADI LAB INVESTIGATION REPORT PEER-REVIEW GUIDE: MIDDLE SCHOOL VERSION

Report By: _____
          ID Number

Author: Did the reviewers do a good job?
   1 2 3 4 5
Rate the overall quality of the peer review

Reviewed By: _____ _____ _____ _____
      ID Number   ID Number   ID Number   ID Number

Section 1: Introduction and Guiding Question	Reviewer Rating			Teacher Score
1. Did the author **provide a context** for the study?	☐ No	☐ Partially	☐ Yes	0 1 2
2. Did the author provide enough **background information** about the phenomenon being studied?	☐ No	☐ Partially	☐ Yes	0 1 2
3. Is the background information **correct**?	☐ No	☐ Partially	☐ Yes	0 1 2
4. Did the author make the **guiding question** clear?	☐ No	☐ Partially	☐ Yes	0 1 2

Reviewers: If your group made any "No" or "Partially" marks in this section, please **explain how the author could improve** this part of his or her report.

Author: What revisions did you make in your report? Is there anything you decided to keep the same even though the reviewers suggested otherwise? Be sure to explain why.

Section 2: Method	Reviewer Rating			Teacher Score
1. Did the author provide a clear description of what he or she did during the investigation to **collect data**?	☐ No	☐ Partially	☐ Yes	0 1 2
2. Did the author describe **how** he or she **analyzed** the data?	☐ No	☐ Partially	☐ Yes	0 1 2
3. Did the author use the **correct term** to describe his or her investigation (e.g., experiment, observations, interpretation of a data set)?	☐ No	☐ Partially	☐ Yes	0 1 2

Reviewers: If your group made any "No" or "Partially" marks in this section, please **explain how the author could improve** this part of his or her report.

Author: What revisions did you make in your report? Is there anything you decided to keep the same even though the reviewers suggested otherwise? Be sure to explain why.

Section 3: The Argument	Reviewer Rating			Teacher Score
1. Did the author provide a clear and complete **claim**? • Does the claim **answer the guiding question**? • Is the claim **acceptable** (is it consistent with the evidence)?	☐ No	☐ Partially	☐ Yes	0 1 2
2. Did the author use **evidence** to support his or her claim? • Is there an **analysis of the data**? • Did the author include **a correctly formatted and labeled graph or table** that uses appropriate metric units (e.g., g, ml, m/s)? • Is the analysis of data **explained**? • Is the evidence **sufficient**?	☐ No	☐ Partially	☐ Yes	0 1 2
3. Did the author include a **justification of the evidence**? • Is a **scientific concept** used to explain why the evidence matters? • Is the justification **acceptable**?	☐ No	☐ Partially	☐ Yes	0 1 2
4. Does the author discuss how well his or her claim agrees with the claims made by other groups and **explain any differences**?	☐ No	☐ Partially	☐ Yes	0 1 2

Reviewers: If your group made any "No" or "Partially" marks in this section, please **explain how the author could improve** this part of his or her report.	**Author:** What revisions did you make in your report? Is there anything you decided to keep the same even though the reviewers suggested otherwise? Be sure to explain why.

Mechanics	Reviewer Rating			Teacher Score
1. Organization: Is each section easy to follow? Do paragraphs include multiple sentences? Do paragraphs begin with a topic sentence?	☐ No	☐ Partially	☐ Yes	0 1 2
2. Grammar: Are the sentences complete? Is there proper subject-verb agreement in each sentence? Are there no run-on sentences?	☐ No	☐ Partially	☐ Yes	0 1 2
3. Conventions: Did the author use appropriate spelling, punctuation, capitalization, and word choice (e.g., *there vs. their, to vs. too*)?	☐ No	☐ Partially	☐ Yes	0 1 2
4. Did the author **use scientific terms correctly** (e.g., *hypothesis* vs. *prediction, data* vs. *evidence*) and **reference the evidence in an appropriate manner** (e.g., *supports* or *suggests* vs. *proves*)?	☐ No	☐ Partially	☐ Yes	0 1 2

Teacher Comments:

Total: _____/40

ADI LAB INVESTIGATION REPORT PEER-REVIEW GUIDE: HIGH SCHOOL VERSION

Report By: _____ Author: Did the reviewers do a good job?   1 2 3 4 5
                    ID Number   Rate the overall quality of the peer review

Reviewed By: _____ _____ _____ _____
                    ID Number   ID Number   ID Number   ID Number

Section 1: Introduction and Guiding Question	Reviewer Rating			Teacher Score
1. Did the author **provide a context** for the study?	☐ No	☐ Partially	☐ Yes	0 1 2
2. Did the author provide enough **background information** about the phenomenon being studied?	☐ No	☐ Partially	☐ Yes	0 1 2
3. Is the background information **accurate**?	☐ No	☐ Partially	☐ Yes	0 1 2
4. Did the author make the **guiding question** explicit and explain how the guiding question is related to the background information?	☐ No	☐ Partially	☐ Yes	0 1 2

Reviewers: If your group made any "No" or "Partially" marks in this section, please **explain how the author could improve** this part of his or her report.

Author: What revisions did you make in your report? Is there anything you decided to keep the same even though the reviewers suggested otherwise? Be sure to explain why.

Section 2: Method	Reviewer Rating			Teacher Score
1. Did the author describe **the procedure** he or she used to gather data and then explain why this procedure was used?	☐ No	☐ Partially	☐ Yes	0 1 2
2. Did the author explain **what data** were collected (or used) during the investigation and why they were collected (or used)?	☐ No	☐ Partially	☐ Yes	0 1 2
3. Did the author describe **how he or she analyzed the data** and explain why the analysis helped answer the guiding question?	☐ No	☐ Partially	☐ Yes	0 1 2
4. Did the author use the **correct term** to describe his or her investigation (e.g., *experiment, observations, interpretation of a data set*)?	☐ No	☐ Partially	☐ Yes	0 1 2

Section 2: Method (continued)

Reviewers: If your group made any "No" or "Partially" marks in this section, please **explain how the author could improve** this part of his or her report.	**Author:** What revisions did you make in your report? Is there anything you decided to keep the same even though the reviewers suggested otherwise? Be sure to explain why.

Section 3: The Argument	Reviewer Rating			Teacher Score
1. Did the author provide a **claim** that answers the guiding question?	☐ No	☐ Partially	☐ Yes	0 1 2
2. Did the author include **high-quality evidence** in his or her argument? • Were the data collected in an appropriate manner? • Is the analysis of the data appropriate and free from errors? • Is the author's interpretation of the analysis (what it means) valid?	☐ No	☐ Partially	☐ Yes	0 1 2
3. Did the author **present the evidence** in an appropriate manner by • using a correctly formatted and labeled graph (or table); • including correct metric units (e.g., m/s, g, ml); and • referencing the graph or table in the body of the text?	☐ No	☐ Partially	☐ Yes	0 1 2
4. Is the claim **consistent with the evidence**?	☐ No	☐ Partially	☐ Yes	0 1 2
5. Did the author include a **justification of the evidence** that • explains why the evidence is important (why it matters) and • defends the inclusion of the evidence with a specific science concept or by discussing his or her underlying assumptions?	☐ No	☐ Partially	☐ Yes	0 1 2
6. Is the **justification of the evidence** acceptable?	☐ No	☐ Partially	☐ Yes	0 1 2
7. Did the author discuss **how well his or her claim agrees with the claims made by other groups** and explain any disagreements?	☐ No	☐ Partially	☐ Yes	0 1 2
8. Did the author **use scientific terms correctly** (e.g., *hypothesis* vs. *prediction, data* vs. *evidence*) and **reference the evidence in an appropriate manner** (e.g., *supports* or *suggests* vs. *proves*)?	☐ No	☐ Partially	☐ Yes	0 1 2

Section 3: The Argument (*continued*)

Reviewers: If your group made any "No" or "Partially" marks in this section, please ***explain how the author could improve*** this part of his or her report.	**Author:** What revisions did you make in your report? Is there anything you decided to keep the same even though the reviewers suggested otherwise? Be sure to explain why.

Mechanics	Reviewer Rating			Teacher Score
1. *Organization:* Is each section easy to follow? Do paragraphs include multiple sentences? Do paragraphs begin with a topic sentence?	☐ No	☐ Partially	☐ Yes	0 1 2
2. *Grammar:* Are the sentences complete? Is there proper subject-verb agreement in each sentence? Are there no run-on sentences?	☐ No	☐ Partially	☐ Yes	0 1 2
3. *Conventions:* Did the author use appropriate spelling, punctuation, paragraphing, and capitalization?	☐ No	☐ Partially	☐ Yes	0 1 2
4. *Word Choice:* Did the author use the appropriate word (e.g., *there* vs. *their, to* vs. *too, than* vs. *then*)?	☐ No	☐ Partially	☐ Yes	0 1 2

Teacher Comments:

Total: _____ /50

IMAGE CREDITS

All images in this book are stock photographs or courtesy of the authors unless otherwise noted below.

Lab 1

Figure 1.1: Andonee, Wikimedia Commons, CC BY-SA 4.0, *https://commons.wikimedia.org/wiki/File:Moon_Phase_Diagram.GIF*

Figure L1.1: (Waxing Crescent) Jay Tanner, Wikimedia Commons, CC BY-SA 3.0, *https://upload.wikimedia.org/wikipedia/commons/7/7e/Phase-048.jpg*; (First Quarter) Jay Tanner, Wikimedia Commons, CC BY-SA 3.0, *https://upload.wikimedia.org/wikipedia/commons/c/c0/Phase-098.jpg*; (Waxing Gibbous) Jay Tanner, Wikimedia Commons, CC BY-SA 3.0, *https://upload.wikimedia.org/wikipedia/commons/6/64/Phase-133.jpg;* (Full) Jay Tanner, Wikimedia Commons, CC BY-SA 3.0, *https://upload.wikimedia.org/wikipedia/commons/0/01/Phase-191.jpg*; (Waning Gibbous) Jay Tanner, Wikimedia Commons, CC BY-SA 3.0, *https://upload.wikimedia.org/wikipedia/commons/3/34/Phase-231.jpg*; (Third Quarter) Jay Tanner, Wikimedia Commons, CC BY-SA 3.0, *https://upload.wikimedia.org/wikipedia/commons/6/6e/Phase-270.jpg*; (Waning Crescent) Jay Tanner, Wikimedia Commons, CC BY-SA 3.0, *https://upload.wikimedia.org/wikipedia/commons/e/ef/Phase-312.jpg*; (New) Jay Tanner, Wikimedia Commons, CC BY-SA 3.0, *https://upload.wikimedia.org/wikipedia/commons/3/36/Phase-358.jpg*

Figure L1.2: SBS, "Solar Eclipse: Myths and Facts," July 23, 2009, *www.sbs.com.au/news/article/2009/07/22/solar-eclipse-myths-and-facts*

Figure L1.3: Tom Ruen, Wikimedia Commons, Public domain. *http://commons.wikimedia.org/wiki/File:Partial_lunar_eclipse_december_10_2011_Minneapolis_TLR.png*

Lab 2

Figure 2.3: Dennis Nilsson, Wikimedia Commons, CC BY 3.0, *https://en.wikipedia.org/wiki/Axial_tilt#/media/File:AxialTiltObliquity.png*

Figure L2.1: Jet Propulsion Laboratory, NASA, "Season's Greetings: NASA Views the Change of Seasons," December 21, 2010, *www.jpl.nasa.gov/news/news.php?feature=2857*

Figure L2.2: Dennis Nilsson, Wikimedia Commons, CC 3.0, *https://en.wikipedia.org/wiki/Axial_tilt#/media/File:AxialTiltObliquity.png*

Figure L2.3: University of Nebraska–Lincoln, NAAP Astronomy Labs—Basic Coordinates and Seasons—Seasons and Ecliptic Simulator *http://astro.unl.edu/naap/motion1/animations/seasons_ecliptic.html*

Figure in checkout question 2: Modified from Rhcastilhos, Wikimedia Commons, Public domain. *https://upload.wikimedia.org/wikipedia/commons/thumb/1/12/Seasons.svg/2000px-Seasons.svg.png*

Lab 3

Figure L3.4: NASA, Public domain. *http://solarsystem.nasa.gov/galleries/earths-orbit*

Figure L3.5: PhET Interactive Simulations, Univeristy of Colorado—Bolder, *https://phet.colorado.edu/en/simulation/legacy/my-solar-system*

Lab 5

Figure 5.1: Margaret W. Carruthers, Wikimedia Commons, CC BY 2.0, *https://commons.wikimedia.org/wiki/File:Triassic_Sedimentary_Strata_Somerset_1.jpg*

Figure L5.1: (a) Luca Galuzzi, Wikimedia Commons, CC BY-SA 2.5, *https://commons.wikimedia.org/wiki/File:USA_10052_Grand_Canyon_Luca_Galuzzi_2007.jpg*; (b) Ryan Lackey, Wikimedia Commons, CC BY 2.0, *https://commons.wikimedia.org/wiki/File:Sedimentary_Rock_Layers_near_Khasab_in_Musandam_Oman.jpg*

Figure L5.2: Peter Halasz, Wikimedia Commons, Public domain. *https://upload.wikimedia.org/wikipedia/commons/thumb/a/a5/Biological_classification_L_Pengo_vflip.svg/399px-Biological_classification_L_Pengo_vflip.svg.png*

Figure above checkout question 1: Kurt Rosenkrantz, Wikimedia Commons, CC BY-SA 3.0, *https://commons.wikimedia.org/wiki/File:Fossils.png*

Lab 6

Figure 6.1: U.S. Geological Survey, Wikimedia Commons, Public domain. *https://commons. wikimedia.org/wiki/File:FigS1-1.gif*

Figure 6.2: U.S. Geological Survey, Wikimedia Commons, Public domain. *https://en.wikipedia. org/wiki/Plate_tectonics#/media/File:Plates_tect2_ en.svg*

Figure 6.3: Modified from Surachit, Wikimedia Commons, GFDL 1.2, *https://en.wikipedia.org/wiki/ Mantle_convection#/media/File:Oceanic_spreading. svg*

Figure L6.1: U.S. Geological Survey, Wikimedia Commons, Public domain. *https://commons. wikimedia.org/wiki/File:FigS1-1.gif*

Figure L6.2: U.S. Geological Survey, Wikimedia Commons, Public domain. *https://en.wikipedia. org/wiki/Plate_tectonics#/media/File:Plates_tect2_ en.svg*

Figure above checkout question 1: U.S. Geological Survey, Wikimedia Commons, Public domain. *https:// en.wikipedia.org/wiki/Plate_tectonics#/media/ File:Plates_tect2_en.svg*

Figure in checkout question 3: Modified from Виктор, Wikimedia commons, CC BY-SA 2.0, *https://commons.wikimedia.org/wiki/File:Outline_ map_of_Central_America.svg*

Lab 7

Figure 7.1: U.S. Geological Survey, Wikimedia Commons, Public domain. *https://en.wikipedia. org/wiki/Plate_tectonics#/media/File:Plates_tect2_ en.svg*

Figure 7.2: Surachit, Wikimedia Commons, GFDL 1.2, *https://en.wikipedia.org/wiki/Mantle_ convection#/media/File:Oceanic_spreading.svg*

Figure 7.4 , Figure L7.4, and figure in checkout question 1: Significantly modified from NordNordWest, Wikimedia Commons, CC-BY-SA-3.0-DE, *https://upload.wikimedia.org/wikipedia/ commons/thumb/8/86/USA_Hawaii_location_map. svg/2000px-USA_Hawaii_location_map.svg.png*

Figure 7.5: John Power, U.S. Geological Survey, Public domain. *https://woodshole.er.usgs.gov/ operations/obs/rmobs_pub/html/alaska.html*

Figure L7.1: (a) NASA image courtesy Jeff Schmaltz, LANCE/EOSDIS MODIS Rapid Response Team at NASA GSFC, Wikimedia Commons, Public domain. *https://commons.wikimedia.org/wiki/File:Aleutian_ Islands_amo_2014135_lrg*.jpg: (b) NASA, Wikimedia Commons, Public domain. *https://commons. wikimedia.org/wiki/File:Himalayas_landsat_7.png*

Figure L7.2: Surachit, Wikimedia Commons, GFDL 1.2, *https://en.wikipedia.org/wiki/Mantle_ convection#/media/File:Oceanic_spreading.svg*

Figure in checkout question 2: Screenshot from Google Maps, *www.google.com/maps/ place/Japan/@34.2521997,130.0725666,4z/ data=!4m5!3m4!1s0x34674e0fd77f192f: 0xf54275d47c665244!8m2!3d36.204824! 4d138.252924*

Lab 8

Figure 8.2 and Figure L8.3: Natural Resources Conservation Service, U.S. Department of Agriculture, Public domain. *www.nrcs.usda.gov/wps/ portal/nrcs/detail/soils/edu/?cid=nrcs142p2_054311*

Figure L8.1: El Guanche, Wikimedia Commons, CC BY 2.0, *https://en.wikipedia.org/wiki/Aeolian_ processes#/media/File:Arbol_de_Piedra.jpg*

Figure L8.2: Roxy Lopez, Wikimedia Commons, CC BY-SA 3.0, *https://commons.wikimedia.org/wiki/ File:Duststorm.jpg*

Lab 9

Figure 9.1 and Figure L9.1: National Park Service, Wikimedia Commons, Public domain. *https:// commons.wikimedia.org/wiki/File:Mississippi_ River_Watershed_Map.jpg*

Figure 9.2: Dboutte, Wikimedia Commons, CC BY 3.0, *https://commons.wikimedia.org/wiki/ File:Coastal_changediagram5.jpg*

Figure L9.2: NASA, Wikimedia Commons, Public domain. *https://en.wikipedia.org/wiki/Mississippi_ River_Delta#/media/File:Mississippi_River_Delta_ and_Sediment_Plume.jpg*

Figure above checkout question 1: Demis Map Server, Wikimedia Commons, CC BY-SA 3.0, *https:// upload.wikimedia.org/wikipedia/commons/c/c2/ Rogue_River_Watershed.png*

Lab 10

Figure L10.1: (Siltstone) Greg Willis, Wikimedia Commons, CC BY-SA 3.0, *https://commons. wikimedia.org/wiki/File:Balls_Bluff_Siltstone_ (4802112634).jpg*; (Shale) Smurfage, Deviant Art, Public domain. *http://smurfage.deviantart. com/art/Shale-Wall-208582121*; (Sandstone) Óðinn, Wikimedia Commons, CC BY-SA 2.5, *https://commons.wikimedia.org/wiki/File:Navajo_ Sandstone.JPG*

Lab 11

Figure 11.2 and Figure L11.2: Natural Resources Conservation Service, U.S. Department of Agriculture, Public domain. *www.nrcs.usda.gov/wps/ portal/nrcs/detail/soils/edu/?cid=nrcs142p2_054311*

Lab 12

Figure 12.1 and Figure L12.2: Atmospheric Infrared Sounder, Flickr, CC BY 2.0, *www.flickr.com/photos/ atmospheric-infrared-sounder/8265072146/sizes/l*

Figure L12.1: (a) the_tahoe_guy, Wikimedia Commons, CC BY 2.0, *https://en.wikipedia.org/ wiki/Lake_Tahoe#/media/File:Emerald_Bay.jpg*; (b) Luca Galuzzi, Wikimedia Commons, CC BY-SA 2.5, *https://en.wikipedia.org/wiki/Glacier#/ media/File:Perito_Moreno_Glacier_Patagonia_ Argentina_Luca_Galuzzi_2005.JPG*; (c) Diamonds [transliteration], Wikimedia Commons, CC BY_SA 3.0, *https://commons.wikimedia.org/wiki/File:Water_ vapor_from_a_pond_in_Owakudani_Valley_2.JPG*

Lab 13

Figure 13.1 and Figure L.13.3: Rob Lavinsky, Wikimedia Commons, CC BY-SA 3.0, *https://upload. wikimedia.org/wikipedia/commons/9/99/Dolomite-Calcite-201604.jpg*

Figure 13.2 and Figure L13.4: Didier Descouens, Wikimedia Commons, CC BY-SA 4.0, *https://upload. wikimedia.org/wikipedia/commons/c/ce/Quartz_ Brésil.jpg*

Figure L13.1: Jim Champion, Wikimedia Commons, GFDL 1.2, *https://commons.wikimedia.org/wiki/ File:Logan_Rock_from_below.jpg*

Figure L13.2: Lucarelli, Wikimedia Commons, GFDL 1.2, *https://upload.wikimedia.org/wikipedia/ commons/e/e3/Carrara_14.JPG*

Lab 14

Figure 14.1: U.S. Geological Survey, Public domain. *www.usgs.gov/media/images/global-copper-map-0*

Figure L14.1: Jonathan Zander, Wikimedia Commons, GFDL 1.2, *https://commons.wikimedia. org/wiki/File:Native_Copper_Macro_Digon3.jpg*

Figure L14.2: Matthew.kowal, Wikimedia Commons, CC BY-SA 4.0, *https://commons.wikimedia.org/ wiki/File:The_Lavender_Open_Pit_Mine,_Bisbee,_ Arizona.jpg*

Figure L14.3: Supercarwaar, Wikimedia Commons, CC BY-SA 4.0, *https://commons.wikimedia.org/wiki/ File:Kennecott_Tailings_Pond_and_Smokestack.jpg*

Figure below checkout question 1: Kbh3rd, Wikimedia Commons, CC BY-SA 3.0, *https:// commons.wikimedia.org/wiki/File:US_copper_ mine_locations_2003.svg*

Lab 15

Figure 15.1: NASA, Wikimedia Commons, Public domain. *https://commons.wikimedia.org/wiki/ File:Air_masses.svg*

Figure L15.1: NOAA, Public domain. *www.lib.noaa. gov/collections/imgdocmaps/daily_weather_maps. html*

Figure in checkout question 2: NOAA, Public domain. *www.lib.noaa.gov/collections/imgdocmaps/ daily_weather_maps.html*

Supplementary materials: U.S. Department of Commerce, Daily Weather Maps, Public domain. *www.wpc.ncep.noaa.gov/dailywxmap/index.html*

Lab 16

Figure L16.1: NASA, Public domain. *http:// earthobservatory.nasa.gov/IOTD/view.php?id=7205*

Lab 17

Figure 17.1 and Figure L17.1: U.S. Environmental Protection Agency, Wikimedia Commons, Public domain. *https://commons.wikimedia.org/wiki/ File:Earth%27s_greenhouse_effect_%28US_ EPA,_2012%29.png*

Figure L17.2: PhET Interactive Simulations, Univeristy of Colorado—Bolder, *https://phet. colorado.edu/en/simulation/legacy/greenhouse*

Lab 18

Figure 18.1: Modified from Columbia University, Global Temperature, *www.columbia.edu/~mhs119/ Temperature*

Figure 18.2: Eli Duke, Wikimedia Commons, CC By-SA 2.0, *https://commons.wikimedia.org/wiki/ File:Antarctica_WAIS_Divide_Field_Camp_10.jpg*

Figure L18.1: NOAA, Public domain. *www.ncdc. noaa.gov/temp-and-precip/state-temps*

Figure L18.2: NASA's Goddard Space Flight Center/ Ludovic Brucker, Wikimedia Commons, Public domain. *https://commons.wikimedia.org/wiki/ File:An_ice_core_segment.jpg*

Lab 19

Figure 19.1: NOAA and National Weather Service, Public domain. *www.srh.noaa.gov/jetstream/ocean/ currents_max.html*

Figure 19.2: Modified from Crates, Wikimedia Commons, GFDL 1.2, *https://commons.wikimedia. org/wiki/File:World_map_blank_without_borders.svg*

Figure L19.1: Modified from Kaboom88, Wikimedia Commons, Public domain. *https://upload.wikimedia. org/wikipedia/commons/c/ca/Blank_US_map_ borders.svg*

Figure in checkout question 3: Modified from AlexCovarrubias, Wikimedia Commons, Public domain. *https://en.wikipedia.org/wiki/File:North_ America_second_level_political_division.svg*

Lab 20

Figure 20.1: NASA, Public domain. *https:// spaceplace.nasa.gov/hurricanes/en/cyclone_map_ large.en.gif*

Figure L20.1: Jacques Descloitres, MODIS Rapid Response Team, NASA/GSFC, Wikimedia Commons, Public domain. *https://commons. wikimedia.org/wiki/File:Hurricane_Isabel_14_ sept_2003_1445Z.jpg*

Figure L20.2: Nilfanion, Wikimedia Commons,

Public domain. *https://commons.wikimedia.org/wiki/ File:Atlantic_hurricane_tracks_1980-2005.jpg*

Two figures in checkout question 1: (Hurricane David) National Hurricane Center, Wikimedia Commons, Public domain. https://commons. wikimedia.org/wiki/File:David_1979_track.png; (Hurricane Andrew) National Hurricane Center, Wikimedia Commons, Public domain. https:// commons.wikimedia.org/wiki/File:Andrew_1992_ track.png

Lab 21

Figure 21.1: Daphne Zaras, Wikimedia Commons, Public domain. *https://upload.wikimedia.org/ wikipedia/commons/1/1a/Dszpics1.jpg*

Figure 21.2: (a) Vanessa Ezekowitz, Wikimedia Commons, GFDL 1.2, *https://commons.wikimedia. org/wiki/File:Meso-1.svg*; (b) Vanessa Ezekowitz, Wikimedia Commons, GFDL 1.2, *https://commons. wikimedia.org/wiki/File:Meso-2.svg* ; (c) Vanessa Ezekowitz, Wikimedia Commons, GFDL 1.2, *https:// commons.wikimedia.org/wiki/File:Meso-3.svg*

Figure 21.3: Dan Craggs, Wikimedia Commons, CC BY-SA 3.0, *https://commons.wikimedia.org/wiki/ File:Tornado_Alley_Diagram.svg*

Figure 21.4: (a) NOAA, Public domain. *www.spc. noaa.gov/wcm/permonth_by_state/March.png*; (b) NOAA, Public domain. *www.spc.noaa.gov/wcm/ permonth_by_state/April.png*; (c) NOAA, Public domain. *www.spc.noaa.gov/wcm/permonth_by_ state/May.png*

Figure L21.1: Ks0stm, Wikimedia Commons, CC BY-SA 3.0, *https://upload.wikimedia.org/wikipedia/ commons/9/93/May_20%2C_2013_Moore%2C_ Oklahoma_tornado.JPG*

Figure L21.2: Maj. Geoff Legler, Oklahoma National Guard, Wikimedia Commons, Public domain. *https:// upload.wikimedia.org/wikipedia/commons/d/d3/ Aerial_view_of_2013_Moore_tornado_damage.jpg*

Lab 22

Figure L22.1: U.S. Environmental Protection Agency, Wikimedia Commons, Public domain. *https:// commons.wikimedia.org/wiki/File:Earth%27s_ greenhouse_effect_%28US_EPA,_2012%29.png*

Figure L22.2: Narayanese and NOAA, Wikimedia Commons, CC BY-SA 3.0, *https://commons.*

wikimedia.org/wiki/File:Mauna_Loa_Carbon_Dioxide_Apr2013.svg

Figure L22.3: Climate Interactive, *C-ROADS, www.climateinteractive.org/tools/c-roads*

Lab 23

Figure L23.2: Kuczora, Wikimedia Commons, CC BY-SA 3.0, *https://commons.wikimedia.org/wiki/File:Hoover_sm.jpg*

Figure L23.3: Hans Hillewaert, Wikimedia Commons, CC BY-SA 3.0, *https://commons.wikimedia.org/wiki/File:Aquifer_en.svg*

Figure L23.4: Shannon, Wikimedia Commons, GFDL 1.2, *https://commons.wikimedia.org/wiki/File:Coloradorivermapnew1.jpg*

Figure L23.5: Arizona State University, *WaterSim*, *https://sustainability.asu.edu/dcdc/watersim*

INDEX

Page numbers printed in **boldface** type refer to figures or tables.

A

Abrasion, 194, 204

Absolute age of rock strata, 122, 133

Acceleration, 81

Acid reactivity of minerals, 312

Advection, 284

A Framework for K–12 Science Education, xi–xiii, **xii**, xvii–xviii

Air Masses and Weather Conditions lab, 364–385

    checkout questions, 383–385

    lab handouts

        argumentation session, 380–381

        getting started, 379

        initial argument, 379–380, **380**

        introduction, 376–378, **376**, **377**, **378**

        investigation proposal, 379

        materials, 378

        report, 381–382

        safety precautions, 378

        scientific knowledge and inquiry connections, 379

        task and guiding question, 378

    teacher notes

        alignment with standards, 373, **373–375**

        content, 364–367, **365**, **366**, **367**

        explicit and reflective discussion topics, 369–372

        implementation tips, 372–373

        materials and preparation, 368–369, **369**

        purpose, 364

        safety precautions, 369

        timeline, 367–368

Aleutian Islands, 170, **170**, **180**

Aphelion, 78, 90, 91, **91**, 94

Apogee, 78, 90, 91, **91**, 94

Aquifers, 543, **544**, 545

Argumentation session, 10–12, **11**, **19**

    See also specific labs

Argument-driven inquiry (ADI) model

    about, xviii–xxiii, 3

    alignment with *CCSS* ELA, **567**

    alignment with *Framework* standards, **557–559**

    alignment with *NGSS* performance expectations, **561–566**

    alignment with NOSK and NOSI concepts, **560**

    investigation proposal, 5, 27, 577, **578–586**

    peer-review guide (high school), **590–592**

    peer-review guide (middle school), **587–589**

    role of teacher in, 18, **19–20**

    stages of, 3–17, **3**

        stage 1: identification of task and guiding question, **3**, 4–5, **19**

        stage 2: designing a method and collecting data, **3**, 5, **19**

        stage 3: data analysis and development of tentative argument, 5–10, **7**, **8**, **9**, **19**

        stage 4: argumentation session, 10–12, **11**, **19**

        stage 5: explicit and reflective discussion, 12–14, **20**

        stage 6: writing the investigation report, 14–15, **20**

        stage 7: double-blind group peer review, 15–17, **16**, **20**

        stage 8: revision and submission of investigation report, 17, **20**

    timelines for labs, **573–575**

Asthenosphere, 146, **147**, **148**, 157, **157**

Atchafalaya River, 218

Atmosphere. *See* Air Masses and Weather Conditions lab; Atmospheric Carbon Dioxide lab

Atmospheric Carbon Dioxide lab, 428–446

    checkout questions, 444–446

    lab handouts

        argumentation session, 442–443

        getting started, 439–441, **440**

        initial argument, 441, **442**

        introduction, 438–439, **438**

        investigation proposal, 439

        materials, 439

        report, 443

        safety precautions, 439

scientific knowledge and inquiry
connections, 441
task and guiding question, 439
teacher notes
alignment with standards, 435,
435–437
content, 428–430, **429**
explicit and reflective discussion
topics, 431–433
implementation tips, 434
materials and preparation, 430, **431**
purpose, 428
safety precautions, 431
timeline, 430
Atmospheric emissions, 335, 349
Atomic composition, 310
Axial tilt and rotation of the Earth, 56, **56**, 68,
69

B
Bed load, 217
Biological classifications, **136**
Breakage of minerals, 312, 322

C
Capacity of rivers, 217
Carbon dioxide. *See* Atmospheric Carbon
Dioxide lab; Carbon Emissions Minimization
lab
Carbon Emissions Minimization lab, 512–533
checkout questions, 530–533
lab handouts
argumentation session, 527–529
getting started, 525–527, **525**
initial argument, 527, **527**
introduction, 522–524, **523**
investigation proposal, 525
materials, 525
report, 529
safety precautions, 525
scientific knowledge and inquiry
connections, 527
task and guiding question, 524
teacher notes
alignment with standards, 517,
518–521
content, 512–513
explicit and reflective discussion
topics, 514–517
implementation tips, 517
materials and preparation, 514, **514**

purpose, 512
safety precautions, 514
timeline, 513–514
Cassini space probe, 78
Cause and Effect
Air Masses and Weather Conditions lab,
364, 370
Extreme Weather Forecasting lab, 490,
495
Human Use of Natural Resources lab,
534, 537
Hurricane Strength Prediction lab, 472,
476
Natural Resources Distribution lab, 332,
338
NGSS crosscutting concept, 569
Sediment Deposition lab, 240, 245
Surface Materials and Temperature
Change lab, 386, 392
Cementation, 240, 252
Checkout questions, 27
See also specific labs
Chemical properties of minerals, 322, **322**
Cirrostratus clouds, 367
Cirrus clouds, 367
Claims, ADI instructional model, 5–10, **7**, **8**, **9**,
47
Clastic sedimentary rocks, 240, 252, **252**
Cleavage of minerals, 312, 322
Climate. *See* Air Masses and Weather
Conditions lab; Atmospheric Carbon
Dioxide lab; Carbon Emissions Minimization
lab; Global Temperature lab; Regional
Climate lab; Surface Materials and
Temperature Change lab
Cold fronts, 365–366, **366**, 377, **377**
Color, 311
Colorado River watershed, 545, **545**
*Common Core State Sta*ndards in English
language arts (*CCCS ELA*)
and ADI model, xiii
Air Masses and Weather Conditions lab,
373–375
alignment with argument-driven inquiry
(ADI) model, **567**
Atmospheric Carbon Dioxide lab, **435–
437**
Carbon Emissions Minimization lab,
518–521
Extreme Weather Forecasting lab,
498–500

Geologic Features Formation lab, **176–179**

Geologic Time and the Fossil Record lab, **130–132**

Global Temperature lab, **414–418**

Gravity and Orbits lab, **85–88**

Habitable Worlds lab, **107–108**

Human Use of Natural Resources lab, **539–542**

Hurricane Strength Prediction lab, **479–480**

investigation report writing, 15

Mineral Characteristics lab, **318–320**

Moon Phases lab, **41–42**

Natural Resources Distribution lab, **342–346**

peer review of investigation report, 17

Plate Interactions lab, **154–156**

Regional Climate lab, **455–457**

Seasons lab, **64–66**

Sediment Deposition lab, **248–251**

Sediment Transport lab, **225–228**

Soil Texture and Soil Water Permeability lab, **270–273**

Surface Materials and Temperature Change lab, **395–397**

Water Cycle lab, **292–295**

Wind Erosion lab, **202–203**

Compaction, 240, 252

Competence, 217

Conceptual models

Air Masses and Weather Conditions lab, 372–373, 378, 379

Moon Phases lab, 32, 37–40, 45, 46

Regional Climate lab, 450–451, 455, 460

Seasons lab, 54, 60–61, 62, 63, 69, 70–71, 76, 77

Sediment Deposition lab, 246–247, 253–254, 255–256

Sediment Transport lab, 223–224

Water Cycle lab, 284–285, **285**, 290–291, 298

Wind Erosion lab, 194, 198–199, 201

Conduction, 386–387, 399

Contact force, 89

Continental crust, 146, **147**

Continental rifting, 149

Convection, 387, 399

Convergent boundaries, 149, **149**, 151, 157, 162

Copper deposits and mining, 332–335, **333**, 347–350, **347**, **348**, **350**, 358–359

Core ideas for labs. *See* Disciplinary core ideas (DCIs)

Creeping process, 194

C-ROADS (Climate Rapid Overview and Decision Support) simulator, 513, 514, 525–526, **525**

Crosscutting Concepts (CCs)

and ADI model, xi–xiii, **xii**, xix, 3, **558**

Air Masses and Weather Conditions lab, 364, 369–370

Atmospheric Carbon Dioxide lab, 428, 431–432

Carbon Emissions Minimization lab, 512, 514–515

Extreme Weather Forecasting lab, 490, 495–496

Geologic Features Formation lab, 168, 172–173

Geologic Time and the Fossil Record lab, 122, 126–128

Global Temperature lab, 408, 411–412

Gravity and Orbits lab, 78, 80–81

Habitable Worlds lab, 102, 104–105

Human Use of Natural Resources lab, 534, 536–537

Hurricane Strength Prediction lab, 472, 475–477

and lab investigations, 23–24

Mineral Characteristics lab, 310, 315–316

Moon Phases lab, 32, 37–38

Natural Resources Distribution lab, 332, 337–339

Plate Interactions lab, 146, 151–152

Regional Climate lab, 448, 452–453

Seasons lab, 54, 60–61

Sediment Deposition lab, 240, 244–245

Sediment Transport lab, 216, 221–222

Soil Texture and Soil Water Permeability lab, 262, 266–267

Surface Materials and Temperature Change lab, 386, 391–392

Water Cycle lab, 284, 288–289

Wind Erosion lab, 194, 198–199, 208

Cyclones. *See* Hurricane Strength Prediction lab

D

Data analysis and development of tentative argument, 5–10, **7**, **8**, **9**, 19

Data *versus* evidence, 106
Daylight length, 57–58, **58**, 70–71
Deflation, 194, 204–205
Deltas and deltaic lobes, 217–219, **218**, 230, **230**, 231, 236–237
Deposition, 252
    See also Sediment Deposition lab
Deposits, 333
Disciplinary core ideas (DCIs)
    and ADI model, xi–xiii, **xii**, xix, 3, **559**
    Air Masses and Weather Conditions lab, 364, 369–370
    Atmospheric Carbon Dioxide lab, 428, 431–432
    Carbon Emissions Minimization lab, 512, 514–515
    Extreme Weather Forecasting lab, 490, 495–496
    Geologic Features Formation lab, 168, 172–173
    Geologic Time and the Fossil Record lab, 122, 126–128
    Global Temperature lab, 408, 411–412
    Gravity and Orbits lab, 78, 80–81
    Habitable Worlds lab, 102, 104–105
    Human Use of Natural Resources lab, 534, 536–537
    Hurricane Strength Prediction lab, 472, 475–477
    and lab investigations, 23–24
    Mineral Characteristics lab, 310, 315–316
    Moon Phases lab, 32, 37–38
    Natural Resources Distribution lab, 332, 337–339
    Plate Interactions lab, 146, 151–152
    Regional Climate lab, 448, 452–453
    Seasons lab, 54, 60–61
    Sediment Deposition lab, 240, 244–245
    Sediment Transport lab, 216, 221–222
    Soil Texture and Soil Water Permeability lab, 262, 266–267
    Surface Materials and Temperature Change lab, 386, 391–392
    Water Cycle lab, 284, 288–289
    Wind Erosion lab, 194, 198–199
Discipline-specific norms and criteria, **7**, 8
Divergent boundaries, 149, **149**, 151, 157, 162
Diversification of life, 122, 124–125, 134, 135–136, 141–143
Dolomite, 310, **310**, 322, **322**

Drainage basins, 216, **216**, 229, **229**
Dust storms, 204, **204**, 206

E
Earth history. *See* Geologic Features Formation lab; Geologic Time and Fossil Record lab; Plate Interactions lab
Earth layers, 146, **147**, **148**, 157, **157**
Earth physical model, 34–35, **35**
Earthquakes. *See* Plate Interactions lab
Earth systems. *See* Mineral Characteristics lab; Natural Resources Distribution lab; Sediment Deposition lab; Sediment Transport lab; Soil Texture and Soil Water Permeability lab; Water Cycle lab; Wind Erosion lab
Eccentricity of orbits, 78, 81, 89–90, **90**, 91, 94
Ecliptic path, 33
Effluent, 546
Electromagnetic waves, 387–388
Elevation, 448, 459
Elliptical orbits, 89
Empirical criteria, 6–7, **7**
Energy and Matter
    Global Temperature lab, 408, 411
    Hurricane Strength Prediction lab, 472, 476
    NGSS crosscutting concept, 569
    Sediment Transport lab, 216, 222
    Soil Texture and Soil Water Permeability lab, 262, 266
    Water Cycle lab, 284, 289
Environmental Protection Agency (EPA) Surf Your Watershed database, 336
Equinoxes, 56–58, **57**, **58**
Erosion, 204
    See also Wind Erosion lab
Estuaries, 230
Evaporation, 284–285, **285**, 297–298, **297**, 304–305
Evidence
    ADI instructional model, 5–10, **7**, **8**, **9**
    data *versus* evidence, 106
    described, 47
Exoplanets, 102–103, 109–110, **109**, 114–115
Explicit and reflective discussion, 12–14, **20**
    See also specific labs
Extinction, 122, 124–125, 134, 135–136, 141–143
    See also Geologic Time and the Fossil Record lab

Extreme Weather Forecasting lab, 490–509
  checkout questions, 507–509
  lab handouts
    argumentation session, 504–506
    getting started, 503–504
    initial argument, 504, **504**
    introduction, 501–502, **501**
    investigation proposal, 503
    materials, 502
    report, 506
    safety precautions, 503
    scientific knowledge and inquiry
      connections, 504
    task and guiding question, 502
  teacher notes
    alignment with standards, 498,
      498–500
    content, 490–492, **490**, **491**, **492**, **493**
    explicit and reflective discussion
      topics, 495–497
    implementation tips, 497–498
    materials and preparation, 494, **494**
    purpose, 490
    safety precautions, 495
    timeline, 494

F
Family (taxonomic), 127, 136–137, **136**
Faunal succession, 123, 134
Forces, 89
      See also Gravity and Orbits lab
Fossil Record. *See* Geologic Time and the
  Fossil Record lab
Fracture of minerals, 312, 322

G
Gallery walk format for argumentation
  sessions, 11, **11**
Geologic Features Formation lab, 168–189
  checkout questions, 186–189
  lab handouts
    argumentation session, 184–185
    getting started, 182–183
    initial argument, 184, **184**
    introduction, 180–182, **180**, **181**
    investigation proposal, 182
    materials, 182
    report, 185
    safety precautions, 182
    scientific knowledge and inquiry
      connections, 184

    task and guiding question, 182
  teacher notes
    alignment with standards, 175,
      175–179
    content, 168–170, **168**, **169**, **170**
    explicit and reflective discussion
      topics, 172–174
    implementation tips, 174–175
    materials and preparation, 171, **171**
    purpose, 168
    safety precautions, 172
    timeline, 170–171
Geologic Time and the Fossil Record lab,
  122–145
  checkout questions, 141–145
  geologic time scale, 123–124, **124**, 126,
    134–135, **135**
  lab handouts
    argumentation session, 138–139
    getting started, 136–137, **136**
    initial argument, 138, **138**
    introduction, 133–135, **133**, **135**
    investigation proposal, 136
    materials, 136
    report, 139–140
    safety precautions, 136
    scientific knowledge and inquiry
      connections, 137
    task and guiding question, 135–136
  teacher notes
    alignment with standards, 130,
      130–132
    content, 122–125, **122**, **124**
    explicit and reflective discussion
      topics, 126–129
    implementation tips, 129–130
    materials and preparation, 125–126,
      126
    purpose, 122
    safety precautions, 126
    timeline, 125
Global source regions of air masses, 364–365,
  365
Global temperature increases, 428–429, **429**,
  438–439, **438**
Global Temperature lab, 408–427
  checkout questions, 426–427
  lab handouts
    argumentation session, 423–425
    getting started, 421–423, **422**
    initial argument, 423, **423**

introduction, 419–421, **420**

investigation proposal, 421

materials, 421

report, 425

safety precautions, 421

scientific knowledge and inquiry
connections, 423

task and guiding question, 421

teacher notes

alignment with standards, 414,
414–418

content, 408–409, **409**

explicit and reflective discussion
topics, 411–413

implementation tips, 413–414

materials and preparation, 410, **410**

purpose, 408

safety precautions, 410

timeline, 409–410

Google Earth, 336

Google Maps, 336

Gradient of river channels, 217, 229

Granite outcrop, 321, **321**

Gravity and Orbits lab, 78–99

checkout questions, 97–99

lab handouts

argumentation session, 94–96

getting started, 92–94, **92**

initial argument, 94, **94**

introduction, 89–91, **90**, **91**

investigation proposal, 92

materials, 91

report, 96

safety precautions, 92

scientific knowledge and inquiry
connections, 94

task and guiding question, 91

teacher notes

alignment with standards, 84, **85–88**

content, 78–79

explicit and reflective discussion
topics, 80–83

implementation tips, 83–84

materials and preparation, 80, **80**

purpose, 78

safety precautions, 80

timeline, 79–80

Greenhouse effect, 408–409, **409**, 419–420,
420, 426, 522–524, **523**

The Greenhouse Effect simulation, 410, 421–
423, **422**

Groundwater, 284–285, **285**, 297, **297**, 543,
544, 545

Guiding question, 4–5, 6, 8, **8**, **19**, 26
See also specific labs

H

Habitable Worlds lab, 102–117

checkout questions, 114–117

lab handouts

argumentation session, 112–113

getting started, 111

initial argument, 111–112, **112**

introduction, 109–110, **109**

investigation proposal, 111

materials, 111

report, 113

safety precautions, 111

scientific knowledge and inquiry
connections, 111

task and guiding question, 110–111

teacher notes

alignment with standards, 107,
107–108

content, 102–103

explicit and reflective discussion
topics, 104–106

implementation tips, 106

materials and preparation, 103, **104**

purpose, 102

safety precautions, 104

timeline, 103

Hardness, 311–312, **311**

Hawaiian Islands, 169–170, **170**, 172–173,
174–175, 180–181, **181**, 182–183, 186,
188–189

Heat, 386, 399

Himalayas, **180**

Hubble Space Telescope, 78

Human impacts. *See* Carbon Emissions
Minimization lab; Extreme Weather
Forecasting lab; Human Use of Natural
Resources lab; Hurricane Strength
Prediction lab

Human Use of Natural Resources lab, 534–
554

checkout questions, 552–554

lab handouts

argumentation session, 549–551

getting started, 547–549, **548**

initial argument, 549, **549**

introduction, 543–546, **543**, **544**, **545**

investigation proposal, 547
materials, 546
report, 551
safety precautions, 546
scientific knowledge and inquiry
  connections, 549
task and guiding question, 546
teacher notes
  alignment with standards, 539,
    539–542
  content, 534
  explicit and reflective discussion
    topics, 536–538
  implementation tips, 538–539
  materials and preparation, 535, **535**
  purpose, 534
  safety precautions, 535
  timeline, 535
Hurricane Strength Prediction lab, 472–489
checkout questions, 487–489
lab handouts
  argumentation session, 485–486
  getting started, 483–484, **484**
  initial argument, 485, **485**
  introduction, 481–483, **481**, **482**
  investigation proposal, 483
  materials, 483
  report, 486
  safety precautions, 483
  scientific knowledge and inquiry
    connections, 485
  task and guiding question, 483
teacher notes
  alignment with standards, 479,
    479–480
  content, 472–474, **472**, **473**
  explicit and reflective discussion
    topics, 475–478
  implementation tips, 478
  materials and preparation, 474–475,
    475
  purpose, 472
  safety precautions, 475
  timeline, 474
Hutton, James, 123, 134

I
Ice cores, 428–429, **429**, 439–440, **440**
Identification of task and guiding question, **3**,
  4–5, **19**
Igneous rocks, 321

Infrared radiation, 408–409, **409**, 419–420,
  420
Intergovernmental Panel on Climate Change
  (IPCC), 513, 524
International Space Station, 78
Investigation design and data collection, **3**, 5,
  19
Investigation design improvement, 38
Investigation proposal
  and ADI model, 5, 27, 577
  comparative or experimental studies,
    581–586
  descriptive studies, **578–580**
  See also specific labs
Investigation report
  ADI instructional model, 14–15, 17, **20**
  peer-review guide (high school), **590–
    592**
  peer-review guide (middle school),
    587–589
  See also specific labs

J
Japanese archipelago, 187
Justification of evidence, 6–10, **7**, **8**, **9**, 47

K
Kepler objects of interest (KOIs), 102–103,
  109–110, **109**
Kinetic energy, 285, 296, 386
Köppen Climate Classification System, 450,
  459

L
Lab investigations, xxiv, 23–28
  alignment with standards, 23–24, 26,
    557–567
  checkout questions, 27
  explicit and reflective discussion topics,
    25–26
  handouts, 26
  implementation hints, 26
  instructional materials for, 26–28
  investigation proposals, 27, 577, **578–586**
  major concepts overview, 25
  materials and preparation for, 25
  peer-review guide and teacher scoring
    rubric (PRG/TSR), 27, **587–592**
  purpose, 24
  safety precautions, xxiv–xxv
  supplementary materials, 28

teacher notes for, 24–26
timelines for, 25, **573–575**
Lake Mead Reservoir, **544**
Lake Tahoe, 296, **296**
Latitude, 448, 459
Layers of the Earth, 146, **147**, **148**, 157, **157**
Lithification, 240, 252
Lithosphere, 146, **147**, **148**, 151, 157, **157**
Loam, 262, **263**, 274, **274**
Lunar cycle, 43–45, **43**, 49–53
Lunar eclipse, 44, **44**
Luster, 311
Lyell, Charles, 123, 134

M
Mantle convection cells, 148, **148**
Marble, 321, **321**
Mass, 388
Mean monthly temperatures at different
    latitudes, 58, **59**
Mesocyclones, 490–491
Mesosphere, 146, **147**, **148**, 157, **157**
Metamorphic rocks, 321
Meteorology, 376
Mid-latitude cyclones, 491–492
Milling, 334, 348–349
Mineral Characteristics lab, 310–331
    checkout questions, 328–331
    lab handouts
        argumentation session, 325–327
        getting started, 324–325
        initial argument, 325, **325**
        introduction, 321–323, **321**, **322**
        investigation proposal, 324
        materials, 323
        report, 327
        safety precautions, 323–324
        scientific knowledge and inquiry
            connections, 325
        task and guiding question, 323
    teacher notes
        alignment with standards, 318,
            318–320
        content, 310–312, **310**, **311**
        explicit and reflective discussion
            topics, 315–317
        implementation tips, 317–318
        materials and preparation, 312–314,
            313
        purpose, 310
        safety precautions, 314–315

timeline, 312
Mine water, 335, 350
Mississippi River delta, 217–218, **217**, 230,
    230
Mississippi River drainage basin, 216, **216**,
    229, **229**
Mohs Hardness Scale, 311–312, **311**
Moon Phases lab, 32–53
    checkout questions, 49–53
    lab handouts
        argumentation session, 47–48
        getting started, 46
        initial argument, 47, **47**
        introduction, 43–45, **43**, **44**
        investigation proposal, 46
        materials, 45
        report, 48
        safety precautions, 45
        scientific knowledge and inquiry
            connections, 47
        task and guiding question, 45
    teacher notes
        alignment with standards, 40, **41–42**
        content, 32–33, **32**
        explicit and reflective discussion
            topics, 37–39
        implementation tips, 39–40
        materials and preparation, 34–36,
            34, **35**, **36**
        purpose, 32
        safety precautions, 36–37
        timeline, 33
My Solar System simulation, 79, 80, 92–93, **92**

N
National Oceanic and Atmospheric
    Administration (NOAA), 368, 379, 450, 474,
    483
National Weather Service Weather Prediction
    Center, 368, 379
Natural Hazards Viewer interactive map, 150,
    153, 158, 171, 174, 182–183
Natural Resources Distribution lab, 332–360
    checkout questions, 358–360
    lab handouts
        argumentation session, 355–356
        getting started, 351–354, **354**
        initial argument, 355, **355**
        introduction, 347–350, **347**, **348**, **350**
        investigation proposal, 351
        materials, 351

report, 356–357
    safety precautions, 351
    scientific knowledge and inquiry
        connections, 354
    task and guiding question, 351
    teacher notes
        alignment with standards, 342,
            342–346
        content, 332–335, **333**
        explicit and reflective discussion
            topics, 337–340
        implementation tips, 340–341, **342**
        materials and preparation, 336–337,
            337
        purpose, 332
        safety precautions, 337
        timeline, 335
Natural resources. *See* Human Use of
    Natural Resources lab; Natural Resources
    Distribution lab
Nature of Scientific Knowledge (NOSK) and
    Nature of Scientific Inquiry (NOSI)
    Air Masses and Weather Conditions lab,
        371–372, 379
    alignment with argument-driven inquiry
        (ADI) model, **560**
    Atmospheric Carbon Dioxide lab, 433,
        441
    Carbon Emissions Minimization lab,
        516–517, 527
    explicit and reflective discussion, 13–14
    Extreme Weather Forecasting lab,
        496–497, 504
    Geologic Features Formation lab, 173–
        174, 184
    Geologic Time and the Fossil Record
        lab, 128–129, 137
    Global Temperature lab, 412–413, 423
    Gravity and Orbits lab, 82–83, 94
    Habitable Worlds lab, 106, 111
    Human Use of Natural Resources lab,
        538, 549
    Hurricane Strength Prediction lab, 477–
        478, 485
    Mineral Characteristics lab, 316–317,
        325
    Moon Phases lab, 38–39, 47
    Natural Resources Distribution lab,
        339–340, 354
    overview of, 570–572
    Plate Interactions lab, 152–153, 159

    Regional Climate lab, 454, 462
    Seasons lab, 62, 71
    Sediment Deposition lab, 246–247
    Sediment Transport lab, 223–224, 233
    Soil Texture and Soil Water Permeability
        lab, 267–268, 277
    Surface Materials and Temperature
        Change lab, 393, 402
    Water Cycle lab, 290, 301
    Wind Erosion lab, 199–200, 208
Next Generation Science Standards (NGSS)
    Air Masses and Weather Conditions lab,
        373
    alignment with argument-driven inquiry
        (ADI) model, **561–566**
    Atmospheric Carbon Dioxide lab, **435**
    Carbon Emissions Minimization lab, **518**
    Extreme Weather Forecasting lab, **498**
    Geologic Features Formation lab, **175**
    Geologic Time and the Fossil Record
        lab, **130**
    Global Temperature lab, **414**
    Gravity and Orbits lab, **85**
    Habitable Worlds lab, **107**
    Human Use of Natural Resources lab,
        539
    Hurricane Strength Prediction lab, **479**
    Mineral Characteristics lab, **318**
    Moon Phases lab, **41**
    Natural Resources Distribution lab, **342**
    overview of, 569–570
    Plate Interactions lab, **154**
    Regional Climate lab, **455**
    Seasons lab, **64**
    Sediment Deposition lab, **248**
    Sediment Transport lab, **225**
    Soil Texture and Soil Water Permeability
        lab, **269**
    Surface Materials and Temperature
        Change lab, **395**
    Water Cycle lab, **292**
    Wind Erosion lab, **202**
Nimbostratus clouds, 367
Non-contact force, 89

O
Ocean currents, 448, **449**, 459
Oceanic crust, 146, **147**
Orbital velocity, 79
Orbits
    described, 78, 89–91, **90**, **91**

orbital period, 81
    See also Gravity and Orbits lab
Original horizontality principle, 123, 134
Overburden, 334, 348, 349
Owakudani valley, 296, **296**

P
Particle size and shape, 253, **253**
Patterns
    Air Masses and Weather Conditions lab,
        364, 370
    Extreme Weather Forecasting lab, 490,
        495
    Geologic Features Formation lab, 168
    Geologic Time and the Fossil Record
        lab, 122, 127
    Habitable Worlds lab, 102, 105
    Mineral Characteristics lab, 310, 315
    Moon Phases lab, 37–38
    Natural Resources Distribution lab, 332,
        338
    NGSS crosscutting concept, 569
    Plate Interactions lab, 146, 151
    Regional Climate lab, 448, 453
    Seasons lab, 61
Peer-review guide and teacher scoring rubric
    (PRG/TSR)
        described, 27
        double-blind group peer review, 15–17,
            16
        peer-review guide (high school), **590–
            592**
        peer-review guide (middle school),
            587–589
Perigee, 78, 90, 91, **91**, 94
Perihelion, 78, 90, 91, **91**, 94
Phases of the Moon, 43–45, **43**, 49–53
Physical models of Earth and Moon, 34–36,
    35, **36**, 44–45, 46
Physical properties of minerals, 322, **322**,
    328–329
Planets. *See* Habitable Worlds lab
Plate Interactions lab, 146–165
    checkout questions, 162–165
    lab handouts
        argumentation session, 160–161
        getting started, 158–159
        initial argument, 159, **160**
        introduction, 157–158, **157**
        investigation proposal, 158
        materials, 158

report, 161
        safety precautions, 158
        scientific knowledge and inquiry
            connections, 159
        task and guiding question, 158
    teacher notes
        alignment with standards, 154,
            154–156
        content, 146–149, **147**, **148**, **149**
        explicit and reflective discussion
            topics, 151–153
        implementation tips, 153–154
        materials and preparation, 150, **150**
        purpose, 146
        safety precautions, 150
        timeline, 150
Porphyry copper deposits, 333, **333**, 347
Powdered mineral color, 311
Prevailing winds, 448, **449**, 459

Q
Quartz, 310, **310**, 322, **322**

R
Radiation, 387–388, 399
Reclaimed water, 546
Regional Climate lab, 448–468
    checkout questions, 465–468
    lab handouts
        argumentation session, 463–464
        getting started, 461–462
        initial argument, 463, **463**
        introduction, 458–460, **458**
        investigation proposal, 461
        materials, 460–461
        report, 464
        safety precautions, 461
        scientific knowledge and inquiry
            connections, 462
        task and guiding question, 460
    teacher notes
        alignment with standards, 455,
            455–457
        content, 448–451, **449**
        explicit and reflective discussion
            topics, 452–454
        implementation tips, 455
        materials and preparation, 451–452,
            452
        purpose, 448
        safety precautions, 452

timeline, 451
Relative age of rock strata, 122, 133–134
Ridge push, 148, **148**
Rivers, 216, 229
    See also Sediment Transport lab
Rock stratum, 122

S

Saffir-Simpson Hurricane Scale, 483, **484**
Saltation, 194
Sandstone, 252, **252**
Sandy clay loam, 262, **263**, 274, **274**
Satellite images, 398, **398**
Satellites, 78–79, 89–91, 93, 97–98
Scale, Proportion, and Quantity
    Atmospheric Carbon Dioxide lab, 428,
      432
    Geologic Time and the Fossil Record
      lab, 122, 127
    Gravity and Orbits lab, 78, 81
    Habitable Worlds lab, 102, 105
    NGSS crosscutting concept, 569
    Plate Interactions lab, 146, 151
    Sediment Transport lab, 216, 222
    Soil Texture and Soil Water Permeability
      lab, 262, 266
    Wind Erosion lab, 194
Science and engineering practices (SEPs)
    and ADI model, xi–xiii, **xii**, xix, 3, **557**
    and lab investigations, 23–24
    See also Next Generation Science
      Standards (NGSS)
Science proficiency, xvii–xviii
Scientific argumentation, 10–12, **11**
Seafloor spreading, 149
Seasons and Ecliptic Simulator, 59, 69,
    70–71, **70**
Seasons lab, 54–77
    checkout questions, 74–77
    lab handouts
      argumentation session, 72–73
      getting started, 70–71, **70**
      initial argument, 71, **71**
      introduction, 67–68, **67**, **69**
      investigation proposal, 70
      materials, 69
      report, 73
      safety precautions, 70
      scientific knowledge and inquiry
        connections, 71
      task and guiding question, 69

    teacher notes
      alignment with standards, 64, **64–66**
      content, 54–58, **55–59**
      explicit and reflective discussion
        topics, 60–62
      implementation tips, 62–63
      materials and preparation, 59, **60**
      purpose, 54
      safety precautions, 60
      timeline, 59
Sedimentary rock layers (strata), 122–123,
    122, 133
Sedimentary rocks, 321
Sediment Deposition lab, 240–261
    checkout questions, 259–261
    lab handouts
      argumentation session, 257–258
      getting started, 255–256, **255**
      initial argument, 257, **257**
      introduction, 252–254, **252**, **253**
      investigation proposal, 255
      materials, 254
      report, 258
      safety precautions, 254–255
      scientific knowledge and inquiry
        connections, 257
      task and guiding question, 254
    teacher notes
      alignment with standards, 248,
        248–251
      content, 240–241
      explicit and reflective discussion
        topics, 244–247
      implementation tips, 247
      materials and preparation, 242–244,
        242–243
      purpose, 240
      safety precautions, 244
      timeline, 241
Sediment-hosted copper deposits, 333, 347
Sediment Transport lab, 216–239
    checkout questions, 236–239
    lab handouts
      argumentation session, 233–235
      getting started, 232–233, **232**
      initial argument, 233, **233**
      introduction, 229–230, **229**, **230**
      investigation proposal, 231
      materials, 231
      report, 235
      safety precautions, 231

scientific knowledge and inquiry
connections, 233
task and guiding question, 231
teacher notes
alignment with standards, 225,
225–228
content, 216–219, **216**, **218**
explicit and reflective discussion
topics, 221–224
implementation tips, 224–225
materials and preparation, 219–221,
220, **221**
purpose, 216
safety precautions, 221
timeline, 219
"Sense-making," 8–10, **8**, **9**
Settling velocity, 241, 253, 259
Shale, 252, **252**
Siltstone, 252, **252**
Slab pull, 148, **148**
Slag, 335, 349
Smelting, 334, 349
Smith, William, 123, 134
Soil moisture, 195
Soil particle types and sizes, 262–263, **262**,
263, 274, **274**, 280
Soil Texture and Soil Water Permeability lab,
262–283
checkout questions, 280–283
lab handouts
argumentation session, 277–279
getting started, 276–277, **276**
initial argument, 277, **277**
introduction, 274–275, **274**
investigation proposal, 276
materials, 275
report, 279
safety precautions, 275
scientific knowledge and inquiry
connections, 277
task and guiding question, 275
teacher notes
alignment with standards, 269,
269–273
content, 262–263, **262**, **263**
explicit and reflective discussion
topics, 266–268
implementation tips, 268–269
materials and preparation, 263–265,
264, **265**
purpose, 262

safety precautions, 265
timeline, 263
Soil types, 195, **195**, 205, **205**
Solar eclipse, 44, **44**
Solar radiation, 408–409, **409**, 419–420, **420**
Solstices, 56–58, **57**, **58**, 75
Space systems. *See* Gravity and Orbits lab;
Habitable Worlds lab; Moon Phases lab;
Seasons lab
Specific heat, 388
Stability and Change
Atmospheric Carbon Dioxide lab, 428,
432
Carbon Emissions Minimization lab, 512,
515
Global Temperature lab, 408, 411
Human Use of Natural Resources lab,
534, 537
NGSS crosscutting concept, 570
Surface Materials and Temperature
Change lab, 386, 392
Water Cycle lab, 284, 289
Wind Erosion lab, 194
Stationary fronts, 365, 367, **367**, 377, **378**
Steno, Nicholas, 123, 134
Strata, 122–123, **122**, 133
Stratification, 122
Stratus clouds, 367
Streak test, 311
Stream flow, 217, 229–230, 236–237
See also Sediment Deposition lab;
Sediment Transport lab
Stream table, 220–221, **221**, 232, **232**
Structure and Function
Mineral Characteristics lab, 310, 315
NGSS crosscutting concept, 570
Sediment Deposition lab, 240, 245
Sunlight and solar radiation, 54–55, **55**, 57–
58, **58**, **59**
Supercells, 490
Superposition, 123, 134
Surface area to volume ratio, 285, 288–289,
291, 298, 304–305
Surface Materials and Temperature Change
lab, 386–407
checkout questions, 405–407
lab handouts
argumentation session, 402–404
getting started, 401, **401**
initial argument, 402, **402**
introduction, 398–399, **398**

investigation proposal, 400
materials, 400
report, 404
safety precautions, 400
scientific knowledge and inquiry
connections, 402
task and guiding question, 400
teacher notes
alignment with standards, 395,
395–397
content, 386–388
explicit and reflective discussion
topics, 391–393
implementation tips, 394–395
materials and preparation, 389, **390**
purpose, 386
safety precautions, 391
timeline, 389
Surf Your Watershed, 336, 340
Suspended load, 217
Suspension, 194
Systems and System Models
Carbon Emissions Minimization lab, 512,
515
Geologic Features Formation lab, 168
Gravity and Orbits lab, 78, 81
Moon Phases lab, 37–40
NGSS crosscutting concept, 569
Regional Climate lab, 448, 453
Seasons lab, 61

T
Tailings, 334–335, 349, **350**
Tectonic plates. *See* Plate Interactions lab
Temperature. *See* Air Masses and Weather
Conditions lab; Global Temperature lab;
Habitable Worlds lab; Seasons lab; Surface
Materials and Temperature Change lab;
Water Cycle lab
Texture, soil texture, 262–263, **263**, 274, **274**
Theoretical criteria, 7–8, **7**
Three-dimensional instruction, xx
"Tool talk," 4
Topography, 448, 459
Tornadoes, 490–491, **490**, **491**, **492**
See also Extreme Weather Forecasting
lab
Transform boundaries, 149, **149**, 151, 157–
158, 162
Transpiration, 284–285, **285**, **297**

Transport processes. *See* Sediment Transport
lab; Wind Erosion lab
Tropical storms, 472–473, **472**, **473**, 481–482,
481, **482**, **484**
See also Hurricane Strength Prediction
lab
Typhoons. *See* Hurricane Strength Prediction
lab

U
Uniformitarianism, 123, 134
U.S. Fish and Wildlife Service Environmental
Conservation Online System (ECOS), 336,
340, 353
U.S. Geological Survey (USGS) Global
Assessment of Undiscovered Copper
Resources, 336, 340–341, 352–353
U.S. Geological Survey (USGS) Mineral
Commodity Summaries, 336, 353

V
Velocity, 81, 216, 229
Viscous drag, 148, **148**
Volcanic eruptions. *See* Plate Interactions lab
Vortexes, 490

W
Warm fronts, 365, 366–367, **367**, 377, **377**
Water column, 243, **243**, 255–256, **255**
Water Cycle lab, 284–307
checkout questions, 304–307
lab handouts
argumentation session, 301–303
getting started, 299–301, **300**
initial argument, 301, **301**
introduction, 296–298, **296**, **297**
investigation proposal, 299
materials, 299
report, 303
safety precautions, 299
scientific knowledge and inquiry
connections, 301
task and guiding question, 298
teacher notes
alignment with standards, 292,
292–295
content, 284–285, **285**
explicit and reflective discussion
topics, 288–290
implementation tips, 290–292
materials and preparation, 286, **287**

purpose, 284
    safety precautions, 286, 288
    timeline, 286
Watershed boundary, 229
WaterSim visualization tool, 534, 535, 546,
    547–549, **548**
Water supply and distribution, 534, 543–546,
    543, 544, 545
        See also Human Use of Natural
            Resources lab
Weather maps, 376, **376**, 383–384
Weather. *See* Air Masses and Weather
    Conditions lab; Atmospheric Carbon
    Dioxide lab; Extreme Weather Forecasting
    lab; Global Temperature lab; Hurricane
    Strength Prediction lab; Regional Climate
    lab; Surface Materials and Temperature
    Change lab
Wentworth scale, 253, **253**
Wind duration, 195
Wind Erosion lab, 194–215
    checkout questions, 211–215
    lab handouts

    argumentation session, 209–210
    getting started, 207–208
    initial argument, 208, **209**
    introduction, 204–206, **204, 205**
    investigation proposal, 207
    materials, 206
    report, 210
    safety precautions, 206–207
    scientific knowledge and inquiry
        connections, 208
    task and guiding question, 206
teacher notes
    alignment with standards, 201,
        202–203
    content, 194–195, **194, 195**
    explicit and reflective discussion
        topics, 198–200
    implementation tips, 200–201
    materials and preparation, 196, **197**
    purpose, 194
    safety precautions, 196–197
    timeline, 195–196
Wind speed, 195